Cyber Physical Computing for IoT-driven Services

Vladimir Hahanov

Cyber Physical Computing for IoT-driven Services

Vladimir Hahanov
Design Automation Department
Kharkov National University of Radio Electronics
Kharkiv, Ukraine

ISBN 978-3-319-85494-6 ISBN 978-3-319-54825-8 (eBook)
DOI 10.1007/978-3-319-54825-8

Softcover re-print of the Hardcover 1st edition 2018

Printed on acid-free paper

This Springer imprint is published by Springer Nature
The registered company is Springer International Publishing AG
The registered company address is: Gewerbestrasse 11, 6330 Cham, Switzerland

Preface

In this book, the basic definitions, models, methods, and algorithms for cyber physical computing, as the entity of the new Internet of Things (IoT) cyberculture of monitoring and management of virtual and physical processes and phenomena, are proposed. A metric for the measurement of processes and phenomena in cyberspace, and also the architecture of logic associative computing for decision making and big data analysis, are offered. An innovative theory and practice for design, testing, simulation, and diagnosis of digital systems, based on leverage of qubit coverage to describe the functional components and structures, are represented. Examples of cyber physical systems for digital monitoring and cloud management of social objects and transport are proposed. An automaton model of cosmological computing is shown. It explains the cyclical and harmonious development of material and energy essence, and also a space–time form of the universe.

The substance is the creation of the theory of actuative metrical human-free cloud management of virtual, physical, and social processes and phenomena based on comprehensive big data and sensor cyber physical digital monitoring to improve human quality of life and help preserve the ecology of the planet.

Why is human-free management important? The human is arguably the peak of planetary evolution, but at the same time, he is the weakest link in the planet's ecosystem: (1) he cannot optimally manage himself, so he should not be admitted to the management of enterprises, technology, and nature; (2) a human is the most vulnerable biological subject and needs constant protection from the external aggressive environment; (3) a human operator is the most unreliable link in the human–machine system and is the cause of 75% of errors leading to accidents and disasters; (4) a human governor is not able to take socially objective solutions to improve the quality of life and preserve the environment, so wars, social upheavals, revolutions, and riots happen; (5) a human official might enact inhumane laws via a democratic majority, often ignorant or incompetent in specific management issues—therefore, democracy and conformism are the antipodes of science and education; (6) a human figurehead might create for himself an immoral, corrupt relationship system, permitting redistribution of material and time resources for the benefit of his entourage; and (7) a human controller is not able to objectively and metrically evaluate the social significance of each individual work.

Theoretically, removing the person from the cycle of monitoring and management means replacing him in this matter by an all-seeing, objective, and fair system, in the aid of which incorruptible cloud computing and cyber physical, scientifically based, and practically verified services are created. Replacing a human manager with computing frees up dollars and human resources to address pressing issues of quality of life and the environment. Computing reliability and accuracy can help avoid social cataclysms and can help to avoid man-made disasters, wars, and accidents.

The goal is to help ensure the morality of social relationships, quality of life, and preservation of the planet's ecology through the creation of the theory of active metrical human-free cloud management of virtual, physical, and social processes and phenomena based on comprehensive big data and sensor cyber physical digital monitoring.

The object of this research is to make virtual, physical, technological, biological, and social processes and phenomena subject to active moral cloud management on the basis of comprehensive cyber physical monitoring.

The essence of this research is an automaton theory of computing for metrical human-free moral cloud management of virtual, physical, and social processes and phenomena based on cyber physical digital monitoring.

Computing is a technological branch of knowledge focused on the theory and practice of exact metrical computer management of virtual, physical, and social processes and phenomena based on exhaustive big data and sensor-driven digital monitoring.

Cyber physical computing is the theory and practice of cloud calculating, with actuative metrical management of virtual, physical, and social processes and phenomena based on exhaustive big data and sensor-driven digital monitoring.

The Internet of Things is a global, scalable, technological platform structures or cyberculture of actuative metrical moral cloud computing management of virtual, physical, and social processes and phenomena based on exhaustive big data and sensor-driven digital monitoring to help ensure quality of life and preserve the planet's ecology.

IoT-driven service is an economically optimal way to meet market needs in the cloud with actuative metrical management of virtual, physical, and social processes and phenomena based on exhaustive big data and sensor-driven digital monitoring to help ensure quality of life and preserve the planet's ecology.

Cyberculture is the development level of social and technological moral relations between society, the physical world, and cyberspace, determined by the implementation of Internet services for precise digital monitoring and reliable metrical management in all processes and areas of human activity, including education, science, life, production, and transport, in order to improve quality of life and help preserve the planet's ecosystems. Cyberculture is a moral metric of computing excellence to achieve social significance and recognition.

Computing is a binary union. A computer system always consists of two interacting components that perform control and execution functions to achieve a specified goal of functioning in cyberspace. Any other system can be represented by a computer interaction of two components. This book's structure includes several investigations in the cyber physical/social field, considering new models of objects and processes within the mainstream called "computing": (1) cyber physical computing for metrical management of physical, virtual, and social processes and phenomena based on digital monitoring; (2) a metric and structures for logical associative computing and decision making due to proposed specific processor structures; (3) processor architectures for big data analysis, leveraging logic register structures for computing; (4) qubit vector or quantum descriptions of functions and structures for designing digital computing systems; (5) qubit vector (quantum) methods for interpretative simulation and testing of digital computing systems leveraging memory-based logic; (6) quantum interpretative simulation of digital systems implemented into cloud service computing; (7) assertion-driven cloud service computing for diagnosing digital hardware/software systems; (8) qubit-driven computing for diagnosis of digital device faults; (9) the smart cyber university—a new model of cloud computing for digital monitoring and management of research and educational processes; (10) cloud service computing transport monitoring and control, leveraging a streetlight-free infrastructure; (11) cosmological computing of the harmonic interaction of the matter–energy essence and the space–time form of the universe; (12) cyber social computing for cloud governance of citizens; and (13) practical conclusions and further computing investigations.

The company Hewlett Packard Enterprise (HPE) has created a prototype supercomputer—"The Machine"—with memory-driven computing architecture based on photonic optical connections, which is 8000 times better than existing computer performance. For calculations, The Machine leverages eight terabytes of memory, created on memristors, which is several tens of times greater than the capacity of modern servers [https://www.labs.hpe.com/next-next/mdc]. The company believes that by using The Machine, it is possible to solve big data–driven problems connected with transportation systems, smart cities, speech and video recognition, security, healthcare, and IoT computing.

The bottleneck of a classical computer is the interaction channel between a slow-speed memory and a high-speed logic processor. Nowadays, memory performance has increased dozens of times, so the communication channel with the logical processor becomes an unacceptable obstacle in the increase of computer architecture performance.

There is only one solution: to move decision making to large memories used for machine learning, thus minimize the need for traditional logic processor from the computer architecture, solving at the same time the bottleneck problem and the interaction of the memory and logic processor (Yervant Zorian, Yerevan, 2007 IEEE EWDTS).

The innovation advantages are visible: (1) simplifying computing architecture, which is reduced to a regular MAT (memory–address–transaction) model; (2) technological programming associated with the processing of big data analytics on the

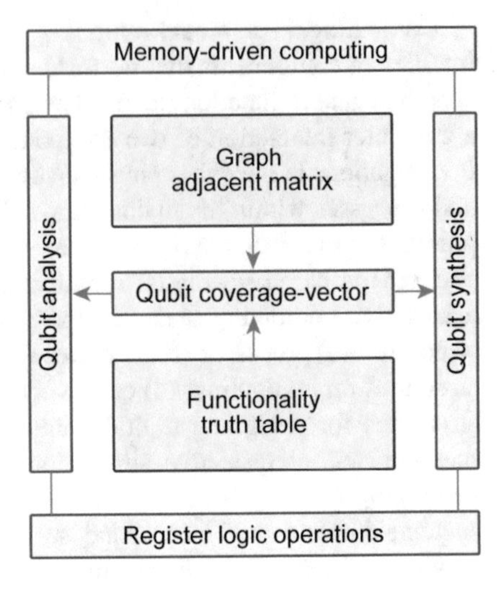

Fig. 1 The qubit coverage vector creates memory-driven computing

memory-driven super computer, free of logic interaction; (3) high-level parallelism of SIMD execution of logic and arithmetic operations, with practically unlimited memory; (4) delivery to the market of new data center technology for processing yottabytes (10^{24}) of information ["yottabyte" means "cosmos"]; and (5) delivery to the IT industry of robust and reliable computing architecture, since any faulty memory component can be readdressed online to the spare area element.

One of the basic proposals is to transform the description of the function and structure to a single one-dimensional format, which makes it technologically easy to solve all functions, synthesize graphs, and analyze problems in memory-driven computing based on qubit coverage vector metrics and logic parallel operation performance (Fig. 1), because two-dimensional register creation, with the corresponding adjacency matrix or truth table, is quite difficult.

The advantages of memory-driven computing, which are obvious to the market, motivate creation of the design and testing theory for a new generation of computing architectures. One of the possible memory-driven trends could be theory and technology, called vector design and testing of digital devices, described by qubit coverage.

The authors greatly appreciate the contributions of the following scientists/colleagues and friends—Stanley Hyduke, Yervant Zorian, Raimund Ubar, Anzhela Matrosova, Vazgen Melikyan, Victor Djigan, Vyacheslav Kharchenko, and Alexander Drozd—which implicitly initiated the research presented in the chapters of this book.

> If you want to live a happy life, tie it to a goal (business, science and education), not to people or objects. (Albert Einstein)

Kharkov, Ukraine Vladimir Hahanov

Contents

1 Cyber Physical Computing 1
Vladimir Hahanov, Eugenia Litvinova, Svetlana Chumachenko, and Anna Hahanova
1.1 Introduction 1
1.2 State of the Art of Cyber Physical Computing 2
1.3 Commitments of Computing Development 4
1.4 MAT Computing for Physical and Virtual Space 9
1.5 Computing of Quantum Teleportation 14
1.6 Quantum Computing 16
1.7 Conclusion 18
References 19

2 Multiprocessor Architecture for Big Data Computing 21
Vladimir Hahanov, Wajeb Gharibi, Eugenia Litvinova, and Alexander Adamov
2.1 Introduction 21
2.2 Nonarithmetic Metrics for Cyberspace Measurement 22
2.3 Structural Solutions for Solving Search Problems 30
2.4 Vector Logical Multiprocessor Architecture 32
2.5 Vector Logical Analysis Infrastructure 35
2.6 Conclusion 39
References 40

3 Big Data Quantum Computing 43
Vladimir Hahanov, Eugenia Litvinova, Svetlana Chumachenko, Tetiana Soklakova, and Irina Hahanova
3.1 Criteria and Structures for Evaluating the Quality of Object Interaction in Cyberspace 43
3.2 Triangular Metric and Topology of Cyberspace 56

3.3 Quality Criterion for Cyber Physical Topology 60
3.4 Quantum Hasse Processor for Optimal Coverage 63
3.5 Conclusion 68
References 68

4 Qubit Description of Functions and Structures for Service Computing Synthesis 71
Ivan Hahanov, Igor Iemelianov, Mykhailo Liubarskyi, and Vladimir Hahanov
4.1 Introduction 71
4.2 Quantum Data Structures 73
4.3 Quantum Description of Digital Functional Elements 78
4.4 Quantum Description of Graph Structures 79
4.5 Qubit Description Analysis of Graph Structures 81
4.6 Modeling Qubit Description of Digital Structures 83
4.7 Synthesis of the Logic Circuit Quantum Vector 85
4.8 Minimization of the Logic Circuit Quantum Vector 86
4.9 Models of Quantum Sequential Primitives 88
4.10 Conclusion 91
References 92

5 Quantum Computing for Test Synthesis 95
Vladimir Hahanov, Tamer Bani Amer, Igor Iemelianov, and Mykhailo Liubarskyi
5.1 Memory-Based Logic 95
5.2 Qubit Test Generation Method 99
5.3 Taking the Boolean Derivatives for *Q* Test Synthesis 109
5.4 Deductive Qubit Fault Simulation Method for Digital Structures 116
5.5 Analysis of the Schneider Circuit 122
5.6 Conclusion 132
References 133

6 QuaSim Cloud Service for Quantum Circuit Simulation 135
Ivan Hahanov, Tamer Bani Amer, Igor Iemelianov, Mykhailo Liubarskyi, and Vladimir Hahanov
6.1 The State of the Art 135
6.2 Model of a Quantum Processor for Binary Simulation 136
6.3 Simulation Method Based on Logic Qubit Coverages 137
6.4 Fault-Free Simulation of Synchronous Circuits 140
6.5 QuaSim: A Cloud Service Simulator of Digital Devices 142
6.6 Conclusion 146
References 147

7 Computing for Diagnosis of HDL Code 149
Vladimir Hahanov, Eugenia Litvinova, and Svetlana Chumachenko
7.1 State of Assertion-Driven Verification 149
7.2 TABA Model for Diagnosis of Faulty HDL Blocks 150
7.3 Design for Diagnosability 155
7.4 Diagnosis Method for a Hardware–Software System 157
7.5 Conclusion 160
References 161

8 Qubit Computing for Digital System Diagnosis 163
Vladimir Hahanov, Svetlana Chumachenko, and Eugenia Litvinova
8.1 Introduction 163
8.2 Qubit Analysis for Digital System Diagnosis 165
8.3 Qubit Modeling of Digital Systems 171
8.4 Repair of Logic 176
8.5 Conclusion 180
References 181

9 Cloud Service Computing: The "Smart Cyber University" 183
Vladimir Hahanov, Oleksandr Mishchenko, and Eugenia Litvinova
9.1 The Smart Cyber University in the Big Data of Cyberspace 183
9.2 Computing as a New Trend in the Cyber Service Market 187
9.3 Motivation and Technological Solutions of a Smart Cyber University 188
9.4 Innovative Services in the Smart Cyber University 191
9.5 Metrics of Employees and University Departments 193
9.6 Leveraging of Cyber Democracy in the University 196
9.7 Conclusion 198
References 199

10 Transportation Computing: "Cloud Traffic Control" 201
Vladimir Hahanov, Artur Ziarmand, and Svetlana Chumachenko
10.1 Introduction 201
10.2 Goal and Objectives of This Investigation 203
10.3 Global Cyber Physical System for Digital Monitoring and Cloud Traffic Control 204
10.4 Register-Based Structure of Smart Car–Driven Traffic Lights 205
10.5 Cyber Physical Monitoring of Transport—Technical Conditions 206
10.6 Formal Model of a Traffic System 207
10.7 Technology of GNSS Navigation 212
10.8 Conclusion 214
References 216

11 Cosmological Computing and the Genome of the Universe 219
Vladimir Hahanov, Eugenia Litvinova, and Svetlana Chumachenko
11.1 Introduction 219
11.2 Cosmological Computing 220
11.3 3D Model of the Universe's Genome 223
11.4 Conclusion 228
References 231

12 Cyber Social Computing 233
Vladimir Hahanov, Tetiiana Soklakova, Anastasia Hahanova, and Svetlana Chumachenko
12.1 Background of Cyber Social Computing 233
12.2 Cyber Social Governance 239
12.3 Cyberculture and the State 242
12.4 Metric Cyber Relationships as the Basis of Management 246
12.5 Conclusion 248
References 250

13 Practical Conclusion 251
Vladimir Hahanov

14 Cyber-Physical Technologies: Hype Cycle 2017 259
Vladimir Hahanov, Wajeb Gharibi, Ka Lok Man, Igor Iemelianov, Mykhailo Liubarskyi, Vugar Abdullayev, Eugenia Litvinova, and Svetlana Chumachenko
14.1 Introduction 259
14.2 Three Main Areas of the Cyber Culture 260
14.3 Examples of Leveraging Top Technologies 263
14.4 Innovation for the Architecture of Quantum Computing 267
14.5 Conclusion 270
References 272

Index 273

Chapter 1
Cyber Physical Computing

Vladimir Hahanov, Eugenia Litvinova, Svetlana Chumachenko, and Anna Hahanova

1.1 Introduction

Our motivation is to create a general picture of computing development in space and time to predict the innovative trends in this field. Every 20 years the technological basis of human development is changed on a global scale. Here computing plays the main role, and in the past 60 years, it has been scaled at three spatial levels: single, network and global. The first level is developing according to the scenario of personal, mobile, and interface computing in details called the embedded brain–computer interface. The second level is defined by the following components: network, cloud and cyber physical computing. The third level includes the Internet, the Internet of Things (IoT), and the Internet of Nature Computing (cyber nature computing). According to the proposed scenario, within the current 20 years (2010–2030), scientists need to solve the following problems: (1) creation of a built-in interface for direct communication of the human brain with a computer; (2) development of cyber physical systems for monitoring and management of all sociotechnological processes and phenomena, including transport and social groups; and (3) creation of a global human brain within the culture of cyber nature computing for accurate monitoring and optimal management of all natural and artificial processes and phenomena.

Cyber physical computing is the main attractor of IT engineering in the twenty-first century. The cyber physical ecosystem leverages big data centers, knowledge, services, and applications focused on monitoring and controlling processes and smart everything in digital physical space with the purposes of providing high

V. Hahanov (✉) • E. Litvinova • S. Chumachenko • A. Hahanova
Design Automation Department, Kharkov National University of Radio Electronics, Ukraine, Nauka Avenue 14, Kharkov 61166, Ukraine
e-mail: hahanov@icloud.com; litvinova_eugenia@icloud.com; svetachumachenko@icloud.com; anna_hahan@mail.ru

V. Hahanov, *Cyber Physical Computing for IoT-driven Services*,
DOI 10.1007/978-3-319-54825-8_1

standards of human life, and saving natural resources, energy, and our world for future generations.

Cyberculture is the level of moral–ethical, social, and technological relationships between society, the physical world, and cyberspace, defined by the introduction of Internet services for accurate digital monitoring and reliable metrical management in all processes and areas of human activity, including education, science, manufacturing, and transportation, in order to improve the quality of human life and preserve the planet's ecosystem.

The essence of the research is technology, methods, and models for sustainable development of computing for metrical human-free management of cyber physical and social processes and phenomena based on digital monitoring.

The goal is to show market-attractive features of the emerging technological pattern associated with active digital human-free management of all cyber physical processes and phenomena based on accurate monitoring, focused on the creation of a green planet and provision of a high quality of human life.

The chapter structure includes the following topics: (1) commitments of cyber physical space development; (2) cyber physical computing structure in time and space; (3) market-oriented attractors in cyber physical computing, supported by global business; (4) MAT computing as a universal and scalable monitoring and management structure of cyber physical and cosmological processes and phenomena; (5) classification of cyber and physical computing; (6) scalable architectures: embedded microsystems, cyber physical systems, IoT platforms; and (7) quantum-like computing and teleportation.

1.2 State of the Art of Cyber Physical Computing

The most significant market-oriented innovations and discoveries are associated with superposition of interdisciplinary research, as well as implementation of the achievements of one technological culture in related neighboring fields of knowledge. Computing models [1, 2] are increasingly being leveraged for monitoring and control of processes and phenomena in all fields of human and nature activities [3–6]. Integration of physical and biosocial culture with cyber technological solutions for monitoring and control leads to an original system (cyber social, cyber physical, bioinformational) of scientific and practical results in traditionally conservative branches of knowledge, such as biology, sociology, technology, education, transport, and industry. Proof of the aforementioned can serve as a global attractive technology, leveraging a scalable computing model [4, 5]: (1) cyber physical systems; (2) IoT; (3) cloud, mobile, service, big data, and quantum computing; and (4) Internet-driven smart infrastructures: enterprise, university, city, and government. In general, computing covers the fields of nano-, micro- and macroelectronics; radio telecommunications; computer hardware–software design; industrial, transport and social engineering; artificial intelligence and cyber control; and computer science and information technology. The essential difference between computing and information technology [7, 8] is associated with active control and management of processes and phenomena in the real and virtual spaces. It puts a sign of correspondence between the concepts of form and essence: the Internet is associated with information technology, and IoT is associated with computing, where the main idea is control or management.

Bernard Marr of *Forbes* has offered nine attractors [9], supporting the market and global business interests in computing: big data, IoT, mobile computing everywhere, cybersecurity, e-assistants or the brain–computer interface, social networks, gamification for business and education, cloud computing, and video communications. In the near future there is expected to be mass implementation in our life of new technologies, forming a virical(**vir**tual-log**ical**) continuum: the Internet of Everything, wearables, smart cars, smart homes, smart cities, 3D printing, quantum computing, robotics, the cloud, big data, the maker movement, and drones. In fact, there is interaction between the virtual and the real world: in 2016 there were more than 5 billion connected devices and 1.4 billion smart phones; in 2020 there will be 50 billion connected devices, 6.1 billion smart phones, and 250 million vehicles controlled without drivers, using cloud traffic cyber services. Already today, Google cars travel about 10,000 miles a week on an innovative urban road infrastructure. More than 10.2 million smart wearables will be produced by 2020. The market for RFID tags will also increase significantly for digital identification of people, objects, and processes from US$11.1 billion to US $21.9 billion. As a result, the financial impact of IoT in 2025 on the world market will be worth US$11 trillion, and the levels of IoT capitalization will be US $4.6 trillion and US$14.4 trillion in the public and private sectors, respectively.

Susan Galer of *Forbes* believes that the cloud service will be the most interesting model of IT business in the next 10 years [10]. By the end of 2017, two thirds of the 2000 hugest global companies will have transformed their activities under digital process monitoring and control. More than 50% of all investment computing (IT) companies will be focused on the creation of technology platforms and services related to cloud, mobile, social business, and big data analytics. By 2020, cloud-based IT capitalization companies will have reached 70% of all software and technology services. In 2018, 22 billion IoT devices will be installed, which will have access to more than 200,000 new IoT applications and services. By 2018, more than 50% of software companies will embed into their applications educational services (1% today), providing just to USA enterprises more than US$60 billion in annual savings. By this time, 50% of all businesses will have created a cloud-based platform for disseminating their own innovations and consumption of external offers. By 2018, 80% of companies in the B2B (business-to-business) format and 60% of B2C (business-to-customer) organizations will have restructured their portals to digital, which will increase the number of customers and clients by 3–4 times. In summary, all companies and organizations—including universities, cities, and countries—will be uncompetitive without integration into cyberspace with preprepared digital legislation.

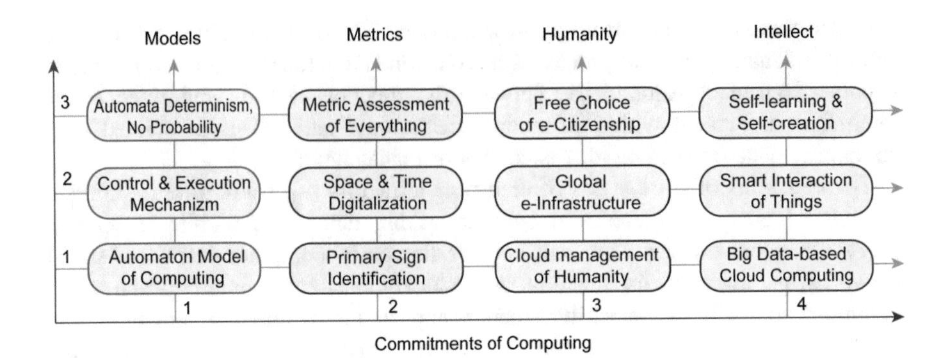

Fig. 1.1 Commitments of computing

1.3 Commitments of Computing Development

The computing model of sustainable development of humanity defined in the metric of the moral–ethical concept calls for us to save the planet and improve the quality of life [1]. There are 12 commitments (Fig. 1.1) as the market attractors, defined in world development computing.

Direction 1 connects with the kind of model leveraging in computing:

(1-1) The *automaton model of computing* is focused on monitoring and controlling all processes and phenomena as the most reliable and easy way to understand and implement. The model has a minimum number of interacting components and signals in the implemented closed-loop deterministic system to achieve some goals. The long life of the automaton model is confirmed not only in the creation of the computer industry but also in the adequate description of computing processes and phenomena in physics, biology, medicine, engineering, sociology, and cosmology.

(1-2) *Control and execution mechanisms* for the implementation of the computing model should not be crossed in the two main composed components. The conditions regulate the strict separation of functions between management and execution mechanisms, which also means noninterference of the first one in the case of the other. Violation of this rule is a typical system error, followed by corruption in social group management.

(1-3) *Automaton determinism*: no probability assumes exact predictability of computing reaction on actuative and initiative signals, excluding the probability from events and processes, as a “fig leaf on the naked body of our ignorance” according to Einstein’s definition. The policies are focused on computing service determinism, excluding the unpredictable probabilistic methods, which significantly reduces the overhead for duplication and redundancy of system components and processes.

Direction 2 operates metrics for measurement processes and objects in cyber physical space:

(2-1) *Primary sign identification* means complete elimination of secondary signs of human identification in cyber physical space based on widespread introduction of e-interfaces for entering primary authenticators (fingerprint, iris, DNA). The conditions allow removing from circulation billions of paper passports and electronic cards that can be faked or lost, which makes it possible to preserve world forests and save up to US$10 billion in manufacturing documents.

(2-2) *Space and time digitalization* of all spatial, cyber physical, biological, and social processes, objects, and transport provides precise control based on digital monitoring and positioning. On a global scale the conditions allow solving the problem of reliable human-free control of all transport to reduce the costs of vehicle license plate production (US$5 billion), infrastructure, road signs, and traffic lights (US$500 billion) by creating virtual traffic management based on e-infrastructure. About 1.2 million people die annually at the present time on the roads of all countries. Digital monitoring of natural phenomena, hurricanes, typhoons, earthquakes, and tsunamis, with warnings by introducing a network of sensors, will provide actuative management of climatic disasters and geopathogenic actions.

(2-3) *Metric assessment of everything* (human-free) or all processes and phenomena, forming an integral criterion of "time–money–quality," allows reaching the purpose of moral and material incentives for socially and environmentally significant projects. Metric estimation and adequate moral and material rewarding for members of social groups and collectives have a significant positive impact on the process of improving the quality of life of people and the planet's ecosystems. Cyberculture is based on the moral–ethical evaluation metric of the social significance of every planet citizen. Implementation of the global market trend in e-infrastructure from the EU Horizon 2020 program is used today for all terrestrial processes and phenomena, including health care e-infrastructure, smart home e-infrastructure, smart university e-infrastructure, smart company e-infrastructure, smart city e-infrastructure, social management and government e-infrastructure, traffic control e-infrastructure, and Internet-driven e-infrastructure for diagnosis and repair of technical objects and computers.

Direction 3 is focused on the computing solution to humanity's problems:

(3-1) *Cloud management of humanity* means application of the God concept, where the God sees and pays on merit, to monitor and control the individual, social groups, and humanity as part of a cloud computing service, excluding a dependent person, as the most unreliable linked element, in the monitoring and control of cyber physical, biological, technical, technological, and social processes and phenomena. These terms allow optimal management by every individual, social groups, and states, and elimination of all the negative effects of human intervention, including wars, social conflicts, corruption, and injustice. The economic effect from the implementation of the concept is measured in tens of trillions of dollars.

(3-2) *Global e-infrastructure* relates to reliable and secure access to the cyber physical space of the planet and the development of appropriate legislation ensuring legitimate online execution of functional responsibilities not linked to the workplace. The conditions allow a global scale to reduce transport traffic, petrol consumption, and carbon dioxide emissions by 20%, thanks to the implementation of the functional responsibilities of employees in the home community. Online work for the IT industry should be seen as an essential step toward the creation of a green planet. It involves employees performing their jobs, not tied to a defined workplace. It has the following advantages over onsite operation: saving companies money due to not creating workplaces and infrastructure maintenance (light, heat, Internet, desks, and chairs); time and cost savings associated with driving from home to the workplace and back again; global saving of time and money on the scale of the city, the country, and the planet by reducing traffic, fuel consumption, and harmful emissions; regional development of cyber infrastructure involving villagers in remote productive process execution; and online work providing villagers with an equal opportunity to earn good money. Work on the result is the main argument for total implementation of online work from the side of companies; constructive employees actually work 24–7 and they are available to the company to communicate with colleagues through e-infrastructure (phone, Skype, e-mail, e-document management, video conferencing); almost all the leading companies in Silicon Valley have introduced this work form. The results of the creative work of employees should be metrically evaluated during the reporting period what is a conditions for the legalization of online work; Awards, moral compensation, and material compensation would be given only for actual design achievements.

(3-3) *Free choice of e-citizenship* for all inhabitants of the planet and elimination of paper-based information from all spheres of human activity is the next step in the green strategy impact. This will allow us to save at least 20% of the world's forests from being cut down to make paper, which will increase the oxygen content in the world and improve its environment. E-citizenship will reduce the migration of citizens and make all countries more attractive places to live in, thanks to the real danger of the leaders of backward states ending up with empty countries due to the possibility of electronic movement of all citizens, with their taxes, to more successful states.

Direction 4 connects with artificial intelligence development and global cyber brain creation:

(4-1) *Big data–based cloud computing* will leverage a planetary network of data centers to solve the problems of information, services, and personal data security, and will eliminate millions of desktops and local servers through use of cost-effective energy-saving tablets to access services and personal virtual offices. Brain–computer interfaces will make it possible to eliminate multiple input devices and go directly to remote image–impulse communication between the human brain and computing terminals. Brain scanning based on external or built-in sensors can prevent criminal or illegitimate human actions that significantly impact the efficiency of the police and special intelligence services.

(4-2) *Smart interaction of things* performs functions in online digital monitoring processes and phenomena in cyber physical network structures for precise and high-speed management of cars, airplanes, robots, and social groups in real time. In addition, smart interaction will be a useful technology for precise monitoring and management of industrial, scientific, and educational processes, sports, and everyday human activities.

(4-3) *Self-learning and self-evolution*: self-creation of an intelligent monitoring and management system in the environment will include allowing it to reproduce itself in the best form for metrical estimation of the effectiveness of functioning by using the integral criteria: time, money, and quality.

Implementation of these attractors allows movement of all the components of the control mechanisms of worldwide processes and phenomena to cloud computing services, which will free up a substantial portion (20%) of humanity from execution of unproductive management processes. To give control to cloud computing means a significant reduction of overhead costs and a way to create a green planet (with a high quality of human life plus a green environment).

The cost of implementing these attractors is about US$50 billion; the money savings from these actions will be no less than US$50 trillion, together with the abovementioned improvement in the quality of every person's life and the renaissance of ecology and a green planet.

All technical problems for further implementation of the above 12 trends have been solved in one degree or another. The main obstacle to the creation of a cyber physical space and attainment of a happy life is the low level of human cyberculture. Therefore, the reachability of human-focused goals is rigidly connected to the raising of people's trust in the reliability, equity, validity, and integrity of computing service monitoring and management.

Three historical periods—called Reflection, Control, and Creation—intersected with four directions—identified as single, network, data, and global computing—present sustainable development of cyber computing in space and time (Fig. 1.2).

There are three smart levels of computing structures:

1. *Reflection*, or only monitoring physical processes and phenomena, represented by single desktop computing, network computing and servers, database computing, and global Internet computing.
2. *Control* and management of physical processes and phenomena based on e-infrastructure, digital monitoring, cloud computing, cyber physical systems and networks, big data computing, and IoT. At the level of the present time, the main actors are the edge gadget, laptop, data center, big data, and cloud service.
3. *Creation* of intelligent cyber physical processes and phenomena under the control of the global cyber brain represented by the following components: brain computing, robotic networks, smart data computing, and the Internet of

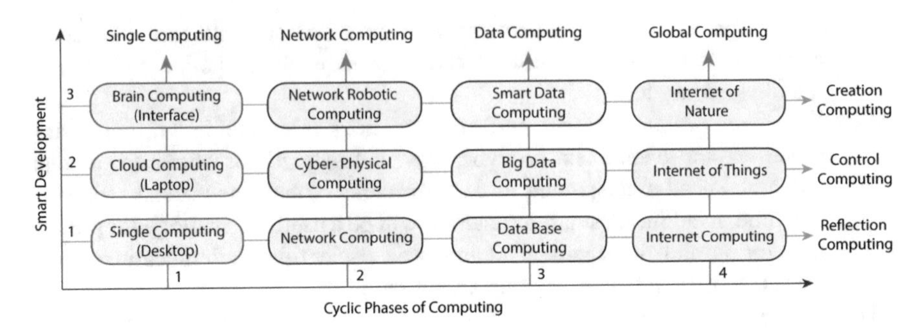

Fig. 1.2 Three levels of computing

Nature. Massive quantum–atomic computing, the brain–computer interface, atomic data center networks, and smart big data networks are expected to appear.

Cyber human computing means the time is not far away when the human brain will be directly connected with cyberspace. Development of computing whose main function is optimal and reliable control of all processes and phenomena on the basis of precise digital monitoring without direct human intervention should be considered only in the interaction of the two worlds—the real and the virtual: (1) in instances where humans cannot competently manage certain aspects of the real world and need to create computing assistance; (2) in instances where, as a perfect mechanism, computing can take over a huge controlling part of technological processes from humans; (3) in order to save humanity from self-destruction, with computing in the near future picking up the role of controlling parts of social processes under its jurisdiction; and (4) where humans and computing are united in the desire for creative change of the cyber physical continuum for peaceful coexistence and mutual combination to form a new concept called a cyberman that can freely leverage the global technology cyberculture directly connected via the brain–computer interface to the cyber physical space in order to optimize the performance of the cyber social role. Computing today is being transformed from an untrusted local structure to a permanent global substance. The human is gradually being transformed from a vulnerable biological subject to the first cyber biological form and then to cyber energy, as a permanent informational and physical reincarnation substance.

As a result of human interaction with the cyber physical world, a new concept called cyberculture has been formed, as a system of social and technological relations between society, the physical world, and cyberspace, with determined implementation of Internet services of digital monitoring and cloud control management in all processes and areas of human activity, including education, science, medicine, manufacturing, and transport, in order to improve human quality of life and protect the planet's ecosystems.

1.4 MAT Computing for Physical and Virtual Space

The goal is to create a general model of Memory–Address–Transaction (MAT) computing as a universal and scalable monitoring and management structure of physical, virtual, and cosmological processes and phenomena, which provides people with opportunities to discover new living space and create market-oriented products and cyber services aimed at improving the quality of human life and protecting the planet's ecosystems [2, 3, 11–15].

The objectives are (1) creation of a universal and scalable model of MAT computing, describing the physical and virtual processes of digital monitoring and exact control; (2) classification of existing computing technologies in real and virtual space to create an overall picture of the cyber physical continuum; (3) quantum process teleportation, as an info-technological expansion of humanity into space; and (4) cyber physical and quantum computing, as a new methodology for describing evolutionary processes of nature and society.

Further modifications have introduced some definitions, aimed in the direction of computing model simplification. Computing means technology of interactive monitoring and management of processes and phenomena in order to achieve the goal of a given program without direct human participation. The basis of computing is transactional interaction addressable data or memory components to achieve a predicted goal. The model of computing contains the components <Memory, Address, Transaction> (Fig. 1.3).

Memory is a substance that can store data or information. The *address* is a substance that determines the structure by the components' coordinates in the virtual space (Fig. 1.3). The computing structure is formed by various substances: gravity, intra- and interatomic interactions (Higgs bosons), electromagnetic fields, galvanic connections, human associations, and addressable interactions. *Transaction* is a goal-oriented process of implementation of the functionality or service based on data reception and transmission between the addressable components of

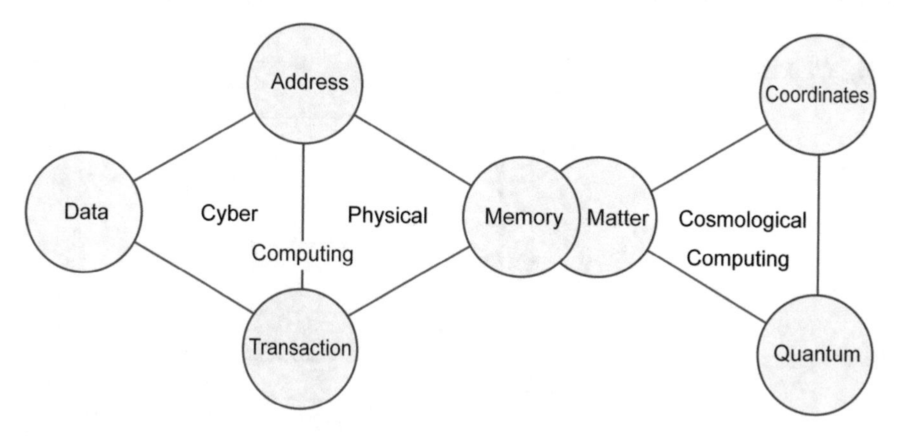

Fig. 1.3 Computing structures: physical, cyber, and cosmological

structured memory. Memory is any physical substance that can store, receive, and transmit data as a virtual entity. Since all matter is composed of atoms and electrons, each of the four forms of matter existence has memory properties.

The physical computing equation defines system symmetric *xor* interaction of three equivalent substances: (Memory – Address – Transaction) $M \oplus A \oplus T = 0$. This means that any component is determined by the interaction of the other two: $M = A \oplus T, A = T \oplus M, T = M \oplus A$. Virtual computing operates on a triad: (Data – Address – Transaction). The cosmological structure (Fig. 1.3) consists of the universal components, their coordinates, and quantum transactions. The universal characteristic computing equation operates on a single addressable transaction (read–write) of data between the source and the receiver [2, 3]:

$$M(Y_i) = Q_i[M(X_i)].$$

Here, $W = <M, Q, X, Y>$ is a structure that contains addressable memory components and nothing more: M is computing memory, Q is memory for quantum description of functions, and X and Y are memory for input and output variable addresses of functional primitives. Naturally, all the components and the memory cells are addressable. At the micro, macro, and cosmological levels, all the physical, social, and virtual processes and phenomena have their own addresses. This means that the processes, phenomena, and components can be replaced and repaired in remote and online mode. Faults in the memory are only the loss in the speed of transaction execution, meantime, any defect in reusable logic makes device to the out of operation. Thus, the computing process is reduced to the transaction on the addressable data in the space–time structure.

Figure 1.4 represents an automaton model of computing focused on interpretative simulation system-on-a-chip (SoC) components. The model can be implemented as a cloud service, so the input stimuli and output responses are associated with the big data of cyberspace. The structure contains only two components. The first (M) is represented by memory for storing all the states of digital devices, including input, internal, and output ones. The second (Q) is intended for

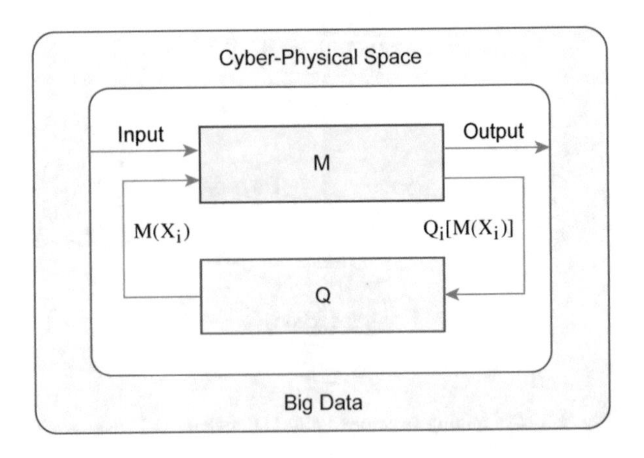

Fig. 1.4 MAT automaton

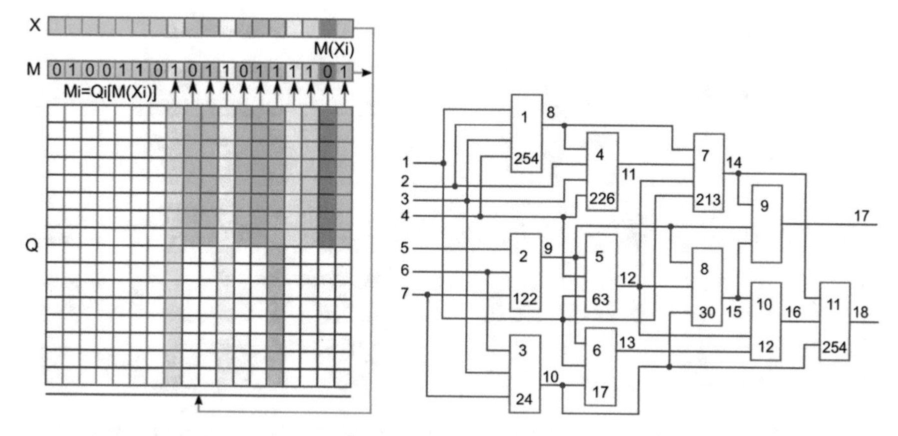

Fig. 1.5 Qubit data structures for circuit simulation

storing descriptions of the structural logical components of the circuit in the form of qubit coverage. The functioning of the model is reduced to two transactions $M(Y_i)$, $M(X_i)$, which carry out address reading and writing data into a structured memory component (M) through the transactional (address) transformation $Q_i[M(X_i)]$, performed by qubit coverage (Q).

Figure 1.5 shows the logic circuit example and data structures that correspond to the MAT automaton. The bottom number of the element corresponds to a decimal form of qubit vector of function. The mentioned data structures are focused on the high performance simulation because of parallel register operations and are very easy to implement in software/hardware cloud services.

The next step is introduced with some pictures connected to the computing leverage. There are seven attractors of computing (Fig. 1.6), without human control, but for people and the protection of the planet, represented in the structural tetrads: logic, structure, addressability, and transaction carrier.

(1) The physical one leverages reusable logic and addressable memory structures, and the carriers are the electrons and bits; (2) the virtual one leverages memory-based logic and addressable data structures, and the carriers are data and big data; (3) the technological one leverages sensor–actuator logic and addressable cyber physical structures, and the carriers are physical signals and data; (4) the biological one leverages the logic of the intellect and brain structures, addressable sensors–actuators, and the carriers are low current spikes; (5) the social one leverages the logic of legislations and cyber social addressable structures, and the carriers are legislation data; (6) the quantum one leverages quantum logic and electron–atom addressable structures, and the carriers are quanta or photons; and (7) the cosmological one leverages harmonic logic and cosmic addressable structures, and the carriers are gravitational and electromagnetic fields.

Thus, the computing picture that is shown allows continuing implementation of the MAT automaton in all fields of humanity's life.

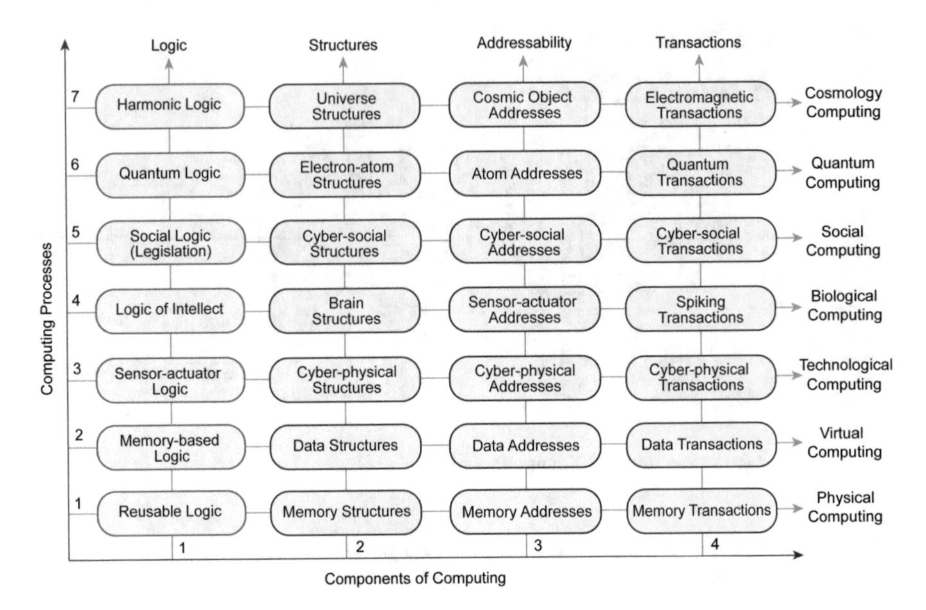

Fig. 1.6 Computing development

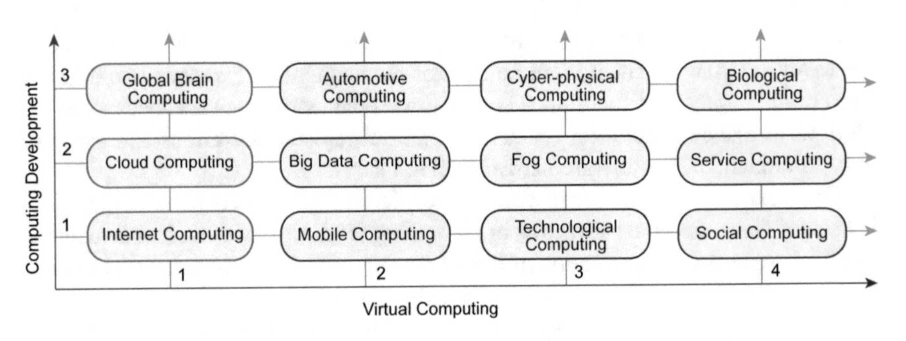

Fig. 1.7 Virtual or cyber computing

Virtual cyber computing manipulates data instead of memory, allocated in cyber space, nowadays presented by attractive structural levels and directions (Fig. 1.7):

1. Internet, mobile, technological, and social computing
2. Cloud, big data, fog network, and service computing
3. Global brain, automotive, cyber physical (IoT), and biological computing

Virtual computing gradually involves in its scope all the physical processes and phenomena that exist today in the world.

Physical computing (Fig. 1.8) is defined by development directions of functional units, concentrated in small and distributed areas of specialized solid-state devices:

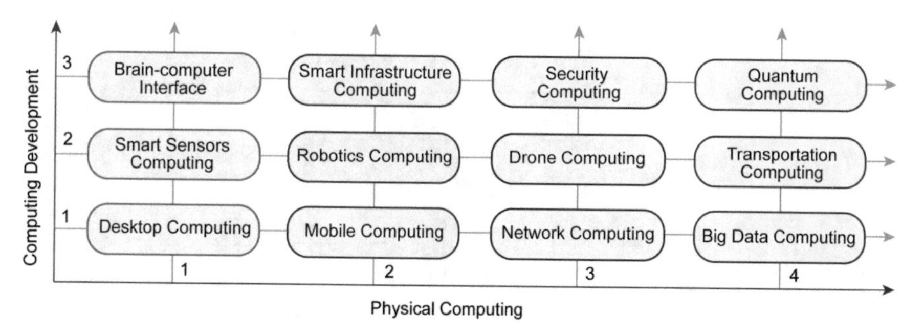

Fig. 1.8 Physical computing

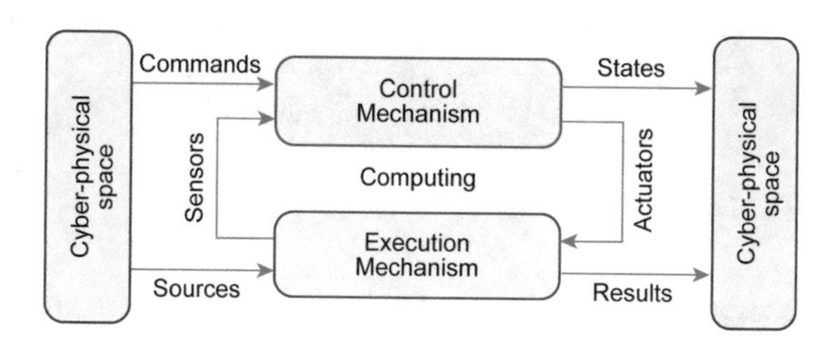

Fig. 1.9 Automaton structure of computing

1. Desktop, mobile, network, and big data center computing.
2. Smart sensor–actuator, robotic, drone, and transportation computing.
3. Brain–computer interface, smart infrastructure, security, and quantum computing. The main idea of solid computing is to create a cyber world of smart things focused on human-free monitoring and control of all processes and phenomena on Earth.

The types of computing, presented above, are covered by the general automaton structure for controlling the physical and virtual processes known to humanity (Fig. 1.9). Here, the main structural component of the computing (control and execution) appears to be memory, represented by any substance that can store information. The operation and development of the system are supported by monitoring and control components, which are represented by M signals of sensors and A signals of actuation. The inputs and outputs of the system are loaded on a common external virtual–physical (virical) continuum, as described by Josh Linkner of *Forbes*. Such a continual definition of cyber physical space applies not only to the world but also to the universe, which has its own program of evolution for the most simple and easy to understand machine computing model. The computing model adequately, perfectly, and simply describes all the processes of functioning and evolution of technical, biological, social, virtual, and cosmological objects and structures based on the use of monitoring and control signals. The automaton structure shown above receives the data and resources from a cyber

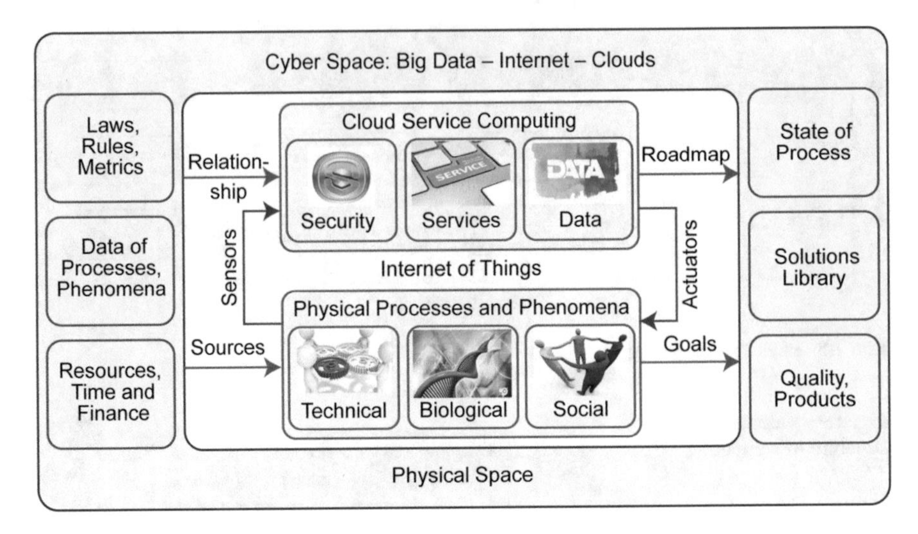

Fig. 1.10 IoT computing

ecosystem, which is also the place to put the products of particular computing. Therefore, such an automaton structure, depending on the purpose, could be a converter of the cyber physical ecosystem to a green planet for humans or a destroyer causing Armageddon.

Based on the automaton structure, IoT is considered an innovative technology for monitoring and control of the physical, biological, and social processes in the real world. IoT gives the opportunity to exclude, step by step, a human as the most unreliable link in the management chain (Fig. 1.10).

1.5 Computing of Quantum Teleportation

Quantum teleportation computing involves transmission across a distance not of objects, but of the relationship between the components of technological processes, which can be reconstructed at the receiving end of the telecommunication channel. Space aliens' appearance on Earth is not so unbelievable. Today confidence is maturing that humanity will be able to continue life on Earth-like planets in the universe by transmitting quantum information as an actuator streaming a description of the structure and algorithms of evolutionary synthesis and development of technical and biological substances. Coming to the favorable environment of some planet, the quantum-encoded genomes could prepare the infrastructure and then grow biological and technical objects in forced or evolutionary ways. The implementation of such disruptive human expansion into the appropriate area of space is real, even at the present level of science and technology. It is important to understand that space missions carrying biological or technical substances on board across interstellar distances would mean a "road to nowhere" for scientists' imagination. Thus, the extension of human life in space is real, and it would happen through the quantum teleportation processes of biotechnical object synthesis in an acceptable planetary environment. Beforehand, it is necessary to solve the technological problem of quantum actuation of forced or evolutionary synthesis of biotechnical substances in Earth conditions. It is easy to assume that more advanced civilizations have already addressed the problem of space expansion of their own life forms and have committed quantum teleportation of biotechnical facilities to Earth by photon delivery of synthesis algorithms into the Earth's environment.

The structural and algorithmic nature of object quantum teleportation is to deliver components' interconnection data to the space point where there are suitable chemical elements and a suitable environment. It is also necessary to bring algorithms that determine the optimal technology of some object structure production. For instance, in Mongolia, one can build a factory for the production of integrated circuits, which requires the company Intel to find one local expert in cyberspace, and after that to teleport to that country finance, culture, relationships, and technologies for the chip production. To create a world-class university (such as Cambridge, Stanford, or MIT), one needs to deliver to a problematic institution only a relationship from leaders of science and education. To create conditions favorable for the population of a country, one should teleport humane-oriented legislation or the Constitutions of the USA, Germany, Norway, and Singapore.

Thus, if there are exactly the same conditions and components, the efficiency and quality of system structures are determined only by relationships, as well as the presence of the same personnel, buildings, and purposes of the university, with system efficiency being defined by relationships (statutes, regulations, and traditions). All of the abovementioned is applicable to the success of cities and countries. It can also be added that in case of failure or errors in the system, one does not need to change the people. That is a stupid case, because humans are the same everywhere. To change the structure of relations between them is problem solving. "We are not looking for the people who made mistakes, but the situations where the mistakes have been possible" (Stanley Hyduke). The quantum or photon transaction of cosmological computing on gravity structures of matter in the forms of stars and planets, with addresses in space, expands the horizons of human life in the universe. The quantum structure of a biological genome penetrates space and creates new forms of life on other space worlds. Humanity is not destined to be forever on the Earth. Gravity, using the Higgs boson, creates structures, and photons make a transaction on them (Fig. 1.3, right). Structural elements of the universe have addresses, represented in terms of space–time coordinates, thanks to gravity. Higgs boson scientists now regard as a solution the bonding of the bricks of the universe in physical form and structure. Gravity is the attraction force generated by the masses of interacting cosmic bodies, creating a structural order in the universe in a short—by space standards—time interval. The Big Bang counteracts gravity, where the strength of the Big Bang is the cause of all the changes in the cycle of universe evolution until the next one.

1.6 Quantum Computing

A computer is the addressable memory structure for targeted transaction (read–write) data to achieve a goal. The universe as a computer consists of gravity-structured matter as memory, which implements the quantum transaction. Otherwise, the gravity-interacting matter structure is able to receive and emit energy flows of photons, which act as sensors and actuators of cosmological computing.

Sensors and actuators are an integral part of the natural and technical computing that performs monitoring and control functions not only in virtual cyberspace but also in the physical four-dimensional world. As part of the planet's cyber physical ecosystems, sensors and actuators, including mobile devices, form fog networks, which practically digitize all processes and phenomena: social, biological, technical, technological, energy, space, and time. Physically existing fog networks interact with cyber ecosystems that more and more are identified with cloud service computing, which forms the cyber brain of humanity. Global structuring of cyber physical space creates the Internet of Things (everything, anything, us) with a three-level hierarchy: cloud services, fog networks, and smart sensors and actuators (Sen-Act). Sen-Act devices are becoming nanoscalable and integrated with digital systems on crystals, which today have options: 10 billion gates on a chip, a 5-micron-thick plate, three-dimensional FinFET transistors, 10- to 2.5-nanometer-resolution connections and components, and p–n transition control with separate electrons. New solutions at the subatomic level for gyroscopes, temperature, acceleration, pressure, positioning, and navigation sensors are expected.

The main disadvantage of a classical computer is its inability to write zero and one signals at one space point. This disadvantage is eliminated by a quantum computer, which is capable of storing two binary states at one point in the Hilbert space, which has a similar physical interpretation. The electron has quantum uncertainty. This means that it can be simultaneously in two points of space. It is also true that the two electrons can simultaneously be at the same point of space. This is its unique advantage. In getting a photon or a quantum, the electron acquires a higher orbit, which can be interpreted as 1 unit. In passing out a photon, the electron falls below the level that can be identified with a zero value. It is hard to imagine a more compact computer than a quantum one that uses the energy level of the electron spin to encode binary states.

Given that all matter consists of atoms and electrons, humans can realize a quantum computer in any of four substances: liquid, solid, gas, or plasma. Plasma computer implementation is preferable because there is a cloud of free electrons in pure form separated from atoms, which need only be structured in a spatially stable computing environment. It would be a useful Higgs boson. Regarding the other three substances, the problem is to grow faster computers from atoms using algorithms based on waste-free additional molecular technology. The scalability of computer design provides an opportunity for the market to make nanocomputers at the cellular level and supercomputers for massive parallel transactions of big data in cyberspace with molecular energy costs. This is our most immediate future. With regard to the past, electrons and atoms are capable of storing information.

Otherwise, any form of matter has a data storage function due to the presence of atoms and electrons. Another thing is that the level of technology and the natural noise do not allow decoding the data about the past of humans, the Earth, and the universe. In addition, it should be recalled that the memory on the matter could be both long term and short term.

Matter and information as memory and data, in the real and virtual worlds, complement each other as a complete system and cannot exist without each other. Matter stores information, as its carrier. Information forms the genome or the evolution algorithm of the material world's components in time and space. The genome of the matter's evolution in time and space is encoded in the matter. The universe is the matter carrier of information, which specifies the algorithm of space development in time and space. The God idea is the pure material, because it is written in the form of matter. The problem of the primacy of matter or ideas, God or the material world, does not make sense, because they cannot exist without each other.

The God idea also puts on the mantle of an automaton computing model. There are sensors that inform the God about the state of the world in all its forms while driving toward the goal as it changes. There are electromagnetic actuators, which influence the destinies of people, cities, and countries, as well as all the Earth and space. There are rules, written and unwritten, on which lives the real world, moving toward the goal as it changes. There are also external sources to serve the execution mechanism of materials to achieve planned results. In fact, nature has chosen us as an actuator for the universe's discovery. But at the same time, we should not think that the God automaton model is a monopoly in terms of the execution mechanism. There are lots of such human mechanisms in the universe, which makes the universe system a multiverse and tolerant of natural and human errors. So nature or the universe gives us a chance to know the God's idea, which gives to humanity infinity in time and space.

The electron is a data carrier; the photon or quantum is a transactor. Each electron in the universe, as well as a point in space, has its own address, which is an order condition of the physical world, which should be used to create new computers based on anything, including the hydrogen in the universe. The future is hidden behind quantum computing close to nature, where the information unit would be identified with electron energy levels or atoms controlled by photons. In the universe, gravity creates the structure of the space computer, where the exchange of data between nano- and macromatter (which is equal to the memory of the cosmological computer) is executed via photons of electromagnetic radiation.

However, today we can leverage the advantage of quantum computing in classical computing processes. In this case, the entanglement is represented by two bits, or two points of physical space. A perfect demonstration of isomorphism between "quantum" data structures and the quantum mechanical notation of Dirac is Cantor algebra [2, 3]. Here, alphabet symbols $A = \{0, 1, X = \{0,1\}, U\}$ are encoded via binary vectors {10, 01, 11, 00}, respectively. The entanglement property is identified with the uncertainty of the state $X = \{0,1\}$, which sets at a mathematical point of space the existence of superposition 0 and 1 at the same time. The superposition property is connected with the logic *or* operations in terms of the Cantor alphabet or corresponding codes. The parallelism property is based on the performance of register operations (*and*, *or*, *not*) over the quantum data structures. The teleportation property uses negation or addition, when each of the two states knows everything about the other at any distance. The operation *xor* ($\bigoplus$) is also able to teleport the state of interrelated bits, because of the axioms $a \bigoplus b = 1, a \bigoplus b = 0$, from which follows $a = 1 \bigoplus b, a = b$. Why is the set theoretic definition of the Cantor alphabet "quantum"? There is an isomorphism between quantum Dirac notations and Cantor algebra, where both have a symbol for binary uncertainty [2, 3].

Qubit or Q coverage [2] is the binary vector form description of functionality, aimed at the parallel organization of computing processes by using the addressability of all primitives of the data structures. The vector qubit form can store the uncertainty as a discrete state, which is defined with parallel logic operations. With the improvement of quantum computer technology and performance, it is necessary to develop new quantum parallel programming methods of hardware–software algorithms for discrete optimization problem solving.

Thus, by using any of the four forms of matter existence, it is possible to build or grow a computer. To do this, it needs a suitable environment and the computer genome (structure and operation algorithm), which can be encoded by electron states or photons. It means that human travel in space is very simple if you send at light speed not matter but the human genome or any other technical or biological object encoded by photons. If such a genome arrives in a favorable environment, it elementarily actuates the growing of people, animals, rockets, computers, plants, or anything. Perhaps this once happened to the Earth's civilization. The paradox of a particle's split or cloning confirms the consistency of teleportation of anything with the help of photons, which remember the electron's condition, which can be restored on another planet. Therefore, it is possible to send life to the solar system's planets not by sending a robot or a person but by sending the photonic structure of the genomes focused on growing objects in a suitable environment. To do this, we just need to solve on Earth the substance synthesis process controlled by photon irradiation in a favorable physical environment.

1.7 Conclusion

The scientific novelty of the results is as follows: (1) world trends in cyber computing as a technological culture aimed at research, design, and implementation of computer systems, networks, and cloud–mobile services for digital monitoring, intelligent analysis of big data, and accurate social management of cyber physical and bioinformatics processes and phenomena in a scalable format of IoT are presented; (2) the automaton model of computing, which formulates and explains technology monitoring and control of processes and phenomena in physical, virtual, and cosmological space, is described; (3) verbal and structural definitions of the main types of computing based on current evolution trends of the planet cyber ecosystem are presented, and the universal computing model MAT (<Memory, Address, Transactions>), which uses three components to create any computational structure in technologically acceptable matter, is formulated; and (4) the directions of quantum human expansion into space and the possibility of similar penetration of alien biotechnical facilities into the Earth's ecosystem are shown, and a computing model that defines the optimal structure of the monitoring and control of scalable processes of different natures—technical, biological, social, virtual, and cosmological—are proposed.

The practical significance of the results is as follows: (1) provision of a methodology of human life expansion in a suitable space by teleportation of object information carried by electromagnetic quanta; (2) introduction of energy-saving additive technologies to create a planet cyber ecosystem through promotion of quantum computing in four forms of matter existence; (3) cyber computing execution of all monitoring systems and evolutionary control of social, physical, biological, and technical processes and phenomena in order to preserve the planet's ecosystems by human exceptions from organizational structures and decision-making technologies; and (4) reduction of computing to memory-based architecture, excluding reusable logic, in order to obtain reliable and online repairable products through addressability and programmability of memory components.

References

1. Hahanov, V. I., Bondarenko, M. F., & Litvinova, E. I. (2012). Structure of logic associative multiprocessor. *Automation and Remote Control. J, 10*, 73–94.
2. Hahanov, V. I., Gharibi, W., Litvinova, E. I., & Shkil, A. S. (2015). Qubit data structure of computing devices. *Electronic modeling. J, 1*, 76–99.
3. Hahanov, V. I., Amer, T. B., Chumachenko, S. V., & Litvinova, E. I. (2015). Qubit technology analysis and diagnosis of digital devices. *Electronic modeling. J, 37*(3), 17–40.
4. Hahanov, V., Litvinova, E., Gharibi, W., Chumachenko, S. (2015). *Big data driven cyber analytic system*. In: 2015 I.E. Int. Congress on Big Data, New York, 2015.
5. Dehorter, N., Ciceri, G., Bartolini, G., Lim, L., del Pino, I., & Marín, O. (2015). Tuning of fast-spiking interneuron properties by an activity-dependent transcriptional switch. *Science. J, 349* (6253), 1216–1220.
6. Merolla, P. A., et al. (2014). A million spiking-neuron integrated circuit with a scalable communication network and interface. *Science. J, 345*(6197), 668–673.
7. Standard 34.003-90 Automated systems. Terms and Definitions.
8. ISO/IEC 38500:2008. Corporate governance of information technology.
9. Bernard Marr. Forbes. 9 Attractors in the Computer Market. 17 'Internet Of Things' Facts Everyone Should Read. http://www.forbes.com/sites/bernardmarr/2015/10/27/17-mind-blowing-internet-of-things-facts-everyone-should-read/
10. Susan, G.F. IDC Releases Top Ten 2016 IT Market Predictions. http://www.forbes.com/sites/sap/2015/11/05/idc-releases-top-ten-2016-it-market-predictions/

11. Bondarenko, M.F., & Hahanov, V.I. (2012). Quantum technologies of brain-like computing structures of academician V.M. Glushkov. In: Proceedings XIX Conference on Automation Control. Kiev.
12. Palagin, A. V., Yakovlev, S., Tikhonov, B. M., & Pershko, I. M. (2005). Architectural and structural organization of computer-class “processor-in-memory”. *Mathematical Machines and systems. J, 3*, 3–16.
13. Palagin, A. V., & Yakovlev, Y. S. (2005). *System means integration of computer technology*. Vinnitsa: Universum.
14. Internet of Things. (2016). IoT Infrastructures. Mandler, B., Barja, J., Mitre Campista, M.E., Cagáňová, D., Chaouchi, H., Zeadally, S., Badra, M., Giordano, S., Fazio, M., Somov, A., Vieriu, R.-L. (Eds.), Second International Summit, IoT 360° 2015, Rome, Italy, October 27–29, 2015, Revised Selected Papers, Part II.
15. Yasuura, H., Kyung, C.-M., Liu, Y., & Lin, Y.-L. (2017). *Smart sensors at the IoT frontier*. Springer International Publishing, New York.

Chapter 2
Multiprocessor Architecture for Big Data Computing

Vladimir Hahanov, Wajeb Gharibi, Eugenia Litvinova, and Alexander Adamov

2.1 Introduction

The main goal is the development of a multiprocessor infrastructure of vector logic computing for searching, decision making, and pattern recognition when analyzing cyberspace using primitive logic operations: *and*, *or*, *not*, *xor*. The specialization of computer architectures, based on logical operations, significantly (×100) increases the performance of big data analysis. The market feasibility of the vector logical multiprocessor (VLMP) is determined by hundreds of logically computable tasks related to data retrieval, pattern recognition, and decision making: (1) synthesis, correction, and analysis of textual frameworks; (2) accurate recognition of images, sounds, and video data; (3) testing, diagnosis, and repair of software and hardware; (4) online testing of knowledge, skills, and habits; (5) comparative identification of a process or phenomenon for accurate decision making; (6) exhaustive information retrieval on the Internet according to a given parameter vector; (7) precise target designation of coordinates and piloting parameters for automatic landing of aircraft in digitized 3D space; and (8) online control of moving objects in time and space in order to avoid collisions. All listed tasks can be solved online using structured big data, structured in tables or matrices. A logical architecture (VLMP) is developed for these data in order to perform parallel searching, recognition, and decision making evaluated on the basis of the nonarithmetic quality criteria described below.

V. Hahanov (✉) • E. Litvinova • A. Adamov
Design Automation Department, Kharkov National University of Radio Electronics, Ukraine, Nauka Avenue 14, Kharkov 61166, Ukraine
e-mail: hahanov@icloud.com; litvinova_eugenia@icloud.com; alex.adamoff@gmail.com

W. Gharibi
Computer Science Department, Jazan University, KSA, Jizan, Saudi Arabia
e-mail: gharibi@jazanu.edu.sa

V. Hahanov, *Cyber Physical Computing for IoT-driven Services*,
DOI 10.1007/978-3-319-54825-8_2

The goal of this chapter is the development of architectural multiprocessor solutions of vector logical computing for a significant increase in the performance of searching, recognition, and decision making based on parallel implementation of register logical operations for the analysis of regular tabular structures of big data in cyberspace without the use of arithmetic instructions.

The goal is achieved through research and development of the following components: (1) creation of a nonarithmetic vector logical metric for evaluating processes and phenomena in cyberspace; (2) development of architectural solutions for the analysis of big data in cyberspace; and (3) creation of a vector logical multiprocessor for solving the problems of data retrieval, pattern recognition, and decision making.

An object for scientific research is the technological culture of data retrieval, discrete pattern recognition, and adequate decision making in cyberspace based on the use of register logical operations and computing architecture for metrical analysis of big data without the use of arithmetic operations.

The state of the art is represented by the following developments: vector logical structures of big data for solving computing problems [1–5]; software–hardware implementation of parallel register calculations for data retrieval, discrete pattern recognition, and adequate decision making in cyberspace [6–9]; and technologies, models, and methods of discrete analysis and synthesis of vector logical objects in cyberspace [10–15].

2.2 Nonarithmetic Metrics for Cyberspace Measurement

The structure of computational processes and the phenomena interacting on a discrete metric and represented by multivalued (binary) vectors of logical variables defines a discrete vector logical space (cyberspace), whose physical carriers are the computer data centers, systems, and networks.

A metric is an aproach of measuring distance in a space–time continuum between processes or phenomena specified by vectors of logical parameters or variables. The distance in cyberspace is determined by the *xor* operation or the symmetric difference between the vectors, which define processes or phenomena. The following concepts are isomorphic to the distance: the derivative (Boolean), the change, and the measure of the difference or proximity of processes or objects to each other. The methods of estimation, comparison, measurement, recognition, testing, diagnosis, and identification create techniques for determining the ratio in the presence of at least one object.

The element of a discrete space is given by the k-dimensional binary ordered set $a = (a_1, \ldots, a_j, \ldots, a_k), a_j \in \{0, 1\}$, where the vector coordinate is defined by the symbols 0 "false" and 1 "true." The concept of a null vector creates a k-dimensional set, where all the coordinates are zero: $a_j = 0, j = \overline{1, k}$.

Let's introduce the β metric of cyberspace, which is defined by the expression:

$$\beta = \overset{n}{\underset{i=1}{\oplus}} d_i = 0,$$

which creates a zero vector for the *xor* sum of the distances d_i between the finite number of objects closed in the cycle. Here the following designations are used: n is the integer number of distances between the space components creating the cycle $D = (d_1,\ldots,d_i,\ldots,d_n)$, or $d(a,b)$ is the distance vector in the cycle between the adjacent components a,b. In other words, the distance between two discrete objects (a and b) forms a vector: $d(a,b) = (a_j \oplus b_j)_l^k$. The norm of the vector value of distance is formed by the Hamming distance between two vectors, which is equal to the number of 1-units in $d(a,b)$. In other words, the β metric of a discrete space always forms the null vector of the *xor* sum of distances in a cycle containing a finite number of points or objects. Convolution of a discrete space into a zero vector is used in diagnosing and repairing digital systems, and also in transmitting data over communication channels. The introduced β metric creates a new definition of cyberspace, which is vector logical, where the *xor* sum of distances between a finite number of components in a cycle is equal to a zero vector. At first the presented metric leverages not elements of the set but the relations between them, which makes it possible to reduce the axioms of identity, symmetry, and transitive triangular closure to one axiom of convolution. It should be noted that the classical definition of the metric of the interaction of several points in a discrete logical space is a particular case of a β metric with $i = 1,2,3$:

$$M = \begin{cases} d_1 = 0 \leftrightarrow a = b; \\ d_1 \oplus d_2 = 0 \leftrightarrow d(a,b) = d(b,a); \\ d_1 \oplus d_2 \oplus d_3 = 0 \leftrightarrow d(a,b) \oplus d(b,c) = d(a,c). \end{cases}$$

Here the transitive closure is analogous to measuring the distance in a metric M space, where the points interact according to the rules:

$$M = \begin{cases} d(a,b) = 0 \leftrightarrow a = b; \\ d(a,b) = d(b,a); \\ d(a,b) + d(b,c) \geq d(a,c). \end{cases}$$

The disadvantage of the axiom of triangular closure of a metric space is that the third side of a triangle cannot be calculated by using the two known sides. A metric β eliminates this drawback; if two distances (sides) are known in the triangle, the third distance or side of it can be calculated:

$$d(a,b) \oplus d(b,c) = d(a,c) \to d(a,b) \oplus d(b,c) \oplus d(a,c) = 0.$$

Five points of space, for example (000111, 111000, 101010, 010101, 110011), form the sides/distances in the pentagon (111111, 010010, 111111, 100110, 110100), where *xor* addition of the last vectors gives the result (000000).

If the vector of a cyberspace component is defined in a multivalued alphabet, then the metric β forms a vector that is equal to $\varnothing$ in all coordinates, which is created based on the symmetric difference of distances between the points forming the cycle:

$$\beta = \mathop{\Delta}_{i=1}^{n} d_i = \varnothing.$$

Here the vector coordinate is defined in the alphabet realizing the power set in the universe of elements of the cardinality $p: a_j = \{\alpha_1, \ldots, \alpha_r, \ldots, \alpha_m\}, m = 2^p$.

The symmetric difference, equal to the symbol of the empty set in all coordinates, determines the equivalence of components or distances between objects. A single coordinate operation $d_{i,j} \Delta d_{i+l,j}$ applied for the four-valued Cantor model $A = \{0, 1, x, \varnothing\}, x = \{0, 1\}$ is given by the Δ table:

Δ	0	1	x	∅
0	∅	x	1	0
1	x	∅	0	1
x	1	0	∅	x
∅	0	1	x	∅

∩	0	1	x	∅
0	0	∅	0	∅
1	∅	1	1	∅
x	0	1	x	∅
∅	∅	∅	∅	∅

∪	0	1	x	∅
0	0	x	x	0
1	x	1	x	1
x	x	x	x	x
∅	0	1	x	∅

a	0	1	x	∅
ã	1	0	∅	x

Three other truth tables form set theoretic operations of intersection, union, and complement, which are useful for analyzing a digital system on a chip (SoC). The power of the alphabet of p primitive symbols is given by the expression $m = 2^p$. For the purposes of using the introduced metric of the cyber physical space, a system of proofs is proposed that specifies a sequential transition from the numerical characteristic of the binary relation of the components to the vector logical criterion for the relation quality of the objects and includes three scalar estimates of their interaction.

Explanations and definitions of the m and A interaction are represented below. The input vector $m = (m_1,\ldots,m_j,\ldots,m_k)$, $m_j \in \{0,1,x\}$ for the analyzed object $A = (A_1,\ldots,A_j,\ldots,A_k)$, $A_j \in \{0,1,x\}$ is determined, which is also described by the parameter vector of the dimension k. The degree of membership of the m vector to the object A is defined as $\mu(m \in A)$. There are five types of vector Δ interaction $m\Delta A$ between the input vector and the object, shown in Fig. 2.1. Fragments of the picture create variants of the interaction between an effective system for searching, pattern

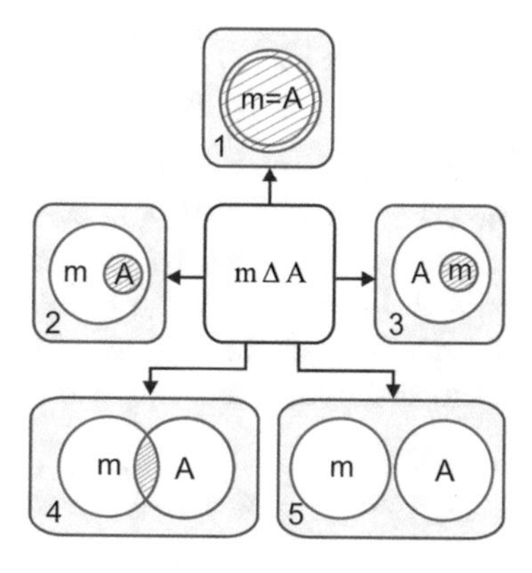

Fig. 2.1 Results of the two vectors' interaction

recognition, and decision making, and the input query vector. In technical diagnosis, the interaction of the test and the digital device is aimed at fault detection and repair.

The integral metric of query quality estimation is a function of the interaction of binary or multivalued vectors $m\Delta A$, which is formed by the average sum of three components: code distance $d(m,A)$, membership function $\mu(m \in A)$, and inverse membership function $\mu(A \in m)$:

$$
\begin{aligned}
&Q = \frac{1}{3}[d(m,A)] + \mu(m \in A) + \mu(A \in m)],\\
&d(m,A) = \frac{1}{n}[n - \text{card}[(i : m_i \cap A_i = \varnothing, i = 1,\ldots,k)]];\\
&\mu(m \in A) = 2^{c\text{-}a}; \mu(A \in m) = 2^{c\text{-}b}; a = \text{card}[(i : A_i = x, i = 1,\ldots,k)];\\
&b = \text{card}[(i : m_i = x, i = 1,\ldots,k)]; c = \text{card}[(i : m_i \cap A_i = x, i = 1,\ldots,k)];
\end{aligned}
$$

The intersection or union is represented by a coordinate set theoretic vector operation. Here and below, all set theoretic operations are defined in the Cantor alphabet $A = \{0,1,x = \{0,1,\varnothing\}$. The normalization of the vector parameters makes it possible to evaluate the interaction of objects in the interval [0,1]. Vectors are equal to each other if the maximum value of each parameter is equal to 1. In the case of a complete mismatch of vectors in n coordinates, a minimal estimate is formed, equal to $Q = 0$. If the intersection is equal to half the power of the vector space A and $m \cap A = m$, then the membership and quality functions are equal:

$$\mu(m \in A) = \frac{1}{2}; \mu(A \in m) = 1; d(m,A) = 1; Q(m,A) = \frac{5}{2 \times 3} = \frac{5}{6}.$$

The parameter Q will have the same value if $m \cap A = A$ and the intersection power is equal to half the power of the vector m space. The entire space of the vector is determined by the function of the number of coordinates ω equal to $x = \{0,1\}$: $q = 2^{\omega}$. If the power of the coordinate-wise intersection card($m \cap A$) is equal to half the powers of the spaces of the components A and m, then the membership functions are determined by the values:

$$\mu(m \in A) = \frac{1}{2}; \mu(A \in m) = \frac{1}{2}; d(m,A) = 1; Q(m,A) = \frac{1}{3} \times \left(\frac{1}{2} + \frac{1}{2} + 1\right) = \frac{2}{3}.$$

If the intersection of two vectors is equal to $\exists i(m_i \cap A_i) = \emptyset$, then the number of common points when intersecting two virtual objects is zero.

The motivation for defining a vector logic criterion for the quality of interaction is to increase the performance of computing processes when calculating the estimate Q of the interaction of the vectors m and A in the analysis of big data by using parallel vector register operations. The previously introduced criterion for the interaction quality Q can be written in the form:

$$Q = d(m,A) + \mu(m \in A) + \mu(A \in m),$$
$$d(m,A) = \text{card}(\{i : m_i \oplus A_i = U, i = 1, \ldots, k\});$$
$$\mu(m \in A) = \text{card}(\{i : A_i = U, i = 1, \ldots, k\}) - \text{card}(\{i : m_i \oplus A_i = U, i = 1, \ldots, k\});$$
$$\mu(A \in m) = \text{card}(\{i : m_i = U, i = 1, \ldots, k\}) - \text{card}(\{i : m_i \oplus A_i = U, i = 1, \ldots, k\});$$
$$U = \begin{cases} 1 \leftarrow \{m_i, A_i\} \in \{0,1\}; \\ x \leftarrow \{m_i, A_i\} \in \{0,1,x\}. \end{cases}$$

The variable $U = 1$ regulates the calculation rules: if the vectors m and A are binary, then the calculations are performed according to the rules of $\oplus$ – operation; in other words, when the vectors m and A have coordinates in the ternary alphabet, then the parameter $U = x$ initiates the operation of the symmetric difference Δ. It should be noted that vector logical operations ($\wedge$, $\vee$, $\oplus$, $\neg$) are isomorphic to set theoretic instructions ($\cap$, $\cup$, Δ, ~). The first component in the formula defines the code distance between the k-dimensional vectors based on the use of the *xor* operation; components 2 and 3 of the formula are designed to calculate the degree of nonmembership of the conjunction to the power of the unit values of each of the two vectors. The quality criterion is zero if the vectors are equal. To exclude arithmetic operations in a vector logical criterion, a logical union of three estimates into one is performed:

$$
\begin{aligned}
&Q = d(m,A) \vee \mu(m \in A) \vee \mu(A \in m) = \\
&= (m \oplus A) \vee \left(A \wedge \overline{m \wedge A}\right) \vee \left(m \wedge \overline{m \wedge A}\right) = \\
&= (m \oplus A) \vee [A \wedge (\bar{m} \wedge \bar{A})] \vee [m \wedge \bar{m} \wedge \bar{A})] = \\
&= (m \oplus A) \vee [(A \wedge \bar{m}) \wedge (A \wedge \bar{A}) \vee (m \wedge \bar{m}) \wedge (m \wedge \bar{A})] = \\
&= [(A \wedge \bar{m}) \vee (m \wedge \bar{A}) \vee (A \wedge \bar{m}) \wedge (A \wedge \bar{A})] \vee (m \wedge \bar{m}) \vee (m \wedge \bar{A})] = \\
&= (A \wedge \bar{m}) \vee (m \wedge \bar{A}) \vee (A \wedge \bar{m}) \vee (A \wedge \bar{A})] \vee (m \wedge \bar{m}) \vee (m \wedge \bar{A})] = \\
&= m \oplus A.
\end{aligned}
$$

The dependency of the vector quality criterion on the power of the alphabet is defined by the formula:

$$
Q' = \begin{cases} m \oplus A \leftarrow \{m_i, A_i\} \in \{0,1\}; \\ m \Delta A \leftarrow \{m_i, A_i\} \in \{0,1,x\}. \end{cases}
$$

⊕	0	1
0	0	1
1	1	0

Δ	0	1	x	∅
0	∅	x	1	0
1	x	∅	0	1
x	1	0	∅	x
∅	0	1	x	∅

Three forms of the interaction of two objects in n-dimensional vector logic space define the quality criterion Q, which forms the dimension and two membership functions. The membership functions are equal to zero if the Hamming distance is equal to zero. In other words, the object interaction is estimated by the values of the membership functions. Increasing the number of units in the criterion decreases the quality of interaction in the corresponding Boolean coordinates. The quality criterion $Q = m \oplus A$ leverages register computing procedures for evaluating decisions based on big data analytics. Vector logical cyberspace should be free of metric distance and scalar quality criteria based on the use of arithmetic operations.

The comparison of the quality criteria by leveraging of the shift left bit crowding (*slc*) register for the n-bit binary vector [6] (Fig. 2.2) is performed in the 1 cycle.

Performing this register procedure forms the value of the quality criterion for vector interaction by the right 1-unit bit number of the crowded sequence of 1 units. Such a number is the proof of a good solution for officials, but not for a computer. The comparative scalar estimate of the binary relation of two or more objects or processes is important to the human-driven decision. Vector estimation is much more comprehensible for the computer that further decides which parameters of the criterion need to be improved. The interaction of two objects defined by the vectors $m = (110011001100)$, $A = (000011110101)$ forms the following quality criterion:

m	1	1	.	.	1	1	.	.	1	1	.	.
A	.	.	.	.	1	1	1	1	.	1	.	1
$Q^* = m \oplus A$	1	1	.	.	.	.	1	1	1	.	.	1
$Q(m,A)$	1	1	1	1	1	1						

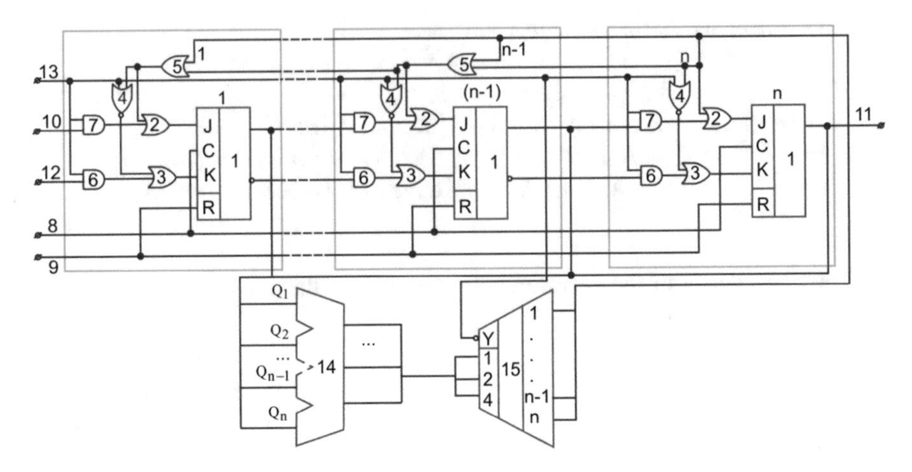

Fig. 2.2 Register for left bit-shifting and compacting

The 1-unit coordinates $Q^* = d(m, A) \vee \mu(m \in A) \vee \mu(A \in m)$ define the variables by which the interaction of vectors does not correspond to the vector logical quality criterion. The compression of the data for the determination of $Q(m{,}A)$ does not result in a loss of the vector estimate. Compression generates scalar criterion estimates for comparing several interactions. Where there are fewer units, there is a better solution. Not only do the table rows show the estimation of vector interaction that is equal to $Q(m, A) = (6/12)$ but also, most importantly, the 1-unit coordinates of the row $Q = d(m, A) \vee \mu(m \in A) \vee \mu(A \in m)$ identify all essential variables, for which there is local low-quality vector interaction.

The method for finding the best solution based on the minimum number of 1-unit coordinates from several alternatives is represented by the computational structure shown in Fig. 2.3 (*top*). The method is based on the execution of operations: (1) entry of unit values in Q for all coordinates of the vector and simultaneous shift left bit crowding in the vector Q_i; (2) comparison of the vectors Q and the current estimate Q_i from the list of solutions to be generated; (3) *and* operation on the contents of vectors ($Q \wedge Q_i$) and subsequent comparison of the result with the vector Q to be able to change it, if Q_i contains a smaller number of units; and (4) a search for the best solution cyclically repeated n times. The best choice between two solutions is presented at the bottom circuit of Fig. 2.3.

To compare the two solutions obtained by logical analysis, compressed quality vectors Q are used; after that the procedure including the following vector operations is performed:

$$Q(m, A) = \begin{cases} Q_1(m, A) \leftarrow \text{or}[Q_1(m, A) \wedge Q_2(m, A) \oplus Q_1(m, A)] = 0; \\ Q_2(m, A) \leftarrow \text{or}[Q_1(m, A) \wedge Q_2(m, A) \oplus Q_1(m, A)] = 1. \end{cases}$$

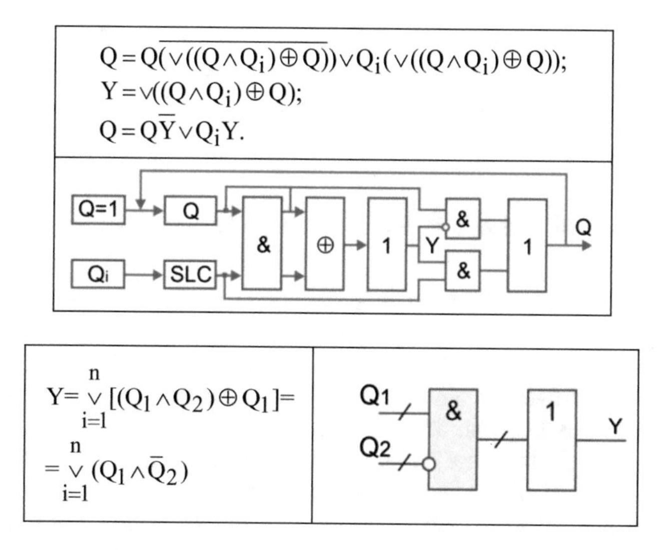

Fig. 2.3 Structures for searching for the best decision

The vector to bit called an *or* operator of devectorization determines a binary bit solution based on applying a logical *or* operation to *n* bits of an essential variable's vector quality criterion. A circuit design for the decision:

$$Q = \begin{cases} Q_1 \leftarrow Y = 0 \\ Q_2 \leftarrow Y = 1 \end{cases},$$

and math process model include three operations, shown in Fig. 2.3 (*bottom*) (Q_1, Q_2 are register variables).

For binary vectors, which have quality criteria, the procedure for choosing the best one based on the $Q(m,A)$ expression is presented below:

$Q_1(m, A) = (6,12)$	1 1 1 1 1 1 . . . . . .
$Q_2(m, A) = (8,12)$	1 1 1 1 1 1 1 1 . . . .
$Q_1(m,A) \wedge Q_2(m,A)$	1 1 1 1 1 1 . . . . . .
$Q_1(m, A) \oplus Q_1(m, A) \wedge Q_2(m, A)$	. . . .
$Q(m, A) = Q_1(m, A)$	1 1 1 1 1 1 . . . . . .

Vector logical criteria for the interaction quality for associative sets enable us to obtain an estimation of the search, pattern recognition, and decision making with high-speed parallel logic operations, which is especially important for critical real-time systems.

So, the scalar estimates are additional numerical information for the person making the decision. The vector criterion for evaluating the interaction of objects in cyberspace makes it possible to find information essential for decision making quickly and in real time, by using logical parallel operations, which is especially important for big data analysis.

2.3 Structural Solutions for Solving Search Problems

The quality metric introduced earlier is aimed at evaluating the interaction between processes and phenomena in cyberspace. One of the possible market-focused problems for which a vector logical criterion should be used is firing for a given target, which has five outcomes: (1) the missile strikes a commensurate target; (2) the unjustifiably large caliber of the missile hits the target, like firing a sparrow out of a cannon; (3) the large target is not hit by the small caliber of the missile; (4) the large-caliber missile partially hits the target; and (5) the launched missile flies past the target. The method for determining the interaction $P(m,A)$ is based on the use of the quality criterion of hit or miss, and also on the correct use of the weapon caliber. The structure of the method for selecting the quasioptimal interaction of the input query m and the set of logical relationships is the following:

$$\begin{array}{l}
P(m,A) = \min Q_i\left(m \mathop{\Delta}\limits_{i=1}^{n} A_i\right) = \vee\left[\left(Q_i \mathop{\wedge}\limits_{j=1,n}^{j\neq i} Q_j\right) \oplus Q_i\right] = 0; \\
Q(m,A) = (Q_1, Q_2, \ldots, Q_i, \ldots, Q_n); A = (A_1, A_2, \ldots, A_i, \ldots, A_n); \\
\Delta = \{\text{and}, \text{or}, \text{xor}, \text{not}, \text{slc}, \text{nop}\}; A_i = \left(A_{i1}, A_{i2}, \ldots, A_{ij}, \ldots, A_{is}\right); \\
A_{ij} = \left(A_{ij1}, A_{ij2} \ldots A_{ijr}, \ldots A_{msq}\right); m = \left(m_1, m_2, \ldots, m_r, \ldots, m_q\right). \\
Q_i = d(m, A_i) \vee \mu(m, \in A_i) \vee \mu(A_i \in m); d(m, A_i) = m \oplus A_i; \\
\mu(m, \in A_i) = A_i \wedge \overline{m \wedge A_i}; \mu(A_i \in m) = m \wedge \overline{m \wedge A_i}.
\end{array}$$

Here, functionality $P(m,A)$ is an analytical model of computation, which gives a minimum quality criterion; the structures of the input data are represented in the form of logically interconnected tables $A = (A_1,\ldots,A_i,\ldots,A_n)$; each of them is given by the rows of the associative table $A = (A_{il},\ldots,A_{ij},\ldots,A_{is})$ of explicit solutions. At that, the row vector $A_{ij} = (A_{ij1},\ldots,A_{ijr},\ldots,A_{nsq})$ is a true statement. The vector $A_{ij} = (A_{ij1},\ldots,A_{ijr},\ldots,A_{nsq})$ contains equivalent variables, which make them invariant for solving the problems of forward propagation and backward implication in $A_i \in A$. A vector A_{ij} is an explicit solution, where the variable is given in a discrete alphabet $A_{ijr} \in \{\alpha_1,\ldots,\alpha_i,\ldots,\alpha_k\} = \beta$. The relation $P(m,A)$ of the input vector query $m = (m_1,\ldots,m_r,\ldots,m_q)$ with the sequence $A = (A_1,\ldots,A_i,\ldots,A_n)$ determines the set of solutions with the choice of the best one by the minimum value of a quality criterion:

$$P(m,A) = \min Q_i[m \wedge (A_1 \vee A_2 \vee \ldots \vee A_i \vee \ldots \vee A_n)].$$

The interaction of a finite number of components among themselves implements the functionality $A = (A_1,\ldots,A_i,\ldots,A_n)$ that is formalized into structures: (1) an associative relational table containing all the explicit logical solutions—in this case, the maximum speed of parallel searching for a solution on the input request is achieved through the organization of associative memory, the implementation of which has high hardware complexity; (2) a graph structure of relationships between standard functional elements represented by truth tables on a smaller number of variables—in this case, the hardware costs are small and the performance of the analysis of the graph structures will be lower than a single associative table; and (3) achievement of a compromise solution by synthesis of a structure of relationships logically related in essential variables between the minimum numbers of primitives, which ensures high performance of the quasiparallel search for solutions for the minimum number of tables composing the graph, and also a low hardware cost—in this case, there is some decrease in the speed of sequential–parallel processing graph component tables. For example, splitting one table into k parts, as a rule, leads to a reduction in the hardware cost or the number of FPGA LUTs [9, 13]. It is known that a memory cell is created using four LUTs. The associative matrix is represented as a square with a side n; at that, the hardware costs $Z(k)$ for the data memory and the analysis time $T(k)$ of the logical graph depends on the number k of table partitions:

$$Z(k) = k \times \frac{1}{4} \times \left(\frac{n}{k}\right)^2 + h = \frac{n^2}{4 \times k} + h, (h = \{n, \text{const}\});$$
$$T(k) = \frac{4 \times k}{t_{\text{clk}}} + \frac{4}{t_{\text{clk}}} = \frac{4}{t_{\text{clk}}}(k+1), (t_{\text{clk}}\text{const}).$$

In this expression, h determines the cost of memory control. Reducing the hardware negatively affects the processing speed of the structural memory. The analysis period of one memory component is set by a cycle of four clock pulses. The number of partitions k increases the number of clock cycles in the consequent loop of the processing memory component of the graph. The ratio $\frac{4}{t_{\text{clk}}}$ specifies the time of data preparation at the input of the computing device and their decoding at the output. The dependence of hardware and time costs for analyzing the graph of memory components on the number of nodes has the form shown in Fig. 2.4.

The efficiency of the processing graph is determined by the number of structure nodes:

$$f[Z(k), T(k)] = Z(k) + T(k) = \left(\frac{n^2}{4 \times k} + h\right) + \left(\frac{4}{t_{\text{clk}}}(k+1)\right).$$

The expression makes it possible to determine the optimal partitioning of the relationship table [11]. For Fig. 2.4 the best partitioning creates the minimum of the function, determined by the value that turns the derivative of the function to zero: $n \times n = 600 \times 600$, $h = 200$, $t_{\text{clk}} = 4$; $k = 4$. The structure of the tables and the introduced quality criterion for the obtained solutions form the theoretical basis for synthesizing multiprocessor architectures of parallel execution of logical vector operations.

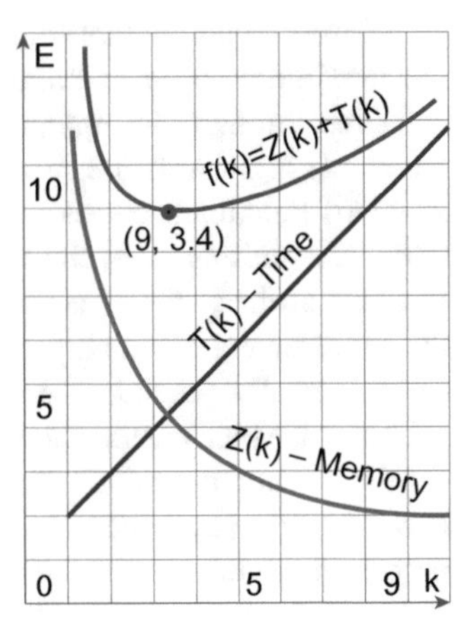

Fig. 2.4 Dependence of hardware and time on the number of partitions

2.4 Vector Logical Multiprocessor Architecture

The analysis of big data is based on the following technology: (1) leveraging of a universal computer for serial processing of information; (2) creation of a parallel PLD processor with a lower clock speed and development cost—here there is the opportunity of reprogramming, similar to software methods for solving analytical problems; and (3) integration of a high-performance ALU with the programmable flexibility of FPGA [8, 13]—a minimal set of instructions, with a simple multiprocessor architecture for parallel computations on a homogeneous logical data structure. The implementation of the multiprocessor in an ASIC chip ensures minimum power consumption, a high clock speed, and a low chip cost for large volumes of manufacturing. These advantages form the spherical architecture of the calculator (Fig. 2.5), consisting of 16 sequencers, each of which is connected to eight neighboring components. The analog of the processor structure is the PRUS device [7].

The main idea of a multiprocessor is parallel execution of simply implementable logical operations and minimization of the time of data transfer between neighboring processor elements. The compiler for the multiprocessor distributes the data between the processor elements, determines the time of the operation, and transfers the results between the components.

A vector logical multiprocessor (VLMP) is a spherical network of logical processors for parallel processing of big data and the minimum time of transactions between network components in the process of their analysis. Primitive multiprocessor circuitry is designed to process large information arrays quickly in hundreds of times, compared with the use of an universal processor.

The vector sequencer, as the base cell of the multiprocessor, is implemented in 200 gates, which means that 4096 cells can be implemented in an ASIC chip. Therefore, the compactness of the digital product has market feasibility for use in such areas of human activity as production management, cybersecurity, cyber physical systems, the Internet of Things (IoT), big data analytics, and recognition

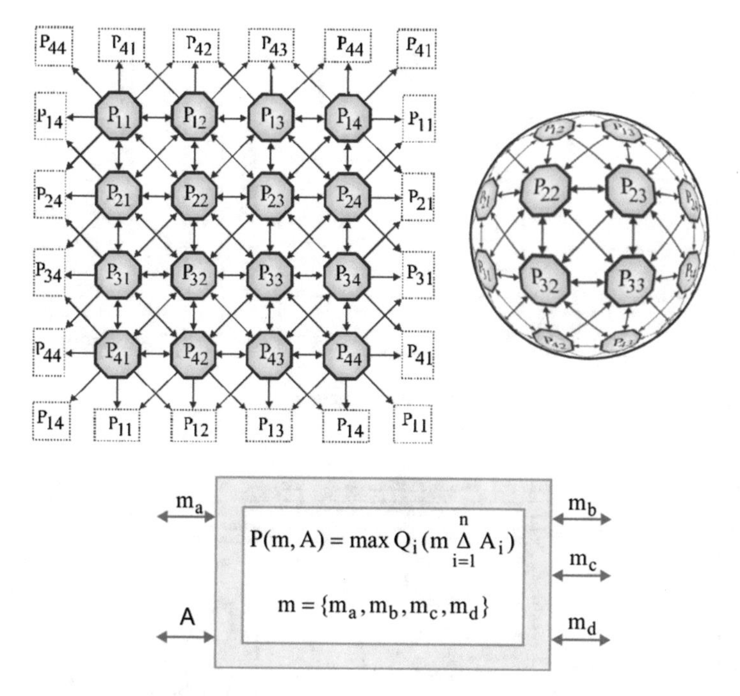

Fig. 2.5 Multiprocessor architecture

and encryption of patterns, sound, and speech. Making or formation of an exact decision based on the analysis of big data using a spherical architecture to perform vector logical operations is the main task of the multiprocessor:

$$P(m, A) = \min Q_i\left(m \mathop{\Delta}_{i=1}^{n} A_i\right), m = \{m_a, m_b, m_c, m_d\}.$$

The interface of the multiprocessor system is formed by the components $\{A, m_a, m_b, m_c, m_d\}$, which can be input and output. Therefore, the structural model of a multiprocessor can be used to perform forward propagation and backward implication in a discrete vector logical space. This is a difference from an automaton model of a computing device with uniquely assigned inputs and outputs. Registers $m = (m_a, m_b, m_c, m_d)$ are used to form a solution in a structure of buffer, input, and output vectors, and also to identify the response quality to the input request.

One of the possible variants of the VLMP architecture is shown in Fig. 2.6. The main matrix component $P = [P_{ij}], (i.j = \overline{1,4})$ contains 16 processors for executing five logical instructions on the data of table A of the dimension $(m \times n)$.

The interface provides data exchange and downloads instructions for processing data in command memory. The control unit starts execution of instructions for data processing and synchronizes the operation of the multiprocessor components. The IP module [8] is designed for servicing all components, fault diagnosis, and repair

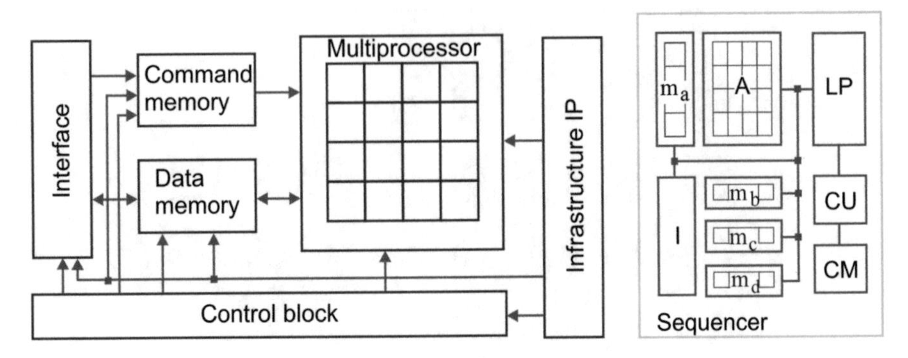

Fig. 2.6 Architecture of VLMP and sequencer structure

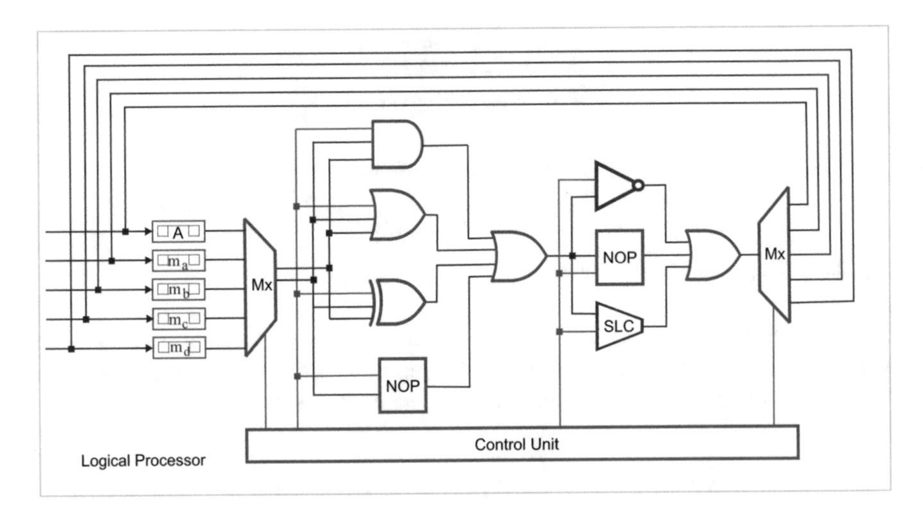

Fig. 2.7 Architecture of a logical processor

of the device. The logical sequencer (Fig. 2.6) is a special processor, containing the logical module LP, the memory matrix A for parallel execution of the basic operations, the vector module m for parallel processing of rows and columns of the A matrix and data exchange, instruction memory for a program of processing information, a control unit CU for logical operations, and a sequencer communication interface I with other VLMP modules.

The *LP* module (Fig. 2.7) implements operations (*and*, *or*, *not*, *xor*, *slc*) for performing algorithms of searching and estimating solutions. The LP block contains a multiplexer for matching operands with a vector logical operator. The result is stored in one of the four operands defined by the address through the multiplexer.

The logical processor contains three binary (*and*, *or*, *xor*) and two unary (*not*, *slc*) instructions. The last ones are performed in the processing cycle of the register data, by additionally selecting one of the three operations *not*, *slc*, *nop*. The execution of only a unary operation is accompanied by the choice of *nop*, which means the data transfers through the repeater to one of the unary operations. Operations in the LP are register or register–matrix. The last ones are necessary for the analysis of the table rows when forming a response to the input m query. The logic calculation block forms the following sequences of operations and operands:

$$C = \begin{cases} \{m_a, m_b, m_c, m_d\}\Delta A_i; \\ \{m_a, m_b, m_c, m_d\}\Delta\{m_a, m_b, m_c, m_d\}; \\ \{\text{not}, \text{nop}, \text{slc}\}\{m_a, m_b, m_c, m_d, A_i\}. \end{cases}$$
$$\Delta = \{\text{and}, \text{or}, \text{xor}\}.$$

The implementation of vector operations of a logical sequencer in a programmable logic chip Virtex 4 (Xilinx) has a computational complexity of 2400 equivalent gates operating at a clock frequency of 100 MHz. The Verilog language design environment was used.

2.5 Vector Logical Analysis Infrastructure

The infrastructure is defined as a set of models, architectural solutions, data structures, and tools for the description, analysis, synthesis, and testing to ensure the functioning quality of a digital product. The model is a set of interrelated components in space and time, describing the process or phenomenon with a specified adequacy to achieve a metrical quality of a solution with given constraints. The limitations are hardware costs, time to market, and yield. A binary logical vector in the discrete Boolean space of essential parameters defines the decision estimation metric. The control and operational automata specify the model of a digital device. Functionality is based on the use of hierarchical technology for creating digital systems with local synchronization of modules and global asynchrony of the entire device [13].

To further detail the architecture of the vector logical processor and sequencers, analytical and structural process models for analyzing the A matrix are further proposed. The first one (Fig. 2.8) defines the set of solutions for the input query m_b, and the second one (Fig. 2.9) is focused to search for the optimal solution among the set of rows found by using the first model.

The second model can also be leveraged separately for fault detection in the digital system.

All the operations executed in both process models are vector. The row analysis architecture (see Fig. 2.8) creates a vector m_a for identifying admissible $m_{ai} = 1$ or conflicting $m_{ai} = 0$ solutions for the input condition m_b per n clock cycles of processing m-bit vectors of the table $A = \text{card}(m \times n)$. The quality of the solution is determined for each interaction of the input vector m_b and the table rows $A_i \in A$ in

$$m_{ai}^{m} = \bar{\vee}\,[(m_b \mathop{\wedge}_{i=1}^{n} A_i) \oplus m_b];$$

$$A_i = (m_b \mathop{\wedge}_{i=1}^{n} A_i).$$

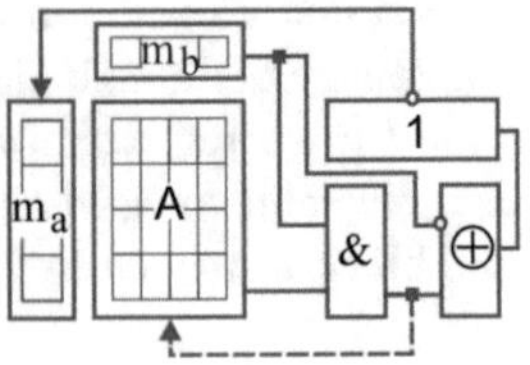

Fig. 2.8 Search for acceptable solutions

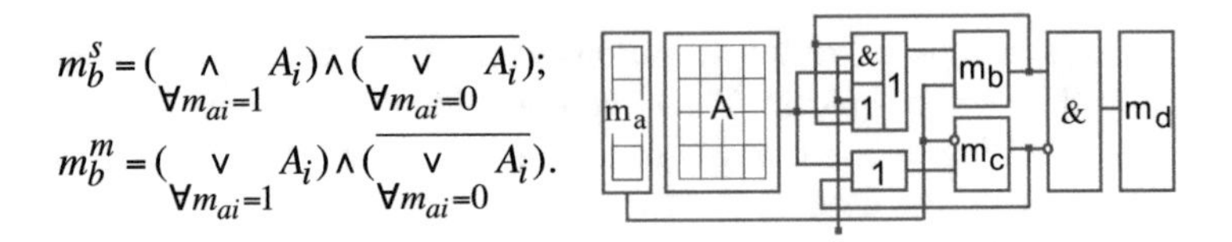

Fig. 2.9 Search for optimal solutions

the disjunction block. The matrix A can be simplified by intersecting it with the input vector based on the logical operation $A_i = \left(m_b \wedge_{i=1}^{n} A_i\right)$ to exclude from the A table inessential for the solution coordinates and vectors marked with a 1 value in the vector m_a.

The method for diagnosing single or multiple faults based on the analysis of rows in the table (see Fig. 2.9) uses a binary output response vector m_a, masking a fault detection table A. The components m_b and m_c are necessary for the accumulation of the results of conjunction and disjunction operations. After that, the logical subtraction of the content of the vector m_c from the register m_b is implemented, and the result is written to the register m_d.

The second equation forms a multiple faults diagnosis, where the element *and* is replaced by the function *or*. To choose the mode, the corresponding variable is used: single or multiple faults. The input condition is formed by a vector m_a that controls the choice of the logical operations (*and*, *or*) for analyzing unit $A_i(m_{ai} = 1) \in A$ or zero $A_i(m_{ai} = 0) \in A$ rows of the A table. Accumulation of unit and zero solutions in the registers A_1, A_0 is performed in n cycles. The vectors of one and zero signals initiate the specified registers: $A_1 = 1$, $A_0 = 0$. Analysis of n rows of the A table is performed per n clock. Then the conjunction operation of the register A_1 and inverted bits of the register A_0 is executed to form a vector m_b, where unit coordinates correspond to the columns of the fault detection table of the digital device, identified by the numbers of faulty blocks to be repaired.

When servicing functional components using the structure of vector logical analysis, it is possible to solve optimization repair tasks of a memory matrix, for example, which has spare rows and columns. To do this, it is necessary to solve the problem of quasioptimal coverage of all the detected faults by the spare

Fig. 2.10 Structure of the method for searching quasioptimal coverage

$$\begin{cases} m_b = (m_b \vee A_i); \\ m_{ai} = \bigvee_{i=1}^{n} [(m_b \vee A_i) \wedge \bar{m}_b]. \end{cases}$$

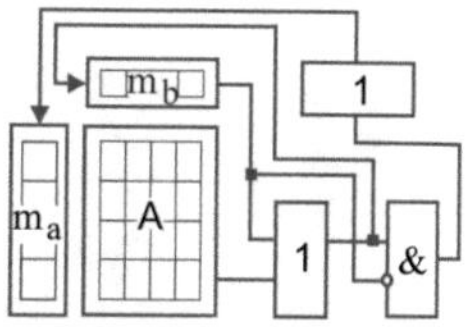

cells. A simple circuit method for obtaining quasioptimal coverage is shown in Fig. 2.10. It has the following advantages: (1) the computational complexity of its implementation depends on the number of vector operations or table rows, $Z = n$; (2) minimum hardware costs are required to store the table and two vectors (m_b, m_a) of intermediate coverage and also to obtain unit coordinates of the result corresponding to the rows of the table, which define the desired coverage; (3) the coverage method does not leverage a traditional division into the search for a kernel and complements; and (4) the block diagram of the solution does not use bit procedures for the cells of rows and columns. The disadvantage is not always the optimal solution that is compensated for by the productivity of vector operations (Fig. 2.10).

The reduction operation at the last stage transforms the binary vector m_a into *a* bit m_{ai} through the use of the *or* operation $m_{ai} = \vee[(m_b \vee A_i) \wedge \bar{m}_b]$. The reduction operation has the following format: <binary operation > <vector>: $\vee A_i$, $\wedge m, \overline{\wedge}(m \vee A_i)$. The inverse vectorization operation is designed to concatenate Boolean variables: $m_a(a, b, c, d, e, f, g, h)$. When searching the coverage, the vectors are zeroed $m_b = 0$, $m_a = 0$. The coverage is accumulated in n cycles in the vector m_a by a successive shift. Bits entered in m_a are formed by the *or* primitive of the reduction after analyzing the result $[(m_b \vee A_i) \wedge \bar{m}_b]$ for the presence of units.

The functional completeness of the diagnostic cycle consists of the repair of faulty memory cells [8] after obtaining quasioptimal coverage of all faulty matrix coordinates by spare cells. The size of the memory module (13 × 15 cells) does not significantly affect the computational complexity of covering ten faulty cells by spare rows and columns (Fig. 2.11).

The optimization problem is solved through constructing a coverage matrix (see Fig. 2.11) of faulty cells ($F_{2,2}$, $F_{2,5}$, $F_{2,8}$, $F_{4,3}$, $F_{5,5}$, $F_{5,8}$, $F_{7,2}$, $F_{8,5}$, $F_{9,3}$, $F_{9,7}$) by spare resources (C_2, C_3, C_5, C_7, C_8, R_3, R_4, R_5, R_7, R_8, R_9), respectively. Columns and rows identify spare components to repair faulty cells. The architecture (Fig. 2.9) allows selection $m_a = \boxed{1111100000}$ or coverage $R = \{C_2, C_3, C_5, C_7, C_8\}$ from three minimal solutions $R = C_2, C_3, C_5, C_7, C_8, \vee C_2, C_3, C_5, C_8, R_9 \vee C_2, C_5, C_8, R_4, R_9$ obtained for repair.

The embedded infrastructure for memory testing, diagnosis, and repair (Fig. 2.12) includes the following components: (1) testing unit under test (UUT) based on the use of the model under test (MUT) to generate a vector of output states m_a, the dimension of which is equal to the length of the test; (2) fault detection using the table *A*; (3) searching for the optimal coverage of faulty cells by spare rows and

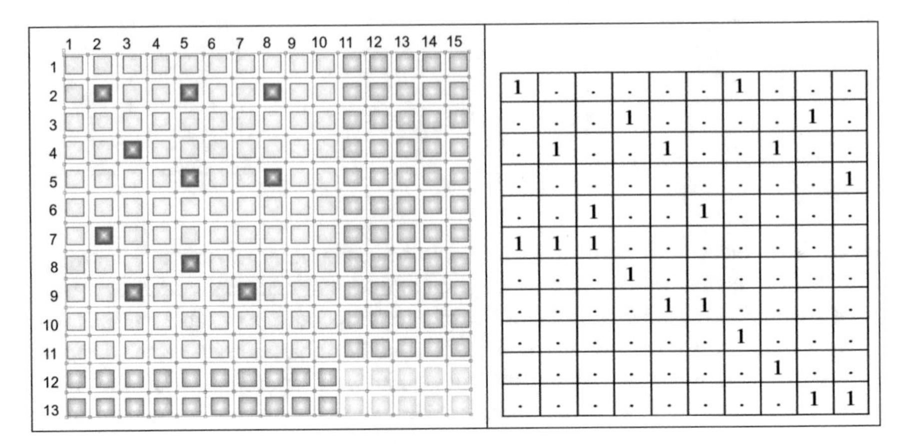

1	.	.	.	.	.	1	.	.	.
.	.	.	1	.	.	.	.	1	.
.	1	.	.	1	.	.	1	.	.
.	.	.	.	.	.	.	.	.	1
.	.	1	.	.	1	.	.	.	.
1	1	1	.	.	.	.	.	.	.
.	.	.	1	.	.	.	.	.	.
.	.	.	.	1	1	.	.	.	.
.	.	.	.	.	.	1	.	.	.
.	.	.	.	.	.	.	1	.	.
.	.	.	.	.	.	.	.	1	1

Fig. 2.11 Memory with spare cells and a coverage table

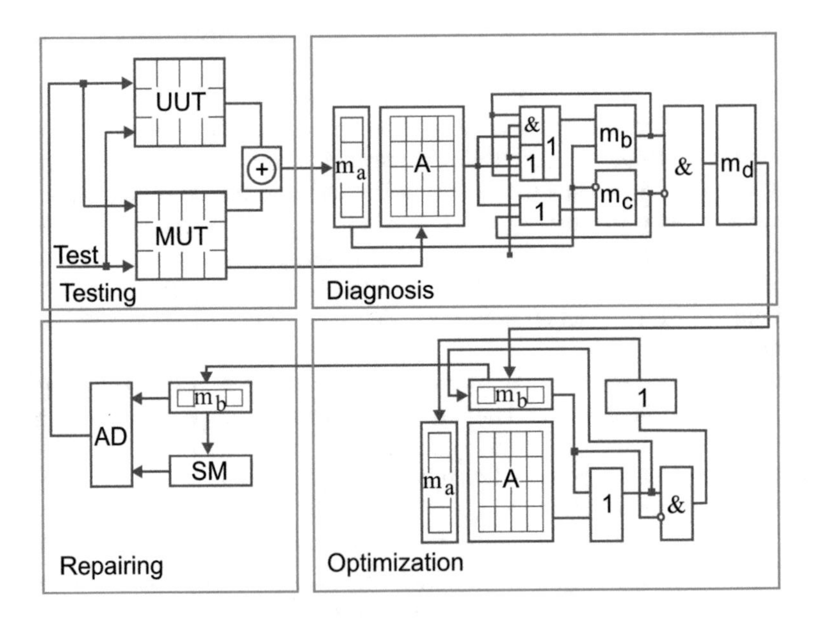

Fig. 2.12 Memory testing and repair architecture

columns using the results of the analysis of the table *A*; and (4) online repair of the memory through readdressing faulty rows and columns (m_a) fulfilled in the address decoder (AD) to the components of the spare memory (SM) [8].

The method of built-in online testing, diagnosis, and repair provides the operability of the digital SoC. The architecture and method can be implemented as a cloud service for computing systems. The proposed methods for analyzing

$$E = F(L,T,H) = \min[\frac{1}{3}(L+T+H)],$$

$$Y = (1-P)^n;$$

$$L = 1 - Y^{(1-k)} = 1-(1-P)^{n(1-k)};$$

$$T = \frac{(1\text{-}k)\times H^s}{H^s + H^a}; \; H = \frac{H^a}{H^s + H^a}.$$

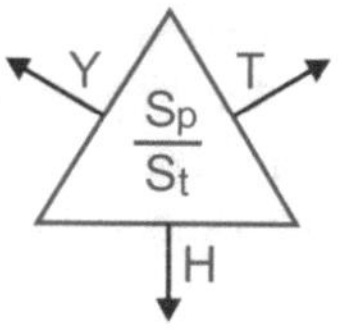

Fig. 2.13 Effectiveness of the prototype

associative tables, and also the metric quality criteria for analytical solutions, make it possible to obtain quasioptimal coverage of faulty software and hardware blocks by fault-free spare components. The proposed method of vector logical computations by using regular matrix structures of big data is the basis for the multiprocessor architecture of searching, pattern recognition, and decision making.

The efficiency of the computational architecture that meets the specialization conditions S_p and the standardization of S_t (Fig. 2.13) is determined by the minimum average of three conflicting parameters: the prototype error level L, the verification and testing time T, and the hardware and software infrastructure redundancy H.

The parameter L is a complement to Y; it determines the yield, depending on the testability of the project k, the probability P for the existence of faulty components, and the number of undetected faults n. The verification time depends on the testability of the project k and the structural complexity of the useful functionality of the digital product divided by the overall complexity of the project in the programming code lines or transistors. Infrastructural redundancy is formed by the dimension of the assertion engine and the boundary scan units divided by the overall complexity of the project. Assertion or scanning redundancy is designed to provide a given depth of online fault diagnosis for the time specified in the specification.

2.6 Conclusion

Software applications are weakly focused on parallel associative–logical technologies for searching, pattern recognition, and decision making in discrete cyberspace [4, 5, 15]. Such applications use a universal system of sequential execution of instructions of an expensive processor with a mathematical coprocessor. Existing hardware specialized logical analysis tools [1, 6, 7] are focused on bit-wise or nonvector processing of information. The innovative approach of parallel vector logic processing of big data eliminates arithmetic operations, which increases the performance and reduces the hardware complexity. The architectural solutions for implementing the described models and methods use a multiprocessor homogeneous structure and a vector logical metric to measure processes and phenomena in cyberspace. The following points determine the scientific novelty:

1. An innovative nonarithmetical metric for measuring the interaction of discrete processes and phenomena in cyberspace, defined by binary or multivalued vectors of essential variables.
2. The method of cyber metric analysis of matrix forms of big data based on parallel vector logical operations and the complete exclusion of arithmetic commands for searching, pattern recognition, decision making, and estimation in cyberspace.
3. A vector logic method for built-in diagnosis of digital SoCs and subsequent repair through defining quasioptimal coverage of faulty components by spares.
4. Practical significance determined by the development of an infrastructure for memory diagnosis and repair in a digital SoC. Further studies are related to the creation of a prototype of a vector logical multiprocessor for the synthesis and analysis of digital systems, searching, pattern recognition, and decision making in cyberspace based on the use of parallel logical operations.

References

1. Bondarenko, M. F. (2004). Brain-like computers. *Radielectronics & Informatics Journal, 2*, 89–105.
2. Cohen A.A. (2004). Addressing architecture for brain-like massively parallel computers. In: Euromicro symposium on digital system design (DSD'04).
3. Kuznetsov, O. P. (1998). Fast brain processes and image processing. *News of artificial intelligence Journal*, 2.
4. Vasiliev, S. N., Zherlov, A. K., Fedosov, E. A., & Fedunov, B. E. (2000). *Intellectual control of dynamic systems*. Moscow: Physics and mathematics.
5. Lipaev V.V. (2006). *Software engineering. Methodological basis*. Textbook. Moscow, Theis.
6. Inventor's Certificate. No. 1439682. 22.07.88. Shift register. Kakurin N.Ya., Hahanov V.I., Loboda V.G., Kakurina A.N.
7. Hyduke, S. M., Hahanov, V. I., Obrizan, V. I., & Kamenyuka, E. A. (2004). Spherical multiprocessor PRUS for solving Boolean equations. *Radioelectronics and Informatics Journal, 4*(29), 107–116.
8. Hahanov, V. I., Litvinova, E. I., & Guz, O. A. (2009). *Digital system-on-chip design and test*. Kharkov: Novoye Slovo.
9. Hahanov, V. I., Hahanova, I. V., Litvinova, E. I., & Guz, O. A. (2010). *Digital system-on-chip design and verification*. Kharkov: Novoye Slovo.
10. Acritas, A. (1994). *Fundamental of computer algebra with applications*. Moscow: Mir.
11. Attetkov, A. V. (2003). *Optimization methods*. Moscow: Bauman Moscow State Technical University.

12. Abramovici, M., Breuer, M.A., Friedman, A.D. (1998). *Digital system testing and testable design*. Comp. Sc. Press, New York.
13. Densmore, D. (2006). A platform–based taxonomy for ESL design. *Design & Test of Computers Journal, 23*(5), 359–373.
14. Malyshenko, Y. V., & Chipulis, V. P. (1986). *Automation of diagnosing electronic devices*. Moscow: Energoatomizdat.
15. Trakhtengerts, E.A. (2009). *Computer methods for the implementation of economic and information management decisions*. SINTEG.

Chapter 3
Big Data Quantum Computing

Vladimir Hahanov, Eugenia Litvinova, Svetlana Chumachenko, Tetiana Soklakova, and Irina Hahanova

3.1 Criteria and Structures for Evaluating the Quality of Object Interaction in Cyberspace

The goal is to develop a universal criterion and structures for evaluating the quality of object interaction in deterministic cyberspace through the use of only logical operations to determine the distances between the processes and phenomena by developing high-performance parallel processors, focused on effectively solving the problems of searching, pattern recognition, and decision making [1–4]. The objectives are the following: (1) development of a universal analytical model to determine binary interaction between processes and phenomena in multivalued logic; (2) development of a nonarithmetic computing parallel structure for metric evaluation of deterministic processes and phenomena, and for making optimal decisions; and (3) examples of the use of computing with logic parallel processors to solve practical problems.

For further understanding of the material, it is necessary to introduce some assumptions and definitions. The input vector $m=(m_1, \ldots, m_j, \ldots, m_k)$, $m_j \in \{0, 1, x\}$ and the object under analysis $A=(A_1, \ldots, A_j, \ldots, A_k), A_j \in \{0, 1, x\}$, represented by a vector too, have the same dimension k. The grade of membership of the m vector to A is designated as $\mu(m \in A)$. There are five types of coordinate set theoretic Δ interaction of two vectors $m\Delta A$. They form all the primitive variants of the responses of the generalized system for searching, pattern recognition, and decision making on the input vector query. In the technological field of knowledge and technical diagnosis, the abovementioned sequence is isomorphic to the route:

V. Hahanov (✉) • E. Litvinova • S. Chumachenko • T. Soklakova • I. Hahanova
Design Automation Department, Kharkov National University of Radio Electronics, Ukraine, Nauka Avenue 14, Kharkov 61166, Ukraine
e-mail: hahanov@icloud.com; litvinova_eugenia@icloud.com; svetachumachenko@icloud.com; tetiana.soklakova@gmail.com; hahanova@mail.ru

V. Hahanov, *Cyber Physical Computing for IoT-driven Services*,
DOI 10.1007/978-3-319-54825-8_3

fault detection, diagnosis, and decision making for repair. These steps of the technological route require a metric for evaluating solutions to select the best option.

The intersection (union) of the vectors is a vector operation based on the corresponding coordinate set theoretic operations. The operations of coordinate-wise intersection and union are defined in the Cantor alphabet $A = \{0, 1, x = \{0, 1\}, \varnothing\}$. The normalization of the parameters allows evaluation of the interaction level of vectors in the numerical interval [0,1]. If necessary to define a maximum value of each parameter as equal to 1, it means the vectors are equal. The minimum evaluation, $Q = 0$, is fixed in the case of a complete mismatch of vectors in all n coordinates. If $m \cap A = m$ and the power of coordinate-wise intersection is equal to half of the power of space of the vector A, the functions of membership and quality are equal:

$$\mu(m \in A) = \frac{1}{2}; \mu(A \in m) = 1; d(m, A) = 1; Q(m, A) = \frac{5}{2 \times 3} = \frac{5}{6}.$$

If vectors m and A are binary at all coordinates, then the variable $U = 1$ and the calculations are performed by using the binary $\oplus$ operation. If vectors m and A are defined in the ternary alphabet, then the variable $U = x$ initiates the calculations, based on the set theoretic symmetric difference Δ operation. Introducing the variable U gives the opportunity to get away from writing two formulae for the criterion, depending on the valuedness of the alphabet for describing coordinates of the interacting vectors. Vector logical operations ($\wedge$, $\vee$, $\oplus$, $\neg$) are isomorphic to the set theoretic ones ($\cap$, $\cup$, Δ, ~). At that set theoretic coordinate, operations corresponding to the logical ones were previously defined in the multivalued Cantor alphabet.

The procedure for calculating the vector criterion depends on the valuedness of the alphabet:

$$Q' = \begin{cases} m \oplus A \leftarrow \{m_i, A_i\} \in \{0, 1\}; \\ m \Delta A \leftarrow \{m_i, A_i\} \in \{0, 1, x\}. \end{cases}$$

For the binary alphabet, the truth table of coordinate *xor* operation has the following form:

$$0 \oplus 0 = 0, \quad 0 \oplus 1 = 1, \quad 1 \oplus 0 = 1, \quad 1 \oplus 1 = 0.$$

In the second case, when the alphabet for describing the coordinates has three values, the calculation of the symmetric difference is carried out in accordance with the Δ-operation.

The quality criterion Q uniquely identifies three forms of interaction between any two objects in the n-dimensional vector logical space: the distance and two functions of membership. If the Hamming distance is not zero, the functions of membership are equal to zero, since in this case the spaces of two vectors are not intersected. Otherwise (the code distance is equal to zero) the interaction of objects is measured by the membership functions. Increasing the number of zeros improves the quality criterion, and increasing the number of units decreases the interaction quality by the respective Boolean variables. The quality criterion $Q = m \oplus A$ agrees with the abovementioned metric for estimating the distance or interaction of objects in vector logical space; it has a trivial computational procedure for evaluating the solutions related to the analysis and synthesis of information objects. In fact, a vector logical space must have no metric distance and numerical quality criteria, including arithmetic operations on scalar values.

To compare quality criteria, it is necessary to determine the number of 1 units in each vector without summation operations. This procedure leverages a shift register [5], which allows executing *slc* (shift left bit crowding) of 1-unit coordinates of the n-bit binary vector in a single clock cycle.

After the *slc* procedure, a number of the right 1 bit of the crowded series of 1s forms a value of the quality criterion for vector interaction. But, in this case, this number is necessary to humans for scalar evaluation of a binary relation as information for comparing the proposed infrastructure in the framework of existing technologies. In practical problems, such evaluation is not necessary for choosing a quasioptimal alternative. The vector evaluation is more convenient for the calculator, which determines the best solution without the participation of a human in this process. For the vectors $m = (110011001100)$, $A = (000011110101)$ the definition of the quality of their interaction by the formulae (8) is presented in the form of the following procedures (zero coordinates are marked by an empty space):

m=	1 1 0 0 1 1 0 0 1 1 0 0
A=	0 0 0 0 1 1 1 1 0 1 0 1
Q*=m⊕A	1 1 0 0 0 0 1 1 1 0 0 1
Q=slc(Q*)	1 1 1 1 1 1

The evaluation of the vector interaction is formed and, most importantly, the 1-unit coordinates of the row $Q^* = d(m,A) \vee \mu(m,A) \vee \mu(m \in A) \vee \mu(A \in m)$ identify all significant variables, for which vector interaction does not meet the quality criterion. The crowding procedure for obtaining $Q(m,A)$ does not mean the loss of informative vector evaluation $Q^* = m \oplus A$. The result of crowding allows choosing the best solution from two or more ones by comparing in parallel the total number of 1 units forming the scalar evaluations of the criteria necessary for a person's beliefs.

As for the formula of cyber choice, when considering practice-focused problems it is important to find the best solution in a finite number of interaction variants $m \oplus A_i$ for its subsequent implementation without arithmetic operations, which decrease the performance of decision making. Suppose, for example, there are two vectors A B, relative to which it is necessary to perform the operations $a = m \oplus A$, $b = m \oplus B$, to evaluate the distance between each of them to the input vector query m:

A= B=	1 0 1 0 0 1 0 1 0 1 0 0 1 0 1 1
m=	1 1 1 1 1 1 0 0
$\mu_a(m \in A) = m \oplus A$	0 1 0 1 1 0 0 1
$\mu_b(m \in B) = m \oplus B$	1 0 1 1 0 1 1 1
$a = slc(\mu_a)$	1 1 1 1 0 0 0 0
$b = slc(\mu_b)$	1 1 1 1 1 1 0 0

A= B=	0 0 0 1 0 0 0 1 0 1 0 0 1 1 0 0
m=	1 1 1 1 1 1 0 0
$\mu_a(m \in A) = m \oplus A$	1 1 1 0 1 1 0 1
$\mu_b(m \in B) = m \oplus B$	1 0 1 1 0 0 0 0
$a = slc(\mu_a)$	1 1 1 1 1 1 0 0
$b = slc(\mu_b)$	1 1 1 0 0 0 0 0

A= B=	1 0 1 0 0 1 1 1 0 1 0 0 1 0 0 1
m=	1 1 1 1 1 1 0 0
$\mu_a(m \in A) = m \oplus A$	0 1 0 1 1 0 1 1
$\mu_b(m \in B) = m \oplus B$	1 0 1 1 0 1 0 1
$a = slc(\mu_a)$	1 1 1 1 1 0 0 0
$b = slc(\mu_b)$	1 1 1 1 1 0 0 0

A technologically simple—and easy to understand and implement—structure for parallel computing of the better variant is proposed. The computational architecture is based on non-numerical logical comparison of two alternative vectors, called (a) and (b), obtained through the leverage of the shift left bit crowding *slc* [5]. After parallel shifting in a single cycle of all 1 units in the registers of the vector quality criterion, which evaluate the interaction of objects in cyberspace, in theory, there are three variants of the ratio of 1 units shown below (the interaction of the previously obtained vectors a and b):

a = b =	1 1 1 1 0 0 0 0 1 1 1 1 1 1 0 0
$a \wedge b$	1 1 1 1 0 0 0 0
$q^b = (a \wedge b) \oplus b$	0 0 0 0 1 1 0 0
$q^a = (a \wedge b) \oplus a$	0 0 0 0 0 0 0 0
$Q^b = \overset{k}{\underset{i=1}{\vee}} q_i^b = 1$	
$Q^a = \overset{k}{\underset{i=1}{\vee}} q_i^a = 0$	w i n n e r

a = b =	1 1 1 1 1 1 0 0 1 1 1 0 0 0 0 0
$a \wedge b$	1 1 1 0 0 0 0 0
$q^b = (a \wedge b) \oplus b$	0 0 0 0 0 0 0 0
$q^a = (a \wedge b) \oplus a$	0 0 0 1 1 1 0 0
$Q^b = \overset{k}{\underset{i=1}{\vee}} q_i^b = 0$	w i n n e r
$Q^a = \overset{k}{\underset{i=1}{\vee}} q_i^a = 1$	

a = b =	1 1 1 1 1 0 0 0 1 1 1 1 1 0 0 0
$a \wedge b$	1 1 1 1 1 0 0 0
$q^b = (a \wedge b) \oplus b$	0 0 0 0 0 0 0 0
$q^a = (a \wedge b) \oplus a$	0 0 0 0 0 0 0 0
$Q^b = \overset{k}{\underset{i=1}{\vee}} q_i^b = 0$	w i n n e r
$Q^a = \overset{k}{\underset{i=1}{\vee}} q_i^a = 0$	w i n n e r

Explanation: A zero value of the Q criterion means a better alternative for consideration, which is then used for comparison with other evaluations, or as a final solution to the problem. The logical structure for the implementation of cyber choice is as follows:

$$\left\langle \begin{array}{c} Q^b = \overset{k}{\underset{i=1}{\vee}} q_i^b \\ Q^a = \overset{k}{\underset{i=1}{\vee}} q_i^a \end{array} \right\rangle \leftarrow \left\langle \begin{array}{c} q^b = (a \wedge b) \oplus b \\ q^a = (a \wedge b) \oplus a \end{array} \right\rangle \leftarrow \langle a \wedge b \rangle \leftarrow \left\langle \begin{array}{c} a \\ b \end{array} \right\rangle \leftarrow \left\langle \begin{array}{c} a = \mathrm{slc}(\mu_a) \\ b = \mathrm{slc}(\mu_b) \end{array} \right\rangle$$

$$\leftarrow \left\langle \begin{array}{c} \mu_a(m \in A) = m \oplus A \\ \mu_b(m \in B) = m \oplus B \end{array} \right\rangle \leftarrow \left\langle \begin{array}{c} m \\ A \\ B \end{array} \right\rangle$$

Register variables (a, b) designated the vectors of crowded to the left 1-unit values are combined and inverted for the simultaneously executing *xor* operations. Then, the results in the form of register states are supplied to the inputs of two logic *or* elements, which form the states of two Boolean variables creating three combinations: 00, 01, 10. A zero value of one of the two variables is the best solution to be selected. Two zero states mean that both solutions are equivalent in terms of preference. A single combination of Boolean variables is not possible. A circuit

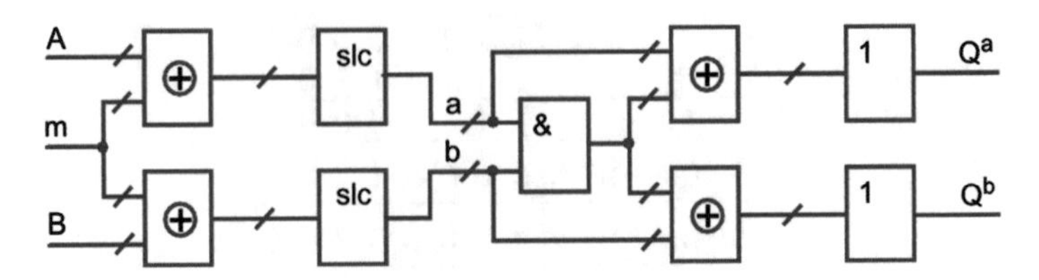

Fig. 3.1 Circuit implementation of the optimal solution

implementation of cyber choice of two alternatives, corresponding to the above logical structure, is shown in Fig. 3.1.

If the best choice of solutions is identified by logical signal 1 (instead of 0), which corresponds to the maximum value of the membership function defined earlier, then a circuit structure for determining the best variant of the two alternatives can be modified by the additional inverters on the outputs of the logic *or* elements, forming $\{Q^a, Q^b\} \in Q$. In this case, the following pairs of states are possible on the output of the digital structure: 10, 01, and 11, where decision making (*a* or *b*) is carried out by a 1-unit value of one of the outputs. The pair of signals 00 on external outputs of the circuit with inverters is impossible. Thus, an accurate search for the requested information in big data can be done based on only the logical operations *and*, *or*, *not*, *xor*, and *slc* without using arithmetic functions, which allows design of high-speed vector logical physical and/or virtual multiprocessors to significantly reduce the time of service execution in cloud applications. The computational procedure of choosing the best variant of the two possible ones is reduced to parallel execution of four vector operations and one logical, as a result of which in one or more outputs a zero value is generated; it identifies the best solution:

$$\begin{aligned} Q^a &= \vee_i \{[\mathrm{slc}(m \oplus A) \wedge \mathrm{slc}(m \oplus B)] \oplus A\}, \\ Q^b &= \vee_i \{[\mathrm{slc}(m \oplus A) \wedge \mathrm{slc}(m \oplus B)] \oplus B\}. \end{aligned}$$

The proposed discrete Boolean metric, vector quality criteria for *xor* interaction of the objects in cyberspace, and non-numerical ranking of the obtained criteria for selecting an object by query create the feasibility of market software–hardware implementation of the metric-driven computing structure in cloud services for analyzing big data.

A vector logical SIMD multiprocessor is characterized by the absence of arithmetic operations, parallel computing distances between the query and information components, and simultaneous determination of the best possible solution by a minimum value of the membership function, which makes it possible to significantly increase the speed of the most accurate data retrieval in big data.

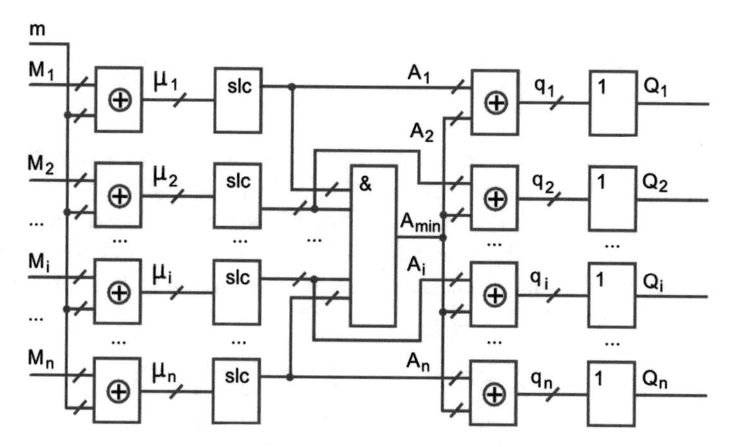

Fig. 3.2 Vector logical multiprocessor

The multiprocessor structure is shown in Fig. 3.2, which includes only logical primitives for performing Boolean and vector operations.

The processor operates as follows: vector query m, consisting of k bits, interacts by an *xor* function with matrix M having n lines or vectors. As a result of the *xor* operation, n membership functions are formed, which define the value of the Hamming distance between the query and each vector row of the matrix M. To estimate distances and to select the best (minimum) interaction, a register *slc* operation is performed, which crowds all 1 units to the left for 1 cycle and makes it possible to mark the minimum distance $m \oplus M_i$ by the last right 1-unit bit number. To determine the vector row number forming the minimum of the membership function, parallel bitwise logical multiplication is performed over all the vectors containing crowded to the left 1 values, which allows calculation of the $A_{\min}$ vector with the minimum number of 1 units. This vector is used to determine the number or index of the vector row of the matrix M, which has the best value of the membership function by performing a vector *xor* operation between $A_{\min}$ and all left shift crowded membership functions A_i $(i = 1, n)$. As a result, the vectors q_i $(i = 1, n)$ are generated, bits of which specify the input values for each of the n logic *or* elements. The output of each *or* element is $A_i \oplus A_{\min}$ equal to one, if there is at least one 1-unit value in the results of the comparison. If there are no such 1 units, the minimum distance between $m \oplus M_i$ is identified by a 0 state of one or possibly several outputs Q_i $(i = 1, n)$. An analytical model for finding the optimal solution in cyberspace by the query vector is based on five parallel logical operations performed sequentially:

$$Q^i = \bigvee_{j=\overline{1,k}} \left\{ \left[\bigwedge_{p=\overline{1,n}} \underset{s=\overline{1,n}}{\text{slc}} \left(m \underset{r=\overline{1,n}}{\oplus} M_i \right) \right] \underset{i=\overline{1,n}}{\oplus} M_i \right\}.$$

The structural model of a vector logical processor corresponding to the analytical model for obtaining an optimum solution is shown in Fig. 3.2.

Three examples illustrating the calculation of output states of a vector logical processor as a response of the matrix M to a query m, consisting of two row vectors, are shown below. The first phase is focused on the formation of vectors with shift left crowded 1 units: $\{A_1, A_2\} \in A$:

$M_1 =$	1 0 1 0 0 1 0 1
$M_2 =$	0 1 0 0 1 0 1 1
$m =$	1 1 1 1 1 1 0 0
$\mu_1(m \in M_1) = m \oplus M_1$	0 1 0 1 1 0 0 1
$\mu_2(m \in M_2) = m \oplus M_2$	1 0 1 1 0 1 1 1
$A_1 = \text{slc}(\mu_1)$	1 1 1 1 0 0 0 0
$A_2 = \text{slc}(\mu_2)$	1 1 1 1 1 1 0 0

$M_1 =$	0 0 0 1 0 0 0 1
$M_2 =$	0 1 0 0 1 1 0 0
$m =$	1 1 1 1 1 1 0 0
$\mu_1(m \in M_1) = m \oplus M_1$	1 1 1 0 1 1 0 1
$\mu_2(m \in M_2) = m \oplus M_2$	1 0 1 1 0 0 0 0
$A_1 = \text{slc}(\mu_1)$	1 1 1 1 1 1 0 0
$A_2 = \text{slc}(\mu_2)$	1 1 1 0 0 0 0 0

$M_1 =$	1 0 1 0 0 1 1 1
$M_2 =$	0 1 0 0 1 0 0 1
$m =$	1 1 1 1 1 1 0 0
$\mu_1(m \in M_1) = m \oplus M_1$	0 1 0 1 1 0 1 1
$\mu_2(m \in M_2) = m \oplus M_2$	1 0 1 1 0 1 0 1
$A_1 = \text{slc}(\mu_1)$	1 1 1 1 1 0 0 0
$A_2 = \text{slc}(\mu_2)$	1 1 1 1 1 0 0 0

The second phase illustrates the formation of the output states of the processor: $\{Q_1, Q_2\} \in Q$, where in the first and second cases there is only one "winner" with zero output, and in the third case both row vectors of the matrix M are the optimal solutions for the query m:

$A_1 =$ $A_2 =$	1 1 1 1 0 0 0 0 1 1 1 1 1 1 0 0
$A\min = M_1 \wedge M_2$	1 1 1 1 0 0 0 0
$q_2 = (M_1 \wedge M_2) \oplus M_2$	0 0 0 0 1 1 0 0
$q_1 = (M_1 \wedge M_2) \oplus M_1$	0 0 0 0 0 0 0 0
$Q_2 = \bigvee_{i=1}^{k} q_i^2 = 1$	
$Q_1 = \bigvee_{i=1}^{k} q_i^1 = 0$	w i n n e r

$A_1 =$ $A_2 =$	1 1 1 1 1 1 0 0 1 1 1 0 0 0 0 0
$A\min = M_1 \wedge M_2$	1 1 1 0 0 0 0 0
$q_2 = (M_1 \wedge M_2) \oplus M_2$	0 0 0 0 0 0 0 0
$q_1 = (M_1 \wedge M_2) \oplus M_1$	0 0 0 1 1 1 0 0
$Q_2 = \bigvee_{i=1}^{k} q_i^2 = 0$	w i n n e r
$Q_1 = \bigvee_{i=1}^{k} q_i^1 = 1$	

$A_1 =$ $A_2 =$	1 1 1 1 1 0 0 0 1 1 1 1 1 0 0 0
$A\min = M_1 \wedge M_2$	1 1 1 1 1 0 0 0
$q_2 = (M_1 \wedge M_2) \oplus M_2$	0 0 0 0 0 0 0 0
$q_1 = (M_1 \wedge M_2) \oplus M_1$	0 0 0 0 0 0 0 0
$Q_2 = \bigvee_{i=1}^{k} q_i^2 = 0$	w i n n e r
$Q_1 = \bigvee_{i=1}^{k} q_i^1 = 0$	w i n n e r

The interest is in the formation of queries in a multivalued alphabet (Cantor alphabet) for describing the variables of the interacting vectors. At first glance, there are problems of counting the distance between the query and the information component of cyberspace that is not binary encoded. But if the multivalued symbols of the power set of primitives involved in the formation of vectors are designated by binary codes ($0 = 10$, $1 = 01$, $X = 11$, $\varnothing = 00$), then the distance "query–component" can be evaluated by the above described procedure of shift left 1-unit crowding:

$M_1 =$	1 0 1 0 0 1 0 1
$M_2 =$	0 1 0 0 1 0 1 1
$m =$	1 X 1 1 X 1 X 0
$\mu_1(m \in M_1) = m \oplus M_1$	∅ 1 ∅ X 1 ∅ 1 X
$\mu_2(m \in M_2) = m \oplus M_2$	X 0 X X 0 X 0 X
$A_1 = slc(\mu_1)$	11 11 11 1 0 0 0 0
$A_2 = slc(\mu_2)$	11 11 11 11 11 11 1 0

$M_1 =$	0 0 0 0 0 0 0 1
$M_2 =$	0 1 1 1 1 1 0 0
$m =$	X X 1 1 X 1 X X
$\mu_1(m \in M_1) = m \oplus M_1$	1 1 X X 1 X 1 0
$\mu_2(m \in M_2) = m \oplus M_2$	1 0 ∅ ∅ 0 ∅ 1 1
$A_1 = slc(\mu_1)$	11 11 11 11 11 11 1 0
$A_2 = slc(\mu_2)$	11 11 1 0 0 0 0 0

$M_1 =$	X 0 1 X 0 X 1 X
$M_2 =$	0 X X 0 X 0 X 1
$m =$	X X 1 1 X X 0 0
$\mu_1(m \in M_1) = m \oplus M_1$	∅ 1 ∅ 0 1 ∅ X 1
$\mu_2(m \in M_2) = m \oplus M_2$	1 ∅ 0 X ∅ 1 1 X
$A_1 = slc(\mu_1)$	11 11 11 0 0 0 0 0
$A_2 = slc(\mu_2)$	11 11 11 11 0 0 0 0

The winners in the three above categories, represented in the last three tables, are vectors, respectively M_1, M_2, M_1, having a minimum number of 1 units, or the maximum number of zero vector coordinates. Thus, there are no fundamental constraints for evaluating the interaction of objects in cyberspace through the use of non-numerical metrics, eliminating arithmetic operations. Moreover, all the distances in the virtual world can be measured using *xor* operation or symmetric difference, which provide a choice of the best solution, based on the vector logical criteria for the interaction quality.

The following two examples, represented by two tables, illustrate the logical operation of a multiprocessor in a multivalued alphabet for describing logical variables, focused on the competence ranking of students. Suppose there is a group of students who pass eight exams in the metric A, B, C, D, which are encoded by the corresponding vectors: 1000, 1100, 1110, 1111. It is necessary to determine the best student who integrally gets the maximum score per session. The calculation results are presented below:

$M_1 =$	A	B	D	D	B	A	C	C
$M_2 =$	C	C	D	A	B	B	A	D
$M_3 =$	B	C	C	B	A	A	C	D
$m =$	A	B	C	A	A	A	A	C
$\mu_1(m \in M_1) = m \oplus M_1$	0000	0000	0001	0111	0100	0000	0110	0000
$\mu_2(m \in M_2) = m \oplus M_2$	0110	0010	0001	0000	0100	0100	0000	0001
$\mu_3(m \in M_3) = m \oplus M_3$	0100	0010	0000	0100	0000	0000	0110	0001
$A_1 = slc(\mu_1)$	1111	111						
$A_2 = slc(\mu_2)$	1111	111						
$A_3 = slc(\mu_3)$	1111	11						
$M_3 =$	w	i	n	n	e	r		

$M_1 =$	A	B	D	D	B	A	C	C
$M_2 =$	C	C	D	A	B	B	A	D
$M_3 =$	B	C	C	B	A	A	C	D
$m =$	A	A	A	A	A	A	A	A
$\mu_1(m \in M_1) = m \oplus M_1$	0000	0100	0111	0111	0100	0000	0110	0110
$\mu_2(m \in M_2) = m \oplus M_2$	0110	0110	0111	0000	0100	0100	0000	0111
$\mu_3(m \in M_3) = m \oplus M_3$	0100	0110	0110	0100	0000	0000	0110	0111
$A_1 = slc(\mu_1)$	1111	1111	1111	1				
$A_2 = slc(\mu_2)$	1111	1111	1111	1				
$A_3 = slc(\mu_3)$	1111	1111	111					
$M_3 =$	w	i	n	n	e	r		

Here in the left table the reference vector m is reduced to the best actual scores obtained by the students for each exam. The right table operates by the reference vector with the theoretically best possible values ($A = 1000$) of knowledge testing. In both cases the integral quality criterion for the session determines the best student with the number 3. To encode all five grades of the Bolognese metric for evaluating knowledge, we propose to use a zero combination too: $A = 0000$, $B = 1000$, $C = 1100$, $D = 1110$, $E = 1111$. In this case, the next two tables give results similar to the previous results to select the best student in the session:

$M_1 =$	A	B	D	D	B	A	C	C
$M_2 =$	C	C	D	A	B	B	A	D
$M_3 =$	B	C	C	B	A	A	C	D
$m =$	A	B	C	A	A	A	A	C
$\mu_1(m \in M_1) = m \oplus M_1$	0000	0000	0010	1110	1000	0000	1100	0000
$\mu_2(m \in M_2) = m \oplus M_2$	1100	0100	0010	0000	1000	1000	0000	0010
$\mu_3(m \in M_3) = m \oplus M_3$	1000	0100	0000	1000	0000	0000	1100	0100
$A_1 = slc(\mu_1)$	1111	111						
$A_2 = slc(\mu_2)$	1111	111						
$A_3 = slc(\mu_3)$	1111	11						
$M_3 =$	w	i	n	n	e	r		

$M_1 =$	A	B	D	D	B	A	C	C
$M_2 =$	C	C	D	A	B	B	A	D
$M_3 =$	B	C	C	B	A	A	C	D
$m =$	A	A	A	A	A	A	A	A
$\mu_1(m \in M_1) = m \oplus M_1$	0000	1000	1110	1110	1000	0000	1100	1100
$\mu_2(m \in M_2) = m \oplus M_2$	1100	1100	1110	0000	1000	1000	0000	1110
$\mu_3(m \in M_3) = m \oplus M_3$	1000	1100	1100	1000	0000	0000	1100	1110
$A_1 = slc(\mu_1)$	1111	1111	1111					
$A_2 = slc(\mu_2)$	1111	1111	1111					
$A_3 = slc(\mu_3)$	1111	1111	111					
$M_3 =$	w	i	n	n	e	r		

If the primitives of the scores are not equivalent by weight in the metric of comparison, they cannot be represented by unitary codes, which perform only the role of element identifiers in the universal set of equivalent primitives. Therefore, to designate the weights in the codes of scores, the factor of the number of units is used. However, the subsequent actions, focused on defining the integral quality criterion of the students' knowledge to select the best one by comparison with the ideal result, are not based on the use of any arithmetic operations, and use only the logical functions. The strategic objective of the linear computational complexity has been formulated as follows: to find a student who has a minimum distance to the well-known ideal result in the form of a reference vector m. An alternative strategy involves a search for the best of the n students by sequentially comparing each of them with each other, which will take much more time because the computational complexity of such a procedure is determined by the equation $(n^2/2) - n$. To rank all the students regarding the ideal result, it is necessary to perform $(n^2/2) - n$ vector logical operations. After determining at each step the best student, the corresponding row of the matrix M (the examination competence of a student group) should be eliminated from further consideration by filling 1 units in the bits of the vector.

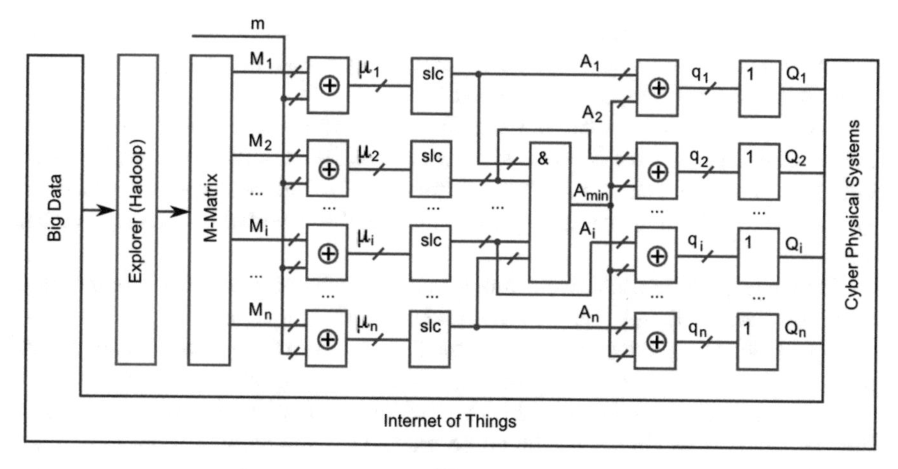

Fig. 3.3 Interaction between the multiprocessor and cyberspace

For efficient operation of the logic multiprocessor, it is necessary to generate the *M* matrix (Fig. 3.3) of possible solutions of the problem, which, in particular, may be the result of usage of the search engine Google (Hadoop) in the cyberspace Internet (big data), used for crude and large extracts, when the number of found information components is in the hundreds or thousands of variants. Then comes the turn of a multiprocessor operation, forming an exact solution to the query m, which should be stored in a structured, specialized part of cyberspace for subsequent reuse. Therefore, the input and output of the logic multiprocessor should be forms of cyberspace: the Internet of Things, big data, and cyber physical systems. The market feasibility of the proposed multiprocessor is the ability to use it to improve the quality and speed of retrieval procedures in big data, and create built-in, automatic, autonomous, and efficient systems for diagnosis and repair, with tools for targeting and pattern recognition. Typical functionality for cyber physical systems using the information space is the generation of multialternative variants of responses to a query in vector logical form of cyberspace components (subjects, processes, or phenomena), which are necessary for the management of social, biological, and non-natural production processes without human intervention.

A direction for future research is "big data image transactions." The image culture of thinking, communication, or transaction means the transfer of an object that describes components from time to space. Is this possible? "Red (100), blue (010), green (001)" is written during a time; if we combine words and replace them with images and eliminate the time, we will get a new but already generated white image (111). The artist Leonardo creates a parallel image of the Mona Lisa by superposition of the sequential visual fragments. But the result does not have the time parameter and so it is valuable. If the picture is partitioned on superposition fragments, it will lose its appeal. The Mona Lisa is indescribable! Any picture is better than textual description. Nevertheless, in the market of electronic technologies, there already exist automatic software applications for the direct synthesis "verbal description–picture" and the inverse analysis "picture–verbal description." It can be started from the voice and/or manual synthesis and analysis of simple geometric figures (triangle, square, circle).

Another direction of research is the "image transactions processor," meaning a processor that creates the bilateral symmetry "image–transaction–image," divided into two types of transactor: "verbal (manual, voice) description–transaction synthesis–image" and "image–transaction analysis–verbal (manual, voice) description."

The language of consecutive symbols used by mankind is imperfect in the time costs associated with information transactions between the transmitter and receiver. Therefore, scientists today have to state the fact that the market needs a disruptor of a new culture for parallel, superposition, creative thinking, communication, generation, transmission, receiving, and perception of information and reality without a time parameter. A cyberspace should be transferred into parallel images to make transactions based on quantum data structures (their properties are superposition, entanglement, parallelism). This allows saving memory, time for learning, transmission, receiving, and perception of information for directly interacting pairs "human–computer," "computer–computer," or "human–human" without the traditional interfaces (keyboard, voice, tactile). History knows analogs in the form of hieroglyphs, cuneiform, or wall paintings, where our ancestors tried to remove the irrelevance of time in describing the facts of the past reality, to minimize the time of attention to get knowledge of the essence, not about the process, which is not very interesting. An image is more effective for perception than a verbal description, so PR actions always operate by photographs with short slogans. Image thinking is compression of the process into a single photographic moment (film—in pictures, words and sentences – in an image; a sequence of logical elements – in information "quantum" primitive—a qubit).

3.2 Triangular Metric and Topology of Cyberspace

The topology of connections or relationships between the components of different natures generates a diversity of system structures. Thus, optimization of the number of components and the number of connections becomes critical to system operation, with an objective and constraints. Further, the idea is advanced that triangular primitives provide efficient transaction structures. A component triad dominates physical and social nature, as one of the most common structures with the properties: (1) the divine triangle unites the Father, Son, and Holy Spirit; (2) the transitivity of triangular closure is the basis of any metric for measuring processes and phenomena; (3) the triangle is the most primitive geometric figure on a plane and a tetrahedron is composed of triangular structures in space; (4) all geometric shapes on a plane can be obtained by superposition of the triangular primitives, and in space through the use of tetrahedrons; (5) all triangle nodes are adjacent, which means there is a minimum path from each node to another one that is absent in any other geometric primitive; (6) a triangle has a minimum cost of interconnections forming a planar structure that can be considered as a primitive unit; and (7) communications made up of triangular structures are optimized by the following criterion: the average path length between any two points on the topology, divided by the cost of its implementation.

Suppose there is a finite number $n \neq 0$ of points in space constituting a cycle, where each of them is given a binary vector of the length k:

$$A = (A_1, A_2, \ldots A_i, \ldots, A_n) = \{(a_{11}, a_{12}, \ldots, a_{1j}, \ldots, a_{1k}), (a_{21}, a_{22}, \ldots, a_{2j}, \ldots, a_{2k}), \ldots,, \ldots, (a_{i1}, a_{i2}, \ldots, a_{ij}, \ldots, a_{ik}), \ldots, (a_{n1}, a_{n2}, \ldots, a_{nj}, \ldots, a_{nk})\}, a_{ij} = \{0, 1\}.$$

The distance between two points is defined as:

$$d_i = d_i(A_i, A_{i+1}) = a_{i,j} \bigoplus_{j=1}^{k} a_{i+1,j}.$$

The metric β of the vector logical binary cyberspace is determined by the zero *xor* sum of the distances d_i between nonzero and a finite number of points in a closed cycle:

$$\beta = \bigoplus_{i=1}^{n} d_i = 0.$$

Metric β cyberspace is a zero *xor* sum of the distances between a finite number of points closed in a cycle. The sum of k-dimensional binary vectors defining the coordinates of points of a cyclical figure is equal to zero. This definition of the metric is based not on a set of elements but on relationships that allow reducing the system of axioms from three to one, and extension of its action to any objects of n-dimensional space.

The metric β of a vector multivalued logic space, where the coordinate is defined in the alphabet constituting a set of all subsets $a_{ij} = \{\alpha_1, \alpha_2, \ldots, \alpha_r, \ldots, \alpha_m,\}$, is a symmetric difference of distances between a finite number of points forming a cycle that is equal to the empty set:

$$\beta = \mathop{\Delta}_{i=1}^{n} d_i = \varnothing.$$

The equality to the empty set of the symmetric difference of the set theoretic interaction emphasizes the equivalence of components (distances) involved in the formation of the equation, where a single procedure used in an analytical four-digit Cantor model is defined by the symmetrical Δ operation, which is transformed in the *xor* function in the binary alphabet:

$$a\Delta b = \{0\Delta 0 = \varnothing, 0\Delta 1 = x, 0\Delta x = 1, 0\Delta\varnothing = 0, 1\Delta 1 = \varnothing,$$
$$1\Delta x = 0, 1\Delta\varnothing = 1, x\Delta x = \varnothing, x\Delta\varnothing = x, \varnothing\Delta\varnothing = \varnothing\}.$$

The introduced metric can be a basis for the quasioptimal cyberspace of the world. It is not associated with the geometric topology of the surface and, in this sense, it is virtual. But to cover the surface and space with telecommunications, it is necessary that the physical basis of cyberspace be majority-voting in order to provide fault tolerance and reliability of communication channels during the occurrence of disasters. An informational structure of cyberspace must be hierarchical and closed, both globally and locally, at any level of the hierarchy. The elementary cell of the space structure has to be triangular. This will provide a significant reduction in the volume of cyberspace information, in the limit—by a third. Does that mean increasing the performance of all transceivers and storage for the information in cyberspace by 33%? This statement is related to the axiom of the transitive closure of a finite number (1,2,3,4,...,n) of points in the informational vector logical space:

1. $d_1 = 0;$
2. $d_1 \oplus d_2 = 0;$
3. $d_1 \oplus d_2 \oplus d_3 = 0;$
4. $d_1 \oplus d_2 \oplus d_3 \oplus d_4 = 0;$
5. $d_1 \oplus d_2 \oplus \ldots \oplus d_i \oplus \ldots \oplus d_n = 0.$

Considering all possible transitive closure for minimizing conditions of data structure recovery, the obvious fact seems to restore the third edge of the triangle by another two well-known ones. In this case, it suffices to transmit the two edges through the channel connection to restore the third one. In this case, the reduction in the volume of the transmitting data is 33%. In other cases, a decrease in the data is less important. For instance, a closed quadrangle requires three edges, transmitted to restore the fourth one, where the benefit is 25%. The triangular planar interpretation of cyberspace has the form shown in Fig. 3.4.

To restore this structure, it is necessary to know already less than 2/3 of the data compiled on two sides of every triangle. It is enough to have only green edges, representing 44% of all the triangles' components. The remaining 56% of the triangle structures can be redefined by the axiom:

$$d_1 \oplus d_2 \oplus d_3 = 0 \rightarrow d_3 = d_1 \oplus d_2.$$

In general, the functional dependence between the restored triangle edge space and the total number n of a triangle structure row is determined by:

$$\eta = \frac{\sum(i+1)}{\sum(3\times i)}\Bigg|_{i=\overline{1,n}} = \frac{1}{3}\left(\frac{2}{n+1}+1\right) = \frac{1}{3}\cdot\frac{2+n+1}{n+1} = \frac{1}{3}\cdot\frac{n+3}{n+1} = \frac{n+3}{3(n+1)}.$$

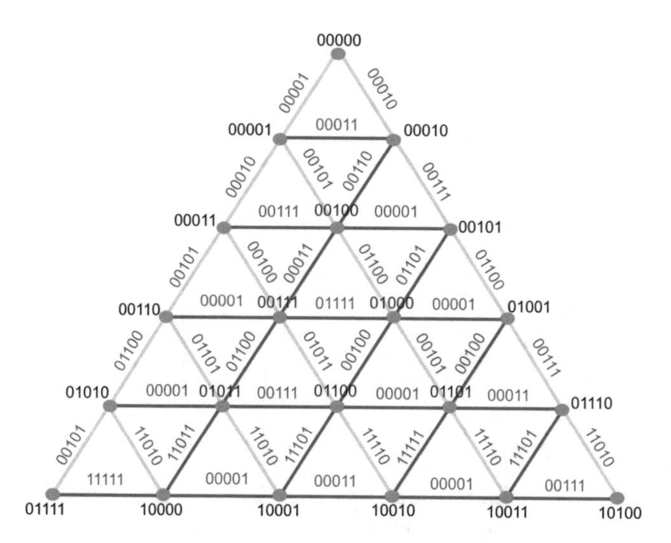

Fig. 3.4 Triangular cyberspace

The transmitted data amount depending on the number n of triangle structure rows in the planar cyberspace is shown in Fig. 3.5. The minimum value of the curve tends to be 33%. This means that for large triangle space data it is enough to transmit a third part of all edges, to restore the full space of the triangular structure.

Triangular metric space is the most economical among the well-known ones. It creates short distances between objects, thanks to the transitive closure. The adjacency of the graph node to the 3–6 points of the metric delivers more advantages in comparison with the "Manhattan" topology in terms of the way of routing between two points. The optimal structure computing system should be made up of triangles. The optimal structure is determined by the ratio between any two points' average path length and the total number of graph edges, forming cost topology. The route length in the triangular metric will always be no worse than in the Manhattan structure, but the total number of edges will be on one diagonal more than in the quadrangle. But this enables direct access to the 3–6 neighbor nodes, which is essential to the reliability of computing. A Manhattan structure with diagonal edges has the same path length as the triangular space but with one edge more. It involves the second diagonal of the quadrangle, which ensures adjacency with 3 planar nodes. Since the triangle is the most primitive figure, forming a planar space, the weight of one edge should not be kept for identification space and will be maximized in the triangle. The triangular system already gives 66% of the structural components, which cannot be described by the system infrastructure creation, because it will be recovered if necessary. All other planar primitives give a smaller benefit in cyber system development. It is very important to improve the reliability of digital systems, having a triangle structure opportunity to recover up to 66% defective components without loss of information.

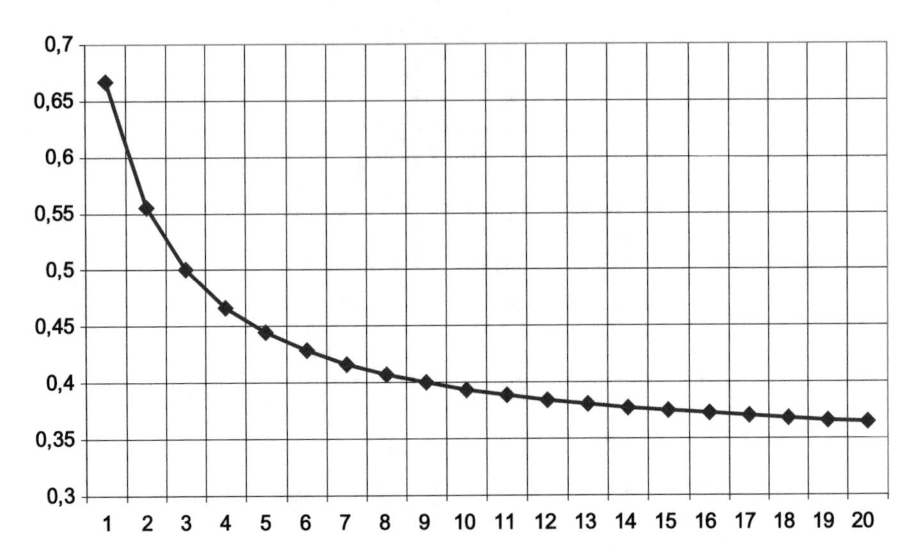

Fig. 3.5 Function of the desired data amount to restore the triangular space

3.3 Quality Criterion for Cyber Physical Topology

The average path length between the components of the digital system topology is the main parameter that affects the speed of the transaction execution between structure nodes. When considering the creation of the topology computing system, it is necessary to determine an integral characteristic as the sum of all distances between each pair of the structural components or corresponding graph vertices given to the number of graph edges. The interest is in the primitive components: a triangle and a tetrahedron. The last one has the unique property where each tetrahedron vertex has three neighbors, while the triangle has the unique property of two adjacent vertices on a planar structure.

To estimate the cost of the cyber physical transaction between the two components of the system graph that contain n vertices, the structure quality criterion is introduced, which is the lengths' sum of the minimal paths min P_i between all pairs of vertices $i = \overline{1, \frac{1}{2}(n^2 - n)}$, divided by the number of topology edges k:

$$L = \frac{\sum_{i=1}^{\frac{1}{2}(n^2-n)} \min P_i}{\frac{1}{2}(n^2 - n)} \times \frac{\frac{1}{2}(n^2 - n)}{k} = \frac{\sum_{i=1}^{\frac{1}{2}(n^2-n)} \min P_i}{k} = \frac{1}{k} \sum_{i=1}^{\frac{1}{2}(n^2-n)} \min P_i.$$

Here, the argument k is the number of edges in the graph, which is a costly part of the topological structure, so it appears in the denominator. But creative leveraging of this parameter can significantly reduce the numerator, which defines a cumulative path length between all pairs of vertices, tending to a minimum.

Thus, the smaller integrated value L leads to higher efficiency of the topological cyber physical structure. Example 1: to determine the quality criteria for the three structural examples with four vertex components (a, b, c, d): $G_1 = \{ab, bc, ad, cd\}$, $G_2 = \{ab, ac, ad, cd, bc, bd\}$, $G_3 = \{ab, bc, ac, ad, cd\}$:

$$
\mathrm{G_1} = \begin{array}{c|cccc} & a & b & c & d \\ \hline a & . & 1 & 1 & . \\ b & 1 & . & . & 1 \\ c & 1 & . & . & 1 \\ d & . & 1 & 1 & . \end{array}
\quad
\mathrm{G_2} = \begin{array}{c|cccc} & a & b & c & d \\ \hline a & . & 1 & 1 & 1 \\ b & 1 & . & 1 & 1 \\ c & 1 & 1 & . & 1 \\ d & 1 & 1 & 1 & . \end{array}
\quad
\mathrm{G_3} = \begin{array}{c|cccc} & a & b & c & d \\ \hline a & . & 1 & 1 & . \\ b & 1 & . & 1 & 1 \\ c & 1 & 1 & . & 1 \\ d & . & 1 & 1 & . \end{array}
$$

Leveraging the above formula, simply calculate the sum of the minimal path lengths for each of the six pairs: $\{ab, ac, ad, cd, bc, bd\}$, connecting $k = (4{,}6{,}5)$ topology edges:

$$
L_1 = \frac{1+2+1+1+2+1}{4} = \frac{8}{4} = 2.0;
$$

$$
L_2 = \frac{1+1+1+1+1+1}{6} = \frac{6}{6} = 1.0;
$$

$$
L_3 = \frac{1+1+1+1+1+1}{5} = \frac{6}{5} = 1.2.
$$

The topology winner is G_2, but this graph is not primitive. So the leader between the remaining two examples will be G_3.

An important feature of the L criterion is its sensitivity, where a slight graph modification by adding one edge can lead to a substantial change in the assessment downward. The market attractiveness of the effectiveness analysis of the cyber physical structure is important not only for digital systems and networks but also for cities, when a transport infrastructure is created. In addition, cyber technology requires a radical change in the organizational structures of state enterprises and organizations, including universities. Introduced quality criteria will significantly simplify the topology of e-documentation flow and vertical subordination of departments and divisions. Already today it should be possible to eliminate unnecessary deans, establishing direct channels of communication between the cyberculture of rector and departments. Also, for many years there has been a drive to remove all auxiliary departments at universities involved in collecting information, transferring these functions to department heads, who in online mode should work with the cloud service, monitoring scientific and educational processes.

Example 2: application of quality criteria to the evaluation of the three graph structures with six vertices and different connection topology, shown in Fig. 3.6:

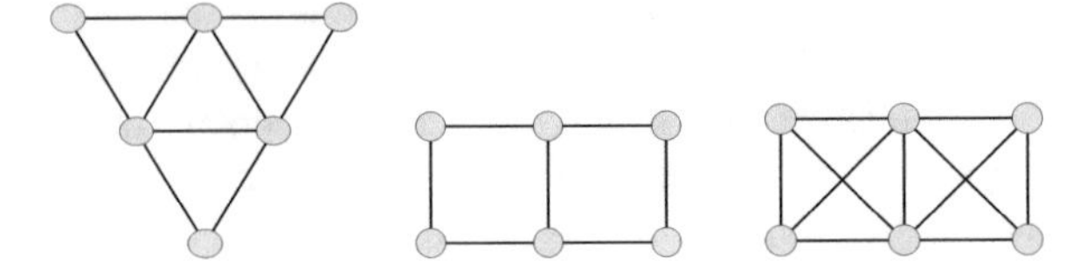

Fig. 3.6 Structures of topological connections for six nodes

$G_1 =$

	a	b	c	d	e	f
a	.	1	1	.	.	.
b	1	.	1	1	.	1
c	.	1	.	.	.	1
d	1	1	.	.	1	1
e	.	.	.	1	.	1
f	.	1	1	1	1	.

$G_2 =$

	a	b	c	d	e	f
a	.	1	.	1	.	.
b	1	.	1	.	1	.
c	.	1	.	.	.	1
d	1	.	.	.	1	.
e	.	1	.	1	.	1
f	.	.	1	.	1	.

$G_3 =$

	a	b	c	d	e	f
a	.	1	.	1	1	.
b	1	.	1	1	1	1
c	.	1	.	.	1	1
d	1	1	.	.	1	.
e	1	1	1	1	.	1
f	.	1	1	.	1	.

There are three graphs with 9, 7, and 11 arcs, respectively. Counting the structure quality criterion according to the L formula represents the following results:

$$L(G_1) = \frac{9 \times 1 + 6 \times 2}{9} = \frac{21}{9} = 2.33;$$
$$L(G_2) = \frac{7 \times 1 + 6 \times 2 + 2 \times 3}{7} = \frac{25}{7} = 3.57;$$
$$L(G_3) = \frac{11 \times 1 + 4 \times 2}{11} = \frac{19}{11} = 1.73.$$

The triangle metric advantages are the following. (1) The obtained topology quality criterion is simple to implement in the cloud service, available for a wide range of users; it is aimed at determining the effectiveness of an old or innovative structure of a computing system, university, social network, company, or transport infrastructure. (2) The criterion shows the obvious advantages of the triangular structure of the interaction of the system components for the service of transactional processes in a cyber physical space. (3) The triangular metric of the space is the most economical. It creates the shortest distances and paths between the coordinates, due to the transitive closure. The triangular urban topology allows a reduction in the traffic routes of between 33% and 66% compared to the Manhattan metric. (4) Implementation of this topology in the structural organization of digital systems on 2D and 3D chips will reduce the length of the connections, increasing the speed by no less than 33%. (5) This topology increases the reliability of computer systems and networks by up to 66%, due to the triangular organization of the structural component relationship, where any failed component has at least two spare neighbors. (6) 3D organization of system components in a triangular tetrahedral topology provides even greater benefits to optimize transactional routes for cyber physical objects, including air transport, global computing, and 3D systems on chips (SoCs). (7) The triangular infrastructure increases the reliability of digital systems, networks, services, the Internet of Things (IoT), cyber physical systems, and the cyber ecosystem by leveraging the opportunity to restore faulty connections in up to 66% of components without loss of information. In other words, the triangular structure of the system means stay in operation with 66% faulty connections. (8) The disadvantage of the triangular topology can be considered the unusual triangular coordinate system for a human, who usually uses the Cartesian metric. The talk is about point identification in the planar structure by leveraging three coordinates that are widely used in global positioning. The Cartesian metric keeps only two coordinates for the exact positioning of objects in the plane.

3.4 Quantum Hasse Processor for Optimal Coverage

Quantum computing architecture is proposed for a significant increase in performance to solve discrete optimization problems [6–9]. Hardware-oriented models of parallel taking the Boolean set of all subsets for the universe of n primitives are described. Hasse processor architecture [9] is focused on solving the problems of coverage, minimizing logic functions, data compression, synthesis, and analysis of digital systems.

The purpose of creation of the Hasse processor is to significantly reduce the time for solving optimization problems by parallel computing of vector logical operations over the set of all subsets of primitive components by increasing the memory for storing matrix data.

The objectives are (1) definition of data structures for taking a Boolean set when solving the problem of covering the columns of a matrix $M = |M_{ij}|, i = \overline{1,m}; j = \overline{1,n}$ by 1-unit row values—in particular, for $m = n = 8$, it is necessary to execute in parallel a logical operation over 256 variants of all possible combinations of the matrix rows that make up the Boolean set, and the processor instruction system includes logical operations (*and*, *or*, *xor*) over vectors with m dimension; (2) development of Hasse processor architecture for parallel computation of $2^n - 1$ combination variants aimed at the optimal solution of the NP-complete coverage problem; and (3) implementation of the Hasse processor prototype in programmable logic and verification of the hardware solution by examples of minimizing logic functions.

As an example, it is proposed to solve the problem of searching the optimal 1-unit coverage of all columns by the minimum number of rows of the matrix M:

M	1	2	3	4	5	6	7	8
a	1	.	.	.	.	1	.	.
b	.	.	1	.	.	.	1	.
c	1	.	.	.	1	.	1	.
d	.	1	.	1	.	.	.	1
e	.	1	.	.	1	.	.	.
f	1	.	1	.	.	1	1	.
g	.	1	.	1	.	.	.	1
h	.	.	1	.	1	.	.	.

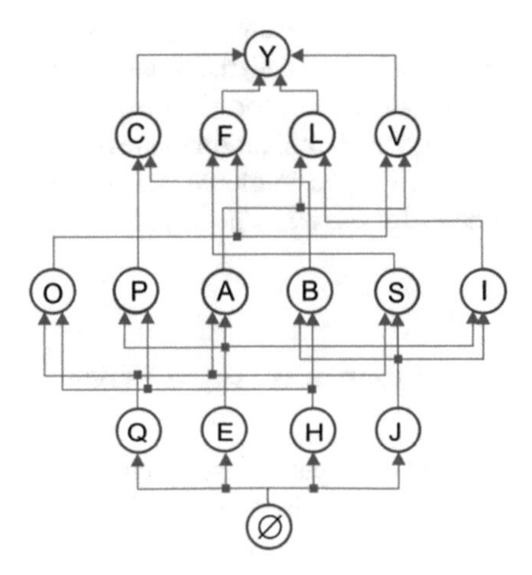

Fig. 3.7 Quantum Hasse structure of the computational process

The task solution is the search that contains 255 combinations. The minimum number of primitive rows that form the coverage is the optimal solution. The Hasse diagram [7–9] is compromise architecture with respect to time and memory. It leverages the previously obtained result to create a more complex superposition of solutions. For each coverage table containing n lines, it is necessary to generate its own Hasse processor for an almost parallel solution of the NP-complete problem. For instance, the four rows of the coverage table create the structure of the Hasse processor shown in Fig. 3.7.

The Hasse processor corresponds to the structural description of a closed alphabet formed by a Boolean set in a universe of four primitives that represent all possible binary transitions of a logical variable in two automaton cycles: $B^*(Y) = \{Q = (00), E = (01), H = (10), J = (11), O = \{Q, H\}, I = \{E,J\}, A = \{Q,E\}, B = \{H,J\}, S = \{Q,J\}, P = \{E,H\}, C = \{E,H,J\}, F = \{Q,H,J\}, L = \{Q,E,J\}, V = \{Q,E,H\}, Y = \{Q,E,H,J\}, U = \emptyset$.

The optimal solutions of the coverage problem for the matrix M, which generates 255 variants of possible combinations, are represented by rows in the DNF form: $C = fgh \vee efg \vee cdf$.

The control algorithm of the computational process for a quantum Hasse structure by an upward analysis of the graph vertices is based on the sequential execution of the following steps:

1. Entering the rows of the matrix in the first level registers $L_i^1 = P_i$ with the subsequent analysis of the coverage quality where each matrix row (primitive) is evaluated by a bit: 1 means coverage presence, 0 means no coverage. If one of the primitives creates the coverage $\wedge_{j=1}^{m} L_{ij}^1 = 1$, the Hasse structure analysis ends. Otherwise, the transition $(r = r + 1)$ to the next higher level of the graph is performed:

$$L_i^1 = P_i \rightarrow \bigwedge_{j=1}^{m} L_{ij}^1 = \begin{cases} 0 \rightarrow n = n + 1; \\ 1 \rightarrow \text{end.} \end{cases}$$

2. Initiating a command for processing to the next level, with consecutive execution of vector (matrix) operations *or*, *and*: $L_i^r = L_{ij}^{r\text{-}1} \wedge_{j=1}^{m} L_{tj}^{r-1}$, $\wedge_{j=1}^{m} L_{tj}^r = 1$ for analysis of the coverage, obtained by superposition of r-level primitives. Here $t = \overline{1, m}, i = \overline{1, m}, r = \overline{1, n}$; n is the number of rows in the coverage table; m is the number of columns in it. If there is a superposition at the current level, creating a coverage indicator 1, processing of all subsequent Hasse levels is not performed. Otherwise, the transition to the next higher level of the structure is performed:

$$L_i^r = L_{ij}^{r-1} \bigwedge_{j=1}^{m} L_{tj}^{r-1} \rightarrow \bigwedge_{j=1}^{m} L_{ij}^r = \begin{cases} 0 \rightarrow r = r + 1; \\ 1 \rightarrow \text{end.} \end{cases}$$

Every operational graph vertex consists of two register variables, which significantly reduces the hardware costs when implementing the Hasse processor. The number of clock cycles for processing the Hasse structure is at worst equal to n. It can also create an algorithm for searching the optimal coverage by bottom-up analysis of the graph vertices. In this case, when the complete coverage in the current level is found, another descent along the structure is necessary to ensure there is no complete coverage on the lower adjacent level. In this case, the solution obtained is optimal. Otherwise, it is necessary to perform a descent to a level where the lower adjacent level will not contain complete coverage. The vertices of the processor structure can have more than one register logical operation. Then the simplest command decoder to activate the logical operations *and*, *or*, *xor*, *not* is created.

Thus, the advantages of a quantum Hasse processor are the leverage of two-input elements for vector logical operations (*and*, *or*, *xor*), which makes it possible to significantly reduce hardware costs through the use of serial–parallel computations and a slight increase in the processing time of all the graph vertices. For each vertex, the coverage criterion is calculated as the presence of all 1 units in the coordinates of the result vector. If the criterion is 1, then all other calculations are not performed, since the Hasse diagram is a strictly hierarchical structure with respect to the number of superpositions in each tier. This means that the best solution is at the lower level of the graph hierarchy. Variants of the same level are equivalent in cost of implementation; therefore the first coverage obtained ($Q = \sum_{i=1}^{n} q_i = n$) is the best solution, which implies stopping the Hasse structure processing algorithm. The serial–parallel analysis cycle of the Hasse graph vertices is determined by the number of levels of the hierarchy or by the number of primitive rows in the coverage table multiplied by the time of analysis of one vertex: $T = \log_2 2^n \times t = t \times n$. In this case, the length m of the row of the coverage table does not affect the speed. Vertex analysis includes two commands: logical (*and*, *or*, *xor*) and the operation of calculating the coverage criterion in the form of a scalar by applying the *and* operation to all bits of the result vector:

$$m_{ir,\ j} = M_{i,j} \vee M_{r,j} \left(j = \overline{1,n}; \{i \neq r\} = \overline{1,m};\right);$$
$$m_{ir}^S = \wedge m_{ir,j} = \wedge\left(M_{i,j} \vee M_{r,j}\right).$$

The hardware costs for the implementation of the Hasse processor depend on the total number of graph vertices and on the number of bits in the row of the coverage table: $H = 2^n \times k \times m$, where k is the hardware implementation parameter of the binary vector logic operation and the criteria for calculating the quality of coverage.

Thus, the high speed of solving the coverage problem is achieved by a significant increase in hardware by $2^n \times k \times m/k \times m \times n = 2^n/n$ times, in comparison with the sequential processing of graph vertices. The Hasse processor provides the optimum between a completely parallel structure of computing processes, where the hardware cost is determined by the number of primitives at each node $H = k \times m \times n \times 2^n$, and a purely sequential computation structure, where the processing time of the Hasse graph is equal to $T = t \times 2^n$ with minimal hardware costs $H^* = k \times m \times n$. Reducing the hardware complexity of the Hasse processor in comparison with the parallel processing of the graph is $Q^H = k \times m \times n \times 2^n / k \times m \times 2^n = n$. Reducing the time of analysis of the Hasse structure vertices, due to hardware redundancy, compared with purely sequential processing of the graph vertices, has the following estimate:

$$Q^T = \frac{t \times 2^n}{t \times n} = \frac{2^n}{n}.$$

The hardware-oriented Hasse architecture of the parallel computation of Boolean set in the universe of n primitive rows is designed to solve the coverage problems, to minimize logic functions, to compress data, and to synthesize and analyze digital systems. A prototype of a quantum Hasse structure implemented in programmable logic for the optimal solution of the coverage problem is used for the cyberspace analysis.

The model of a quantum Hasse device is developed in the Verilog language. The elementary cell of the processor consists of two register gates (Fig. 3.8). The register element (*or*) performs a logical operation between two vectors, forming a vector result too. The register primitive (*and*) performs a convolution of all bits of the vector, forming the bit identifier of the optimal coverage presence.

A fragment of the quantum Hasse processor scheme for solving the coverage problem is shown in Fig. 3.9. Here is a scheme of generating solutions for six levels of the Hasse structure, which contains primitives that implement the *or* operation.

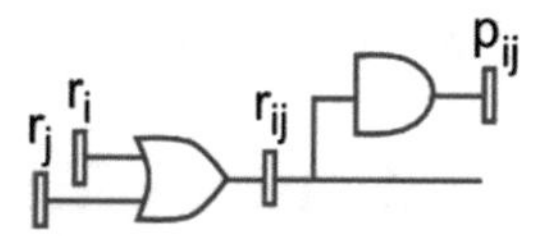

Fig. 3.8 Elementary cell of the quantum processor

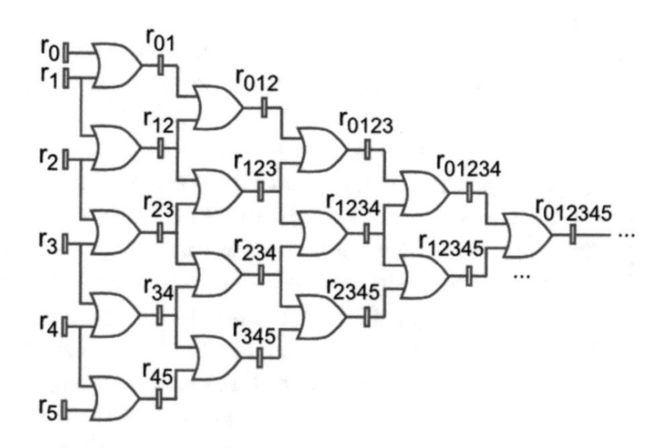

Fig. 3.9 Fragment of the RTL-level scheme

Each element in the scheme has an output *and* gate for analyzing the quality of the coverage.

The implementation of the computing device is based on the FPGA Xilinx xc3s1600e-4-fg484, where the main parameters are presented below:

Map-Report Logic Utilization:
Number of Slice Flip Flops: 2286 out of 29.504 7%;
Number of 4 Input LUTs: 2715 out of 29.504 9%;
Total Number of 4 Input LUTs: 2715 out of 29.504 9%;
Number of Bonded IOBs: 321 out of 376 85%;
Number of BUFGMUXs: 1 out of 24 4%;
Timing Parameters of Project: Tclk_to_clk = 4.672 ns; Tclk_to_pad_max = 11.552 ns;
Period = max {Tclk_to_clk; Tclk_to_pad_max}; Period = 11.552 ns;
Fclk = 86.5 MHz.

The hardware and software implementation of the quantum Hasse processor is used in the educational process to minimize the logical functions and solve the coverage problems in binary tables with a dimension of 100×100 elements. In this case, segmentation of a large table into fragments is used, which are considered as initial data for a quantum Hasse architecture.

3.5 Conclusion

A new cyber informational model for analyzing big data has been developed; it is focused on the use of cloud services, cyber physical systems, and parallel virtual multiprocessors with a minimal set of vector logical operations for precise information retrieval based on the proposed power set metric and vector logical quality criteria, which allows the making of gradual classifications and the ordering of chaotic big data in the framework of the planet.

The practical value of the proposed model lies in the need to restructure cyberspace through replacement of the concept of amorphous big data by a semantically classified information infrastructure of useful data for management of cyber physical processes. Directions for the formation of technological big data culture for a gradual increase in the level of useful information from 0.4% to 10% through competency of a big data infrastructure of cyberspace are proposed.

A new model of a vector logical SIMD multiprocessor is offered, which is characterized by the absence of arithmetic operations, parallel computing distances between the query and information quanta, and simultaneous determination of the best of n possible solutions by the minimum value of the membership function, which makes it possible to significantly increase the speed of the most precise information retrieval in big data.

A criterion for quality evaluation of computing structures is represented; it is intended for choosing the best topology and takes into account the sum of all paths between pairs of nodes and the number of edges in the graph. The advantages of triangular cyberspace for the creation of reliable and high-performance cyber physical systems and networks, which stay in operation with a large number of failed connections, are shown.

An innovative model of quantum computing is proposed to solve the coverage problem, which is characterized by the leverage of the Hasse structure for significantly increasing the performance of vector logical operations on big data in the form of tables.

References

1. Alkhatib, H., Faraboschi, P., Frachtenberg, E., Kasahara, H., Lange, D., Laplante, Ph., Merchant, A., Milojicic, D., Schwan, K. (2014). IEEE CS 2022 report. IEEE Computer Society.
2. Mayer-Schönberger, V. (2013). *Big data: A revolution that will transform how we live, work*. Edition: Mann, Ivanov and Ferber.
3. Demchenko, Y., de Laat, C., Membrey, P. (2014). *Defining architecture components of the big data ecosystem*. In: International conference on collaboration technologies and systems (CTS), 2014.
4. Grolinger, K., Hayes, M., Higashino, W.A., L'Heureux, A., Allison, D.S., Capretz, M.A.M. (2014). *Challenges for MapReduce in big data*. In IEEE World Congress on Services (SERVICES), 2014.

5. Bondarenko, M. F., Hahanov, V. I., & Litvinova, E. I. (2012). Structure of logic associative processor. *Automation and Remote Control Journal, 10*, 71–92.
6. Hiroshi, I., & Masahito, H. (2006). *Quantum computation and information. From theory to experiment*. Springer.
7. Hahanov, V.I., Hahanova, I.V., Litvinova, E.I., Guz, O.A. (2010). *Design and verification of SoC*. Verilog & System Verilog. Kharkov. Novoe Slovo.
8. Hahanov, V. I., Litvinova, E. I., Chumachenko, S. V., & Guz, O. A. (2011). Logical associative processor. *Electronic Modeling Journal, 1*, 73–90.
9. Hahanov, V.I., Hahanova, I.V., Guz, O.A., Abbas, M.A. (2012). *Quantum models for data structures and computing*. In: TCSET Proceedings. Lvov. Slavske.

Chapter 4
Qubit Description of Functions and Structures for Service Computing Synthesis

Ivan Hahanov, Igor Iemelianov, Mykhailo Liubarskyi, and Vladimir Hahanov

4.1 Introduction

The market feasibility of the emulation of quantum computing methods to create computing structures (CS) in cyberspace is based on the use of qubit data models, focused on parallel solving of the problems of designing, testing, and discrete optimization with a substantial increase in memory. Without going into the details of the physical foundations of quantum mechanics related to the nondeterministic interactions of atomic particles [1–3], the notion of the qubit is further used as a binary or multivalued vector for joint and simultaneous definition of a power set (a set of all subsets) of states in a discrete area of cyberspace, based on linear superposition of unitary codes, focused on parallel use of methods for analyzing and synthesizing components of cyberspace. In the rapidly developing theory of quantum computing, the state vectors form a quantum register of n qubits, forming a unitary or Hilbert space H [4, 5], the dimension of which has a power dependence on the number of qubits $\text{Dim } H = 2^n$.

The main goal is substantial improvement of the quality and reliability of high-performance computing by leveraging the addressability of circuit data structures, allowing online repair, as well as increasing the speed of modeling and simulation techniques, and of testing and verification methods for complex digital systems on chips (SoCs), by reducing the dimension of the functional primitive description.

The objectives of this chapter are as follows: (1) creation of an automaton model based on qubit graph structure description; (2) synthesis and analysis of qubit

I. Hahanov (✉) • I. Iemelianov • M. Liubarskyi • V. Hahanov
Design Automation Department, Kharkov National University of Radio Electronics, Ukraine, Nauka Avenue 14, Kharkov 61166, Ukraine
e-mail: ivanhahanov@icloud.com; igor@itdelight.com; mlyubarskyy@gmail.com; hahanov@icloud.com

V. Hahanov, *Cyber Physical Computing for IoT-driven Services*,
DOI 10.1007/978-3-319-54825-8_4

models of digital primitives: logic elements, flip-flops, registers, and counters; and (3) synthesis and optimization of quantum models of digital circuits.

The feasibility and state of the art are as follows:

1. The modern digital SoC contains more than 94% memory and only 6% new or reusable logic, which creates 90% of challenges related to test synthesis, verification, testing, fault diagnosis, and repair [6, 7]. Of course, the performance of reusable logic is much higher than the access to the memory, but most computational processes are focused on exchange data in the memory structure components. Therefore, the benefits of reusable logic in real computing architectures for processing big data are compensated for by timing costs (about 90%) associated with the memory component transaction.
2. The development and implementation of the memory-driven processor provide similarity of the circuit structure and types of functional primitives. That allows developer to make the designing, manufacturing, and functioning more convenient and faster, including verification, embedded testing, and diagnosis. The most important thing is to provide online repair through leveraging of on-chip universal addressable spare memory components [6, 7].
3. Modeling of the circuit for the processes of design and verification of computing systems based on addressable component models makes this procedure technologically very simple through regular data structures and single transactions between the memory elements, as well as being faster, due to the possibility of parallel quantum-like processing of large memory arrays of the same cell type [3–5, 7, 8, 11].
4. Replacement of logical elements by addressable memory components results in slightly increased energy consumption, which is really considered the "price to pay" for the abovementioned significant advantages related to computing product reliability and increased yield, reducing the costs of designing and manufacturing, as well as allowing remote and online human-free repair. However, energy-saving solutions for computing processes in the memory [6, 9, 10, 12] suggest that the energy consumption would not be increased at all. The main feature of the data organization in a classical computer is addressable bits, bytes, or other components. Addressability creates a problem of a processing association for the unaddressed elements of a set, which do not have an order by definition. The solution is a processor where the universal set of n unitary coded primitives uses a superposition for the formation of a power set $|B(A) = 2^n|$ for all possible states [8].

The main idea is to develop quantum-like data structures, which are characterized by the properties of superposition, entanglement, and parallelism, for creating high-performance deterministic computers focused on high-speed analysis of processes and phenomena in cyberspace. The tasks are the following: (1) development of power set data structures that are isomorphic to the qubit structures in nondeterministic quantum computers; (2) use of the power set data structures for a compact description of computational processes based on the use of logical operations; and (3) provision of examples of the compact description and analysis of digital components.

Leveraging the emulation of quantum computing methods when creating a virtual (cloud) computer (VC) in cyberspace is based on qubit data models, focused on parallel discrete optimization problems with a significant increase in memory [1–6]. Without considering the physical foundations of quantum mechanics related to nondeterministic interaction of atomic particles [7, 10–12], we use the notion of the qubit as a binary or multivalued vector for joint and simultaneous definition of the power set of states of a discrete cyberspace area based on linear superposition of unitary codes, focused on parallel usage of methods for analyzing and synthesizing components of cyberspace. In the rapidly developing theory of quantum computing, the state vectors form a quantum register of n qubits, forming a unitary or Hilbert space H, the dimension of which has a power law dependence on the number of qubits Dim $H = 2^n$.

The innovation of an approach for the design of the VC is governed by the advent of cloud services, which are dedicated virtual computer systems, distributed in space and implemented in hardware or software. The programmer is not always interested in how and where the codes of software applications are implemented, and he has no need to know the theory for designing a digital automaton on the gate level or register transfer level and use of specific components of computer devices (flip-flops, registers, counters, multiplexers). Any component can be functionally represented in a vector form of the truth table, implemented by using memory. Traditionally implemented logic functions are not considered. It partially reduces performance, but considering that 94% of an SoC is memory, the remaining 6% can be implemented in memory components that will not be critical for the majority of software and hardware applications. Therefore, for programming effective virtual computers, a theory based on two components of a higher abstraction level (memory and transaction) can be used.

4.2 Quantum Data Structures

The feature of data organization in a classical computer is that each bit, byte, or other component has its own address. Therefore, there is the problem of effective processing of an association (finite symbol alphabet) of equal elements that do not have an address order by definition, e.g., the set of all subsets. A solution could be based on a processor, where the unit cell is the image or pattern of the universe of n unitary coded primitives using superposition to form a power set $|B(A)| = 2^n$ of all possible states of a cell in the form of the set of all subsets [4, 6]. In fact, it turns out that for efficient processing of the set of all subsets, the vector of the addressable cells or bits is introduced; the power set of states is formed from its zeros and ones. Otherwise, the set theoretic culture should be reduced to the address-focused data structures of a modern computer to significantly improve performance when executing set theoretic operations. Thus, the power set vector is the addressable representation of the power set of states as the set of all subsets of unitary encoded elements for executing logic operations in parallel instead of coordinate-wise set theoretic procedures.

There is a certain analogy between the vector data structures of the power set and the qubit of the quantum computer. The quantum qubit concept as an elementary structure for storing information in a quantum computer that enables the superposition of states $|\psi\rangle = \alpha\cdot|0\rangle + \beta\cdot|1\rangle$ can be put in one-to-one correspondence with the power set of states in a classical computer [4–6], forming, for example, Cantor's alphabet $A^k = \{0,1,X,\varnothing\}, X = \{0,1\}$. Here the two first primitives of the alphabet are unitarily encoded by the following vectors (ordered binary sequences): $0 = (10)$ and $1 = (01)$. The codes of the other two symbols can be obtained as derivatives by using the superposition or union operations (logical adding) $(10)\vee(01) = (11)$, as well as the intersection operation (logical multiplication) $(10)\wedge(01) = (00)$, which generates four states of the power set: $\{(10),(01),(11),(00)\}$ [4–6] in the form of binary vectors or tuples.

To prove the correctness of the use of the adjective "qubit" for models of digital units, it is necessary to compare the algebra of set theory and linear algebra by the example of Cantor's algebra $A^k = \{0,1,X,\varnothing\}$. The first two symbols are primitive or primary elements, which are not decomposed into components. The third symbol is derived in the set theory and defined by the superposition or union of symbols–primitives: $X = 0\cup 1$. In a Hilbert space, where the linear algebra operates by the primitives $\alpha\cdot|0\rangle$ и $\beta\cdot|1\rangle$ of a qubit, the third symbol is formed by superposition of two components of the state vector: $|\psi\rangle = \alpha\cdot|0\rangle + \beta\cdot|1\rangle$. The complete analogy is obvious; it can be applied to form the fourth symbol (the empty set) that is absent in the qubit (linear) algebra by using the inverse function of the superposition. In the set theory it is the intersection operation, which forms an empty set by the primitive symbols $\varnothing = 0\cap 1$. In a Hilbert space, such an operation is the scalar (inner) product or Dirac function (symbol) $< \alpha|\beta >$ [1, 5], which has a geometric interpretation in the following form: $<\alpha|\beta> \approx |\alpha| \times |\beta| \times \cos \angle(\alpha,\beta)$. If the projection α,β of a quantum state vector is orthogonal (this is true), the following result appears: $<\alpha|\beta> = |\alpha| \times |\beta| \times \cos \angle(\alpha,\beta) = |\alpha| \times |\beta| \times \cos \angle 90^{\circ} = 0$. The scalar product of orthogonal vectors is zero. This result is an analog of the empty set symbol when performing the intersection operation of the algebra of sets. Thus, the correspondence table of the set algebra and linear algebra presented below confirm the property of isomorphism between the symbols of the power set and states of the qubit vector obtained using isomorphic operations of union—superposition and intersection—with the scalar product:

Boolean A^k =	0	1	$X = 0\cup 1$	$\varnothing = 0\cap 1$
Qubit $\|\psi\rangle$ =	$\|0\rangle$	$\|1\rangle$	$\alpha\|0\rangle + \beta\|1\rangle$	$\alpha\|0\rangle\|\beta\|1\rangle$

So, the data structure "power set" can be considered as a deterministic (not probability) image or analog of the quantum qubit in the algebra of sets (logic), the primitive elements of which are unitarily encoded by binary vectors (tuples) and characterized by the properties of superposition, parallelism, and entanglement. This enables use of the proposed model for improving the performance of the analysis of digital units using classical computers, as well as quantum computers without substantial modification, which will appear for 5–10 years in the market of electronic technology.

In Kurosh [4], the possibility of emulating the classical computing processes in quantum computers is considered. But, given the “reversibility” of the conformity of these algebras, inverse transformation—emulation of some advantages (superposition, parallelism, uncertainty) of quantum computing based on classical processors using the power set as the base model of data structures for interpretative description and modeling of digital systems—is further proposed.

In the market for electronic technology, competition exists between the variants of idea implementation [1–6]:

1. Software implementation of the project is associated with the synthesis of the interpretative model of software functionality or hardware programmable logic devices based on FPGA and CPLD. The advantage is technological design modification; the disadvantage is low performance of the digital system.
2. Hardware implementation is focused on the use of compiling models in the development of software applications or the implementation of the project as VLSI chips [6]. The advantages and disadvantages of hardware implementation are inverse in regard to software implementation: high performance and inability to modify.

Taking into account the basic variants of the idea, implementation of quantum data structures focused on improving the performance of flexible models of software or hardware implementation of the project are proposed.

A quantum description of the structure of digital systems is represented below. There are some definitions. The qubit (n qubit) is the particular vector form of the unary coding of the n primitives universum allowing us to define the power set of the states 2^{2^n} leveraging 2^n binary variables. For instance, if $n = 2$ then the 2-qubit vector defines 16 states based on four variables. If $n = 1$ then the qubit vector defines four states on the two primitives universum (10) and (01) by leveraging two binary variables (00,01,10,11) [4]. The superposition allows the simultaneous existence of 2^n states in the vector, designated as functional primitives. The qubit (n qubit) allows the execution of parallel logic operations instead of element-wise set theoretic operations to significantly accelerate the algorithms of synthesis and analysis of discrete computing systems.

Further, the qubit is identified with an n qubit or a binary vector, for ease of understanding of the presented material. Since quantum computing is related to qubit data structure analysis, we will continue to use the definition “quantum” to identify particular technology that uses three points of quantum mechanics: parallelism of binary vector processing, the states’ superposition, and entanglement. The qubit analog when specifying a binary vector of a logic function is Q coverage or a

Q vector [4] as a unified superposition form of output states, corresponding to the address codes of the logic element input variables.

A qubit in a digital system is considered as a specific form of a structural primitive, which is invariant regardless of the hardware/software implementation technology of functionality. Moreover, the synthesis of digital systems based on qubit structures is not tied rigidly to Post's theorem, which determines five classes of the existence of a functionally complete basis [5]. At the proposed quantum abstraction level, an n qubit vector gives more opportunities for defining any n input functional element where the number of subsets is $|B(A)| = 2^{2^n}$, which certainly contains all the functionalities meeting the five classes of Post's theorem [5]. The structural component qubit vector of the digital circuit $Q^* = (X, Q, Y)$ includes input and output variables, as well as qubit vector Q defining the logic function $Y = Q(X)$, where the dimension is defined by the exponential dependence of the input line number $k = 2^n$.

A practically focused novelty of qubit simulation is to replace the truth tables of components of digital units by vectors of output states. Applying such transformations to the logic gates can be simply demonstrated. Let the functional primitive have the following binary coverage (truth table):

P =

X_1	X_2	Y
0	0	1
0	1	1
1	0	1
1	1	0

which can be transformed by unitary encoding of input vectors through the use of a two-phase alphabet [6], originally designed for a compact description of all the possible automaton variables (Fig. 4.1).

Here there are symbols, their unitary and binary codes (for example, Q = 00–1000) for describing two adjacent states of automaton variables. The proposed alphabet structurally defines the power set (the set of all subsets) of states by using the universe of four primitives: $Y = \{Q, E, H, J\}$. A unitary code corresponds to the format of a vector comprising two qubits, which form 16 symbols of a two-phase alphabet. Using the last one, any coverage of a two-input logic primitive can be represented by two cubes, or even one, given that they are mutually inversed:

$$P = \begin{array}{|c|c|} \hline 00 & 1 \\ 01 & 1 \\ 10 & 1 \\ 11 & 0 \\ \hline \end{array} = \begin{array}{|c|c|} \hline Q & 1 \\ E & 1 \\ H & 1 \\ J & 0 \\ \hline \end{array} = \begin{array}{|c|c|} \hline V & 1 \\ J & 0 \\ \hline \end{array} = \begin{array}{|c|c|} \hline 1110 & 1 \\ 0001 & 0 \\ \hline \end{array} \rightarrow \begin{array}{|c|} \hline 1\ 1\ 1\ 0 \\ \hline \end{array}$$

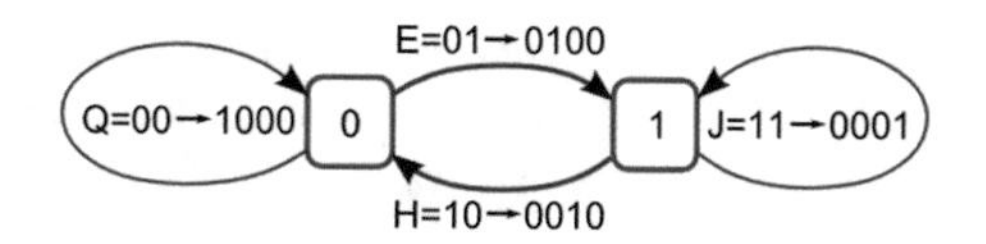

$Q=00\to1000$	$E=01\to0100$	$H=10\to0010$	$J=11\to0001$
$O=\{Q,H\}=$ $=\{00,10\}\to1010$	$I=\{E,J\}=$ $=\{01,11\}\to0101$	$A=\{Q,E\}=$ $=\{00,01\}\to1100$	$B=\{H,J\}=$ $=\{01,11\}\to0011$
$S=\{Q,J\}=$ $=\{00,11\}\to1001$	$P=\{E,H\}=$ $=\{01,10\}\to0110$	$C=\{E,H,J\}=$ $=\{01,10,11\}\to0111$	$F=\{Q,H,J\}=$ $=\{00,10,11\}\to1011$
$L=\{Q,E,J\}=$ $=\{00,01,11\}\to1101$	$V=\{Q,E,H\}=$ $=\{00,01,10\}\to1110$	$Y=\{Q,E,H,J\}=$ $=\{00,01,10,11\}\to1111$	$\varnothing=00\to0000$

Fig. 4.1 Two-phase alphabet of automaton variables and interpretation of the symbols

The qubit vector is used to simulate the components of digital units and two mutually inverse cubes—for the synthesis of tests for one-dimensional activation of circuit variables.

At first, all couples are encoded with symbols of the two-phase alphabet, and then the union of the first three cubes is performed according to the rule of minimizing the co-edge operator [5]: the vectors differing in one coordinate are combined into one. Then the qubit vectors, corresponding to these symbols, encode the resulting coverage of two cubes. To simulate fault-free behavior, just one cube (zero or unit) is enough, as the second one is always the complement of the first one. Consequently, for example, a unit cube forming the output 1 allows elimination of the bit of the primitive output state, thereby reducing the dimension of the cube or the primitive model up to the number of addressable states of the element, where the address is the vector composed of the binary values of the input variables; the vector defines the state of the primitive output. Due to the triviality it does not make sense to show that, by analogy, any truth table can be led to qubit functionality in the form of a vector of output states of the logic gates with n inputs. In addition, the two-phase alphabet is used for a compact description of transition tables of a finite automaton, which significantly reduces the analysis time of such models. For example, a complete directed graph of the 16 transitions on four nodes {00,01,10,11} is minimized in one cube:

{*QQ* = (00-00), *QE* = (00-01), *EE* = (00-11), *EQ* = (00-10), *QJ* = (01-01), *EJ* = (01-11), *QH* = (01-00), *EH* = (01-10), *JJ* = (11-11), *HJ* = (11-01), *JH* = (11-10), *HH* = (11-00), *JQ* = (10-10), *HQ* = (10-00), *JE* = (10-11), *HE* = (10-01)} = *YY*.

To summarize in defining the *n* qubit, it should be noted that its essence is different from the classical byte or bit by superpositional structuring of the binary vector that is capable of simultaneously storing n states (symbols) of the functionality in the power set with $|B(A)| = 2^n$ primitives, and performing in parallel logical operations with set theoretic data in the vector format. For example, the operation of the symmetric difference of the subsets $A = \{a,b,c,d,e,f\}$ and $B = \{a,c,f,g,h,k\}$ is performed in parallel in one clock cycle of an *xor* operation if each element is defined by a unitary code and subsets are defined by the corresponding vectors, which in this case are qubit operands to compute the symmetric difference:

9 – qubit	a	b	c	d	e	f	g	h	k
A =	1	1	1	1	1	1	0	0	0
B =	1	0	1	0	0	1	1	1	1
A⊕B	0	1	0	1	1	0	1	1	1

The abovementioned gives rise to the definition of a structured vector as an n qubit as it is characterized by the properties of computing parallelism, superposition, and uncertainty (entanglement—there are 6 units in each vector) of states, which occur in quantum data structures.

In conclusion, a new model of computing a discrete automaton in the form of qubit data structures is proposed. It is characterized by the transactional interaction of memory components [13, 14], representing combinational and sequential elements implemented in the form of qubit or "quantum" primitives. Qubit data structures allow creation of parallel virtual computers for effectively solving the problems of big data analysis without arithmetic instructions, and provide high performance of cloud-focused processors.

4.3 Quantum Description of Digital Functional Elements

The qubit (n qubit) is the vector logical form of a unitary coding of the power set based on n primitives for a compact description of 2^{2^n} states by using 2^n binary variables. Four variables, when $n = 2$ qubits, define 16 states. The qubit with $n = 1$ defines four states on the 2-primitive universal set (10) and (01) by using two binary variable codes (00,01,10,11) [7]. A superposition in the binary vector of 2^n states, considered as elements, is allowed. The qubit gives an opportunity to leverage parallel vector logical instructions instead of element-wise set theoretical operations. It means an increase in the processing performance of digital system analysis. The analog of the qubit is the binary vector of the logic function, Q coverage or the Q vector [8]. The structural component $Q^* = (X,Q,Y)$ of the digital circuit is described by the input, qubit vector Q, and output, defining a function $Y = Q(X)$. The dimension of the Q vector corresponds to the exponential function of the input line number $k = 2^n$. The novelty of the qubit form is replacing the truth tables of functional elements with disordered rows by the vectors of ordered states of the outputs. If the functional primitive of a digital circuit has a binary coverage, then it can be associated with a qubit or Q coverage: $Q = (1110)$:

P =

X_1	X_2	Y
0	0	1
0	1	1
1	0	1
1	1	0

→ Q = (1110).

The advantages of the *n* qubit are the ability to simultaneously perform logical operations on the vector format of set theoretic data. For example, an *xor* operation on $A = \{a,b,c,d,e,f\}$ and $B = \{a,c,f,g,h,k\}$ is performed in parallel in one cycle if each element is represented by unitary code and each subset is represented by vectors, which are the qubit operands:

9 – qubit	a	b	c	d	e	f	g	h	k
A =	1	1	1	1	1	1	0	0	0
B =	1	0	1	0	0	1	1	1	1
A ⊕ B =	0	1	0	1	1	0	1	1	1

4.4 Quantum Description of Graph Structures

The quantum *Q* vector or qubit can and should be used to represent graph structures. Function and structure are two mutually convertible forms of the description of an entity. The above qubit vector or quantum perfectly fits into the technology of programmable logic devices. , which use addressable memory elements. Given a certain isomorphism of the functions and structures, it is easy to load into memory the second component of the discrete entity description. To do this, it is necessary to represent an oriented graph in the form of the nodes *V*, encoded by the binary vectors and arcs *A* (directed transitions between them):

$$G = (V,A), V = (V_1, V_2, \ldots, V_i, \ldots V_n), A = \left(A_1, A_2, \ldots, A_j, \ldots, A_{n^2}\right).$$

In this case, a directed graph leverages the table of addresses (codes) of all the nodes (sources and drains) representing the Cartesian product. Each pair of nodes $\left(V_i, V_j\right), i,j = \overline{1,n}$ is assigned 1 if there is an arc between them ($V_i \rightarrow V_j$), and 0 if there is no such connection. Thus, the quadratic table or adjacency matrix for the graph becomes a vector of the dimension n^2, which can be located in an addressable memory element. Such a vector for describing the structure is no different from the description of any discrete functionality and is applicable to all formal methods of analysis, modeling, testing, verification, and synthesis of discrete components and systems. Therefore, a single universal data format is created in the form of a qubit or quantum vector to represent the structure of the computing device and functional

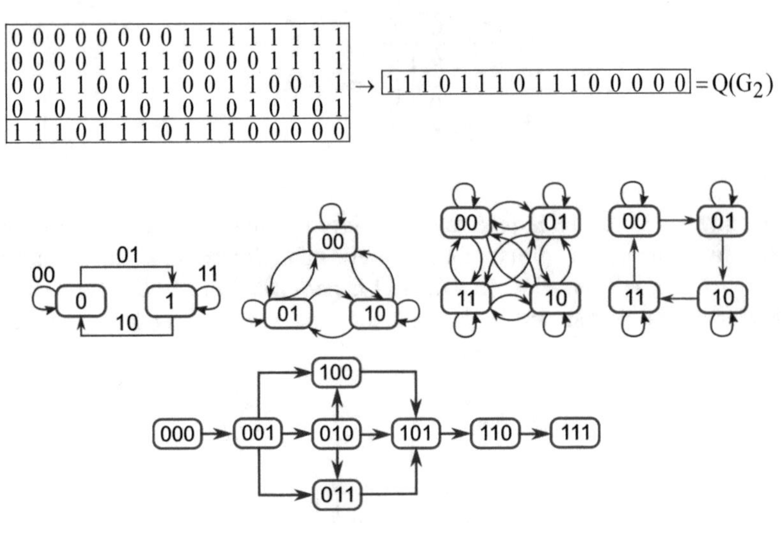

$$\begin{array}{|cccccccccccccccc|}
0&0&0&0&0&0&0&0&1&1&1&1&1&1&1&1\\
0&0&0&0&1&1&1&1&0&0&0&0&1&1&1&1\\
0&0&1&1&0&0&1&1&0&0&1&1&0&0&1&1\\
0&1&0&1&0&1&0&1&0&1&0&1&0&1&0&1\\
\hline
1&1&1&0&1&1&1&0&1&1&1&0&0&0&0&0
\end{array} \rightarrow \boxed{1110111011100000} = Q(G_2)$$

Fig. 4.2 Examples of primitive graphs and their qubit description (G1–G5)

descriptions of all its components. The following are examples of qubit description of the graph structures shown in Fig. 4.2.

The graph G1 is represented by two nodes, which are encoded as 0 and 1, respectively. Between the nodes, there are four transitions, which are designated by the vectors 00, 01, 10, 11. These pairs of binary digits constitute the addresses of memory cells in which the symbols 1 are written because each transition takes place in the graph:

$$\begin{array}{|cc|c|}
\hline
0&0&1\\ \hline
0&1&1\\ \hline
1&0&1\\ \hline
1&0&1\\ \hline
\end{array} \rightarrow \begin{array}{|c|}
\hline
1\\ \hline
1\\ \hline
1\\ \hline
1\\ \hline
\end{array} \rightarrow \boxed{1|1|1|1} = Q(G_1).$$

The absence of a transition in the column is identified by the symbol 0 in the corresponding memory cell, the address of which is composed of the node codes (source and drain). Three nodes represent the graph G2, which means that there are two bits to encode its nodes. The truth table of the graph (it is rotated for the horizontal increment of addresses in order to save space) also has zero coordinates, because all transitions connected to theoretically possible node 11 are absent:

If we use all four nodes and build on them a complete transition graph (G3), then all qubit cells containing 16 coordinates will be equal to 1:

```
|0 0 0 0 0 0 0 0 1 1 1 1 1 1 1 1|
|0 0 0 0 1 1 1 1 0 0 0 0 1 1 1 1|
|0 0 1 1 0 0 1 1 0 0 1 1 0 0 1 1|→|1 1 1 1 1 1 1 1 1 1 1 1 1 1 1 1|=Q(G3)
|0 1 0 1 0 1 0 1 0 1 0 1 0 1 0 1|
|1 1 1 1 1 1 1 1 1 1 1 1 1 1 1 1|
```

Four nodes also represent the graph G4, but it does not have all the possible transitions which means the presence of zero coordinates in the qubit vector:

```
|0 0 0 0 0 0 0 0 1 1 1 1 1 1 1 1|
|0 0 0 0 1 1 1 1 0 0 0 0 1 1 1 1|
|0 0 1 1 0 0 1 1 0 0 1 1 0 0 1 1|→|1 1 0 0 0 1 1 0 0 0 1 1 1 0 0 1|=Q(G4)
|0 1 0 1 0 1 0 1 0 1 0 1 0 1 0 1|
|1 1 0 0 0 1 1 0 0 0 1 1 1 0 0 1|
```

Thus, the qubit form of the graph representation will be more effective if the number of nodes in the structure tends toward a power of two. The qubit should be seen as an implicit and compact form of graph representation, which is unique and technologically can be simply converted into a graphic image based on analysis of the contents of the addressable memory cells (binary vector).

The nodes of the graph G5 are encoded by three bits because the number of nodes $n = 8$. The address of each qubit vector cell contains 6 bits—twice the number of bits for encoding the node, and the length of a qubit vector is represented by the expression $q = 2^{2\times\log_2 n}$. In this case, the number of qubit bits is equal to $q = 64$, the numbers of coordinates in which the units are 1, 10, 11, 12, 19, 20, 21, 29, 36, 46, and 55, and the qubit vector has the following form:

$Q(\mathrm{G}_5) = (010000000111000000111000000010000001000000000100000000)$.

Thus, the length of the qubit vector depends only on the number of nodes in a graph. If the number of arcs is small compared to the complete graph, which contains n^2 ordered connections, the efficiency E of the use of the qubit vector—the ratio of the number of unit coordinates of the qubit card (1) to its overall length $E = \mathrm{card}(1)/2^{2\times\log_2 n}$ —will be low. An example is the graph G5: $E = 11/2^{2\times\log_2 8} = 0,172$.

4.5 Qubit Description Analysis of Graph Structures

The physical nature of the qubit is the memory (binary vector), in which each addressable cell stores a value of zero or one. Its logical essence is a tuple consisting of three components (A, B, C), where A is the first part of the memory cell address represented in binary code of a source node, B is the second part of the address corresponding to a drain node, and C is the identifier of truth $C(AB) = 1$ or falsity $C(AB) = 0$ of the concatenated sentence: "There is an arc from node A to node B."

As a one-dimensional form of the adjacency matrix, a qubit vector is focused on solving all problems of discrete modeling, synthesis, analysis, and optimization. But the question is to find the class of problems for which the use of the qubit description and address-based procedures, excluding the enumeration, gives technological or economic benefits: (1) synthesis (modeling) of the qubit system model based on the qubit descriptions of the components; (2) analysis or modeling (simulation) of the qubit model of the structure based on given input conditions: searching all paths from the input or output node; (3) searching for the minimum or maximum path in the graph based on modeling of the initial conditions; (4) covering all nodes of the graph by the minimum number of paths for design and testing of SoC components; and (5) searching for faults in computational structures based on backtracking an incorrect output response.

Modeling of the graph by using the qubit is reduced to performing a single procedure—reading the state of the qubit vector cell by using the address obtained by concatenating the codes of the source and drain nodes: $M(A,B) = Q(A*B)$. This procedure answers the question of whether there is an arc (AB) in the graph. The purpose of the modeling procedure can be the following: (1) searching the maximum or minimum path in the graph; (2) defining all paths from the input node; and (3) calculating all the successor nodes for a given output. If a system includes a set of interconnected graphs, then the modeling will use the following expression: $M_i(A,B) = Q_i(A*B)$.

A quantum format of the graph representation can be used for efficient parallel execution of elementary algebraic operations. There are algebra–logical operations with qubit vectors of the same dimension: conjunction (intersection of graphs), disjunction (addition of graphs), exclusive *or* (graph comparison), and inversion (complement to the graph).

It is actually manipulation with arcs of qubit vectors of the same length in the vector form. They may be obtained if all the graphs will be reduced to one universal set of nodes, which generates the same number of potentially possible arcs. Examples of algebra–logical operations on graphs given by qubit vectors are presented in the following table:

G1	0	1	1	0	0	1	1	1	0	0	0	1	1	0	0	0
G2	0	0	1	1	0	0	0	1	1	1	0	0	1	0	1	0
And(G1, G2)	0	0	1	0	0	0	0	1	0	0	0	0	1	0	0	0
Or(G1, G2)	0	1	1	1	0	1	1	1	1	1	0	1	1	0	1	0
Xor(G1, G2)	0	1	0	1	0	1	1	0	1	1	0	1	0	0	1	0
NotG1	1	0	0	1	1	0	0	0	1	1	1	0	0	1	1	1

The operation *xor* shows the distance between the graphs, expressed as the number of units corresponding to the Hamming distance. In this example, the distance between two graphs is equal to d(G1,G2) = 8. If two or more of the graphs have a different number of nodes, to perform algebra–logical operations, it is necessary to reduce the node codes of all the graphs to the same dimension. This procedure creates the same address space for the codes (A, B) of oriented arcs of all graphs. The dimension of all qubit vectors will be the same, which is a necessary and sufficient condition for algebra–logical manipulation of them.

The problem of reduction or normalization of all graph structures to the same dimension actually means recoding of existing graph nodes in the new coordinate system. If the solution to the normalization problem is related to the subsequent comparison of the graphs, in order to determine their degree of similarity (difference), it is necessary to carry out synthesis of the arc codes approximating all the graphs to the structure of the maximum of them.

This task (as well as normalization, pattern recognition, clustering, and classification of graphs) is extremely important for the theory, has independent practical interest, and is not considered here. However, four quantum vectors, represented below, describe the first four graphs (see Fig. 4.2) reduced to the format of four nodes and two-bit vectors (the spaces in the table correspond to the zero coordinates):

G1	1			1									1			1
G2	1	1	1		1	1	1		1	1	1					
G3	1	1	1	1	1	1	1	1	1	1	1	1	1	1	1	1
G4	1	1				1	1				1	1	1			1
And(G3, G1) = G1	1			1									1			1
And(G3, G2) = G2	1	1	1		1	1	1		1	1	1					
And(G3, G4) = G4	1	1				1	1				1	1	1			1
Xor(G2, G4) = 7			1		1				1	1		1	1			1

The operation of logical multiplication shows that in the above format consisting of four nodes, all three structures (G1, G2, and G4) are a subset of the graph G3. The Hamming distance between the second and fourth graphs (the last row of the table) is 7 out of 16. The qubit form is also suitable for describing the logical functionality that makes it a universal structure for use in the design, synthesis, and analysis of modern specialized computing systems.

4.6 Modeling Qubit Description of Digital Structures

The procedure for functional modeling with the Q vector of logic functionality is reduced to writing the state of the Q vector bit in the output Boolean variable Y, the address of which is formed by the concatenation of the input variables:

$$Y = Q(X) = Q\left(X_1{}^*X_2 \ldots {}^*X_j \ldots {}^*X_k\right).$$

To simulate digital circuits, the M vector of line states is introduced as an analog of the Q vector specifying the values of the internal variables of the system. It connects the Q vectors of logical primitives (Q primitives) in the structure by using numbering of the input and output variables of each functional element [6]. The procedure for processing the functional element is described by the expression:

$$M(Y) = Q[M(X)] = Q\left[M\left(X_1{}^*X_2\ldots{}^*X_j\ldots{}^*X_k\right)\right].$$

Given through numbering of Q primitives, the universal procedure for modeling the current i element will have the following format:

$$M(Y_i) = Q_i[M(X_i)] = Q_i\left[M\left(X_{i1}{}^*X_{i2}\ldots{}^*X_{ij}\ldots{}^*X_{ik_i}\right)\right].$$

An algorithm for modeling digital systems is significantly simplified and reduced to the procedure of address generation, which makes it possible to increase the speed of interpretive modeling in $2^n - 1$ times by replacing the truth tables of primitives with Q vectors describing only output states. Furthermore, if the logic element has n inputs, the number of rows of the truth table is equal to 2^n, which means its dimension is equal to $d = 2^n \times n$. Given that the length of the Q vector for each logical primitive is 2^n, the gain in the memory for storage and processing is as follows:

$$r = \frac{2^n \times n}{2^n} = n.$$

The combinational circuit (Fig. 4.3) contains six primitives and three different logic elements [6]. A qubit of the graph and Q vectors to define the logic primitives correspond to this circuit. Thus, the graph structure Q(G1: A = 000, 7-001, 8-010, 9-011, B-100, C-101), and also all functional elements Q(A)-Q(8) are shown in the form of qubit coverage or vectors, which require significantly less memory than the adjacency matrix and the truth table of logic elements.

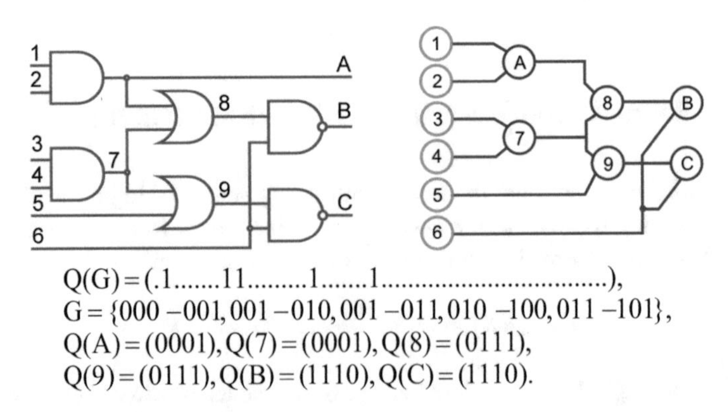

Fig. 4.3 Graph of a digital circuit

4.7 Synthesis of the Logic Circuit Quantum Vector

The qubit of the combinational circuit is a vector of output states on an ordered set of input words identified with the addresses of the memory cells [15–17] of the vector. Synthesis of the Q vector (coverage) of the circuit structure (without truth tables of logic elements) based on the primitives defined by Q vectors is reduced to obtaining a generalized qubit vector by a Cartesian product of a logical operation on the bits of the qubit vectors. The Cartesian procedure for the superposition of two four-digit qubits by using a logical operation *or* (*and*, *xor*) is represented in the following table:

$\vee,\wedge,\oplus$	b(0)	b(1)	b(2)	b(3)
a(0)	c(0) = a(0) v b(0)	c(1) = a(0) v b(1)	c(2) = a(0) v b(2)	c(3) = a(0) v b(3)
a(1)	c(4) = a(1) v b(0)	c(5) = a(1) v b(1)	c(6) = a(1) v b(2)	c(7) = a(1) v b(3)
a(2)	c(8) = a(2) v b(0)	c(9) = a(2) v b(1)	c(10) = a(2) v b(2)	c(11) = a(2) v b(3)
a(3)	c(12) = a(3) v b(0)	c(13) = a(3) v b(1)	c(14) = a(3) v b(2)	c(15) = a(3) v b(3)

Examples of using the logical superposition of two qubits for obtaining the Q vectors of the circuit structures: $c_1 = (a_1 \wedge a_2) \vee (b_1 \vee b_2), c_2 = (a_1 \wedge a_2) \wedge (b_1 \vee b_2), c_3 = (a_1 \wedge a_2) \oplus (b_1 \vee b_2)$
are represented in the following table:

a(and) = b(or) =	0001 0111
c_1 = a(and) $\vee$ b(or)	0111 0111 0111 1111
c_2 = a(and) $\wedge$ b(or)	0000 0000 0000 0111
c_3 = a(and) $\oplus$ b(or)	0111 0111 0111 1000

Q coverage for three circuits is built, consisting of three elements each, where two logical primitives are combined through superposition by a third element (*or*, *and*, *xor*). This results in three vectors, each with a dimension of 16 bits. The computational complexity of the procedure for synthesizing Q coverage of the combination system is the product of the Q vector lengths of primitives p, included in it: $\eta = \prod_{i=1}^{p} \mathrm{card}(Q_i)$. The synthesis of the circuit Q coverage is a more difficult problem. Input lines and reconvergent fan-outs are connected by wires (by the variable a_2): $c = (a_1 \wedge a_2) \vee (a_2 \vee a_3)$. In this case, after the synthesis of the circuit Q coverage, it is necessary to perform its verification to detect the possibility of the existence of conflicting addresses for variables a_2 to minimize the Q vector by subsequently excluding these addresses from consideration; this procedure reduces the dimension of the Q coverage up to card$(Q) = 2^q$, where q is the total number of input circuit variables:

Q=	0111 0111 0111 1111
a1 =	0000 0000 1111 1111
a2 =	0000 1111 0000 1111
a2 =	0011 0011 0011 0011
a3 =	0101 0101 0101 0101

=

Q=	0111 0111 0111 1111
a1 =	0000 0000 1111 1111
a2 =	00xx xx11 00xx xx11
a2 =	00xx xx11 00xx xx11
a3 =	0101 0101 0101 0101

=

Q=	0111 0111
a1 =	0000 1111
a2 =	0011 0011
a2 =	0011 0011
a3 =	0101 0101

=

Q=	0111 0111
a1 =	0000 1111
a2 =	0011 0011
a3 =	0101 0101

The procedure for the synthesis of Q coverage includes the following steps: (1) the correspondence table is built for the address and digits of the circuit Q vector; (2) the conflicting coordinates in two rows a_2 are marked by symbols x; (3) all the columns containing these symbols are then excluded from the table; (4) two identical rows a_2 are obtained, which are combined into one row; and (5) this gives us the Q vector of a combinational circuit, but with a much smaller dimension. The advantages of the proposed Q method for synthesizing computing devices are the compact description of Q vectors and the high-speed simulation of logic element addresses, as well as creating the conditions for developing a market-feasible "quantum" theory for SoC design, based on the use of a qubit vector of the structural components.

4.8 Minimization of the Logic Circuit Quantum Vector

Use of Q coverage performs the synthesis of the circuit quantum vector by using Q coverage through minimization or reduction of the dimension of the Q vector by eliminating nonessential variables. Usually essential variables depend on the galvanic connections of input and internal lines of the digital device, which impose constraints related to inconsistency of signals on communication lines. Therefore, the rule of address space minimization is to remove the address codes, which create inconsistency in the connected variables. Suppose we have the Q vector of a circuit and its address space where the variables b,c,d (a,b,c) are wire connected. Below, the tables of transformation or minimization of the address space in order to obtain a reduced Q vector are represented:

Q=	0111 0111 0111 1111
a=	0000 0000 1111 1111
b=	0000 1111 0000 1111
c=	0011 0011 0011 0011
d=	0101 0101 0101 0101

Q=	0xxx xxx1 0xxx xxx1
a=	0000 0000 1111 1111
b=	0xxx xxx1 0xxx xxx1
c=	0xxx xxx1 0xxx xxx1
d=	0xxx xxx1 0xxx xxx1

=

Q	0101
a=	0011
b=	010

Q=	0111 0111 0111 1111
a=	00xx xxxx xxxx xx11
b=	00xx xxxx xxxx xx11
c=	00xx xxxx xxxx xx11
d=	0101 0101 0101 0101

=

Q	0111
a=	0011
d=	0101

Firstly, it should be noted that we can see mirrored axis symmetry with signal inversion on the coordinates of the address space, which creates the property described by the following expression: $L \oplus R = 1 \rightarrow L_{ij} \oplus R_{ij} = 1$. This fact should be used to reduce the dimension of the analyzed space by two times with a corresponding reduction in the computational complexity of the synthesis problem of quantum vector functionality of the digital circuit. Secondly, the number of different interactions of q input variables associated with wire connections of the various combinations of input lines is determined by the functional dependence, the boundary values of which are in the range card(Q) = [2^q–3^q]. Nevertheless, there is an effective procedure for minimizing the dimension of the Q vector by detecting contradictions in column codes in coordinates (A_{ij}), which correspond to wired w variables by the j parameter. This procedure can be sufficiently performed at half the address space card(Q) = $2^q/2$, and the rest of the conflicting columns are removed in accordance with the mirrored reflection of the numbers of those columns, which were removed from the first half of the table of address codes:

$$\{Q_i, Q_{2^q-i}\} = \varnothing \leftrightarrow \left(\wedge_{j=1}^{w} A_{ij}\right) \oplus \left(\vee_{j=1}^{w} A_{ij}\right) = 1, \;\; i \leq 2^q/2.$$

If in the column Ai there is a group of w-connected variables, for which the conjunction of their states is zero, and the disjunction has a value of one, then the i column and its mirrored reflection ($2^q - 1$) are removed from the address space A. This will automatically lead to the exclusion of two received $\varnothing$ coordinates (in the tables they are marked by the symbol x), corresponding to these columns, from the Q vector.

Really, there is also the symmetry of space for vectors of Hamming distances obtained through *xor* interaction between adjacent rows of the table of the address space, for which the superposition of the left and right sides gives the result $L \oplus R = 0 \rightarrow L_{ij} \oplus R_{ij} = 0$:

Q=	0111 0111 0111 1111
a ⊕ b	0000 1111 1111 0000
b ⊕ c	0011 1100 0011 1100
c ⊕ d	0110 0110 0110 0110
d ⊕ a	0101 0101 1010 1010

$= (L,R);\ (L \oplus R) =$

Q=	0111 0111 0111 1111
a ⊕ b	0000 0000
b ⊕ c	0000 0000
c ⊕ d	0000 0000
d ⊕ a	0000 0000

$\leftrightarrow (L = \bar{R})$

Is it advisable to minimize the logic function described by the quantum vector? One answer is that minimization of Q vectors for obtaining normal or bracket logic forms has no practical significance; it is important only in significantly reducing the dimension of the vector of functional description. It can only be a consequence of defining the insignificance of some input (address) variables. However, there is a problem of partitioning the quantum vector into parts of smaller dimensions, which is related to the implementation of the functionality in the structural components of the LUT FPGA [15–17]. In this case, partitioning of the Q vector into two equal subvectors $Q = (L, R)$ is performed; the subvectors are combined into structural and address organization of the functionality via multiplexer $Q = (\bar{a} \wedge L) \vee (a \wedge R)$. If the multiplexing variable $a = 0$, the functionality Q is formed by the cells of the left L vector; otherwise, if $a = 1$, the value of the function Q is formed by the bits of the right R vector. Partitioning and implementation algorithms of complex logic functions are available in each system for synthesis, simulation, and verification of SoC components.

4.9 Models of Quantum Sequential Primitives

The following are the axioms of a quantum (memory-only-based) processor: (1) there is nothing in a quantum processor but addressable memory; (2) the computational process is represented by the only universal transaction between addressable memory components $M_i = Q_i[M(X_i)]$; (3) the transaction is a universal procedure of reading and writing data to the nonempty set of addressable memory elements; (4) all memory components are repaired online, due to their pseudogalvanic address-connected connectivity; (5) combinational logic elements (reusable logic), as well as sequential components, are implemented in the memory elements; (6) the union of all components in a computing system is performed by (digital) identification of wire connections of input–output variables of the circuit components forming the simulation vector, which stores the states of all the essential lines of the digital system; (7) all components of the quantum model of the digital system $W = \langle Q,M,X \rangle$, including functional modules, simulation vectors, and vectors of input variable addresses, are reprogrammable online, so they are repairable online; and (8) a primitive of a digital system has the format $W = \langle Q,Y, X \rangle$ because a separate element does not have the connections and vector M to create the system from the individual components.

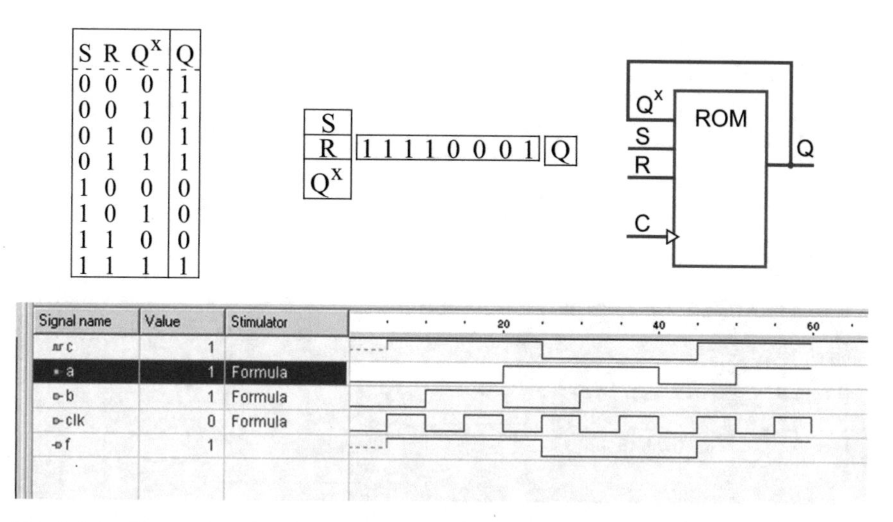

Fig. 4.4 Implementation of an SR flip-flop on a memory element

According to the introduced quantum model, the descriptions of the sequential primitives (flip-flops, registers, and counters) can be represented by Q coverage or qubit vectors, which have pseudovariables to define the internal state. For example, a functional description of the SR flip-flop is transformed into the quantum primitive, specified by Q coverage; then it is implemented in the addressable memory element FPGA, verification diagrams of which are represented in Fig. 4.4.

The truth table of the flip-flop is represented in the form of the output state vector $Q(S, R, Q^X) = (11110001)$, which is written in the element of the constant memory with three address inputs, the synchronization signal, and the feedback that connects the output of the memory element and an address input. HDL implementation of the synthesized SR flip-flop by leveraging the Active-HDL 9.1 design system (Aldec, Inc., Henderson, NV, USA) and verification results confirm the correctness of the proposed circuit solution.

Another example relates to the synthesis of a synchronous DV flip-flop on the element of the constant memory. The truth table of the flip-flop is transformed into a vector of output states $Q(D, V, Q^X) = (01000111)$, which is written in the memory element with three address inputs, the synchronization signal, and the feedback connecting the output of the memory primitive and an address input. All of these components, including timing diagrams of HDL code verification for the DV flip-flop model, are shown in Fig. 4.5.

Fig. 4.6 shows models of two sequential primitives, such as a two-digit register and counter. Their difference lies in the definition of two outputs, the states of which are formed by the same set of input variables.

The register for variables (C,D,$Q1$,$Q2$,$Y1$,$Y2$) realizes a shift to the right of information from the input D by the digits (D-$Y1$-$Y2$) when R = 1, and saving

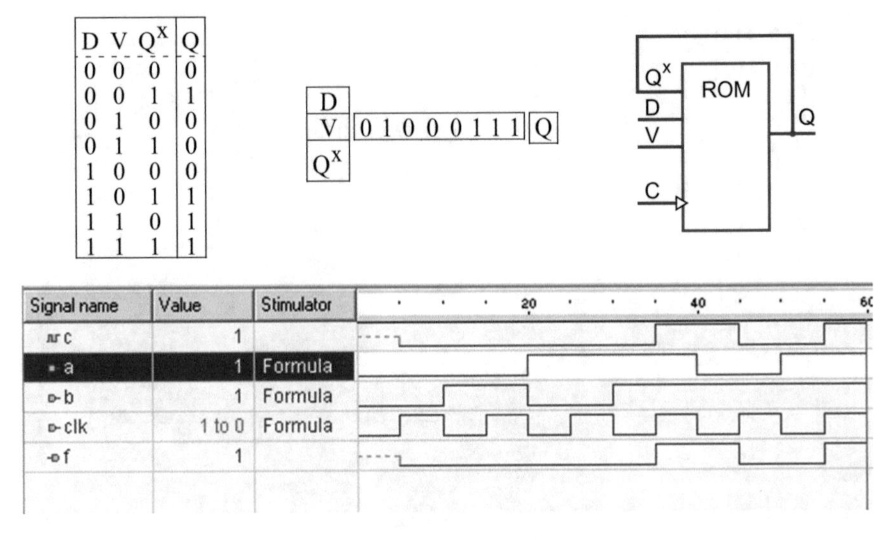

Fig. 4.5 Implementation of a DV flip-flop on a memory element

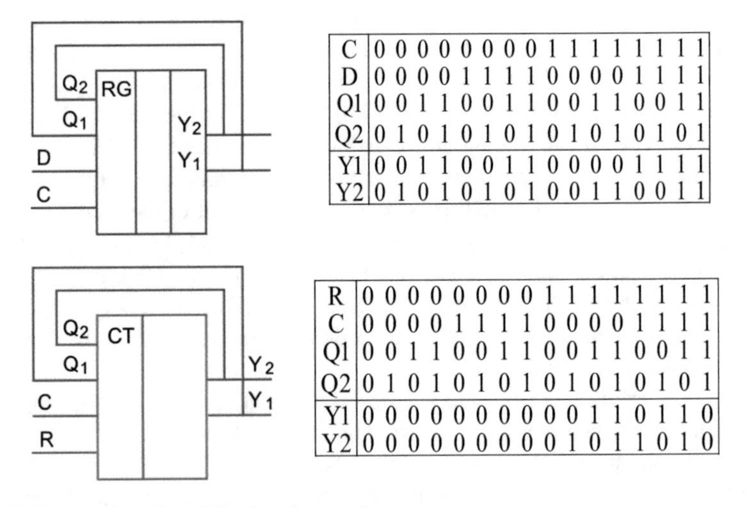

Fig. 4.6 Memory-based models of register and counter

data when $C = 0$. A counter defined on the variables $(R,C,Q1,Q2,Y1,Y2)$ realizes the function of an increment for the digits $(Y1,Y2)$ when $RC = (11)$, and storing information when (R or $C = 0$). Thus, to implement the two-bit register or counter, we need two 16-bit memory elements operating synchronously on the same inputs:

Inputs	Q vector	Output
C		
D	0 0 1 1 0 0 1 1 0 0 0 0 1 1 1 1	Y1
Q1	0 1 0 1 0 1 0 1 0 0 1 1 0 0 1 1	Y2
Q2		

Inputs	Q vector	Output
R		
C	0 0 0 0 0 0 0 0 0 0 1 1 0 1 1 0	Y1
Q1	0 0 0 0 0 0 0 0 0 1 0 1 1 0 1 0	Y2
Q2		

Here, each quantum model is represented by two vectors, where each of them forms a function for a digit in a register or counter as the state of the vector cell obtained when generating an address A by using input variables $\{Y1, Y2\} = A(C, D, Q1, Q2)$, $\{Y1, Y2\} = A(R, C, Q1, Q2)$, correspondingly. Primitive simulation is reduced to a trivial procedure for forming the address where an output state of a primitive is written as the contents of a quantum vector cell.

So, there are no any barriers for step-by-step replacing the truth table and Boolean algebra for doing practical transition to the Q vector description of digital primitives focused into addressable memory-only structures. It is now high time to develop and investigate the innovative theory of Q vector synthesis and analysis of SoC design. Such a theory allows simplification of a lot of problem decisions connected to verification, testing, diagnosis, repair, fault simulation, and hazard detection for complicated discrete systems.

4.10 Conclusion

A new technology for analysis and synthesis of SoC components is suggested. The proposed data structures and methods are determined by the following features: (1) leverage of memory cells as a background for the interaction between the execution and control automata; (2) a synthesis method for SoC components, leveraging the superposition of qubit vector descriptions of the functionalities implemented in the memory cells, which creates opportunities to create high-performance tools for modeling, simulation, testing, and verification, and to simplify the procedures for real and virtual computer system design; (3) original data structures for simulation of digital SoCs, which allow simplification of the design and testing algorithms and improvement of their performance due to functional element addressability and parallel processing of all components; and (4) computing structure and process implementation in cloud services based on the leverage of an addressable memory-driven qubit automaton, which allows creation of a functional infrastructure for testable design to improve SoC yield due to online repair of system components.

The proposed qubit models for describing the structure of digital systems and functions of the components are characterized by a compact description of the adjacency matrix and truth tables in the form of Q coverage, with unitary encoding of the input states allowing increases in the performance of hardware and software for modeling computing devices due to addressable implementation of the analysis of qubit vectors.

The innovative idea of quantum computing is to transfer from computational procedures of byte operands specifying a single solution in a discrete space to logical register parallel processes with qubit operands, which simultaneously form the power set of solutions. This makes it possible to determine new ways of creating high-performance computers for parallel analysis and synthesis of structures and computing services.

A new qubit form of graph description is proposed, which is characterized by a compact description of all oriented arcs, with vector representation of an interconnected structure that makes it possible to significantly increase the speed of methods for analysis and modeling through execution of parallel logical operations.

The high applicability of the qubit (quantum) form for description of computing components creates a background for effective solving of a wide range of discrete optimization problems, synthesis and analysis, recognition and diagnosis, and simulation and testing.

The theoretical and practical significance of the proposed method for analysis and synthesis of digital system components are as follows:

1. Creation of a qubit-driven processor based on memory-only elements that make it uniform in its structure and types of functional primitive description. This delivers market-oriented technological convenience for all design processes, manufacturing, and execution, including SoC verification, embedded testing, and fault diagnosis [15–17]. The main thing is online repair through leverage of universal addressable spare memory cells.
2. Use of qubit memory-only-based models for digital component and structure description, focused on increasing the yield, as well as improving the reliability of computing systems and reducing the costs of design and manufacturing.

Future research will be focused on the development of high-speed methods and algorithms for synthesis, analysis, visualization, normalization, pattern recognition, clustering, and classification of graph-functional structures and systems through the use of a universal metric (such as Manhattan) for reduction and normalization of graphs in a single coordinate system.

References

1. Nielsen, M.A., & Chuang, I.L. (2010). *Quantum computation and quantum information*. Cambridge University Press.
2. Whitney, M.G. (2009). *Practical fault tolerance for quantum circuits*. PhD dissertation. University of California, Berkeley.

3. Nfrfhara, M. (2010). *Quantum computing. An overview*. Higashi-Osaka: Kinki University.
4. Kurosh, A.G. (1968). *The course of higher algebra*. Moscow: Publishing House Nauka.
5. Gorbatov, V.A. (1986). *Basics of discrete mathematics*. Moscow: Higher School.
6. Hahanov, V.I., Litvinova, E.I., Chumachenko, S.V., et al. (2012). *Qubit model for solving the coverage problem*. In: Proc. of IEEE East-West Design and Test Symposium, 2012.
7. Hahanov, V.I., Gharibi, W., Litvinova, E.I., Shkil, A.S. (2015). Qubit data structures of computing devices. *Electronic Modeling Journal*, *1*, 76–99.
8. Hahanov, V.I., Bani Amer, T., Chumachenko, S.V., Litvinova, E.I. (2015). Qubit technology for analysis and diagnosis of digital devices. *Electronic modeling Journal*, *37*(3), 17–40.
9. Hahanov, V., Gharibi, W., Iemelianov, I., Shcherbin, D. (2015). *"Quantum" processor for digital systems analysis*. In: Proceedings of IEEE East-West Design & Test Symposium, 2015.
10. Metodi, T., & Chong, F. (2006). *Quantum computing for computer architects. Synthesis Lectures on Computer Architecture*. Morgan & Claypool.
11. Stig, S., & Suominen, K.-A. (2005). *Quantum approach to informatics*. John Wiley & Sons, Inc.
12. Hahanov, V., Bani Amer, T., Hahanov, I. (2015). *MQT-model for Virtual Computer Design*. In: Proc. of Microtechnology and Thermal Problems in Electronics (Microtherm), 2015.
13. Zorian, Y., & Shoukourian S. (2013). *Test solutions for nanoscale systems-on-chip: Algorithms, methods and test infrastructure*. In: Computer Science and Information Technologies (CSIT), 2013.
14. Zorian, Y., & Shoukourian, S. (2003). Embedded-memory test and repair: Infrastructure IP for SoC yield. *IEEE Design & Test of computers Journal*, *20*(3), 58–66.
15. Dugganapally, I.P., Watkins, S.E., Cooper, B. (2014). *Multi-level, memory-based logic using CMOS technology*. In: 2014 I.E. computer society annual symposium on VLSI (ISVLSI), 2014.
16. Yueh, W., Chatterjee, S., Zia, M., Bhunia, S., Mukhopadhyay, S. (2015). A memory-based logic block with optimized-for-read SRAM for energy-efficient reconfigurable computing fabric. IEEE transactions on circuits and systems II. *Express Briefs Journal*, *62*(6), 593–597.
17. Matsunaga, S., Hayakawa, J., Ikeda, S., Miura, K., Endoh, T., Ohno, H., Hanyu, T. (2009). *MTJ-based nonvolatile logic-in-memory circuit, future prospects and issues*. In: Design, Automation & Test in Europe Conference & Exhibition, 2009.

Chapter 5
Quantum Computing for Test Synthesis

Vladimir Hahanov, Tamer Bani Amer, Igor Iemelianov, and Mykhailo Liubarskyi

5.1 Memory-Based Logic

The concept of address execution of logical operations implemented in the look-up tables (LUTs) of programmable logic devices (PLDs) provides a potential opportunity to create only on-chip address space, applicable for built-in self-repairing of all digital circuit components [1–3]. The motivation for address space creation for all components is supported by the distribution of logic and memory on a chip, shown in Fig. 5.1, where the chip after 2020 will be represented by 1% logic and 99% memory.

The trend toward an increase in memory leads to the ability to build in repair of failed components through spare logic cells. The problem of autonomous correction of faulty logic elements (self-repair) is associated with the lack of their addresses. It can be solved by means of creation of flexible addressable connections between logic elements placed in the memory. Additionally, the memory should contain an algorithm for processing logic elements. If a fault is detected in an addressable logic element, a built-in self-repair system (BISR) restores the functionality by readdressing to the spare analog. The problem of improving the quality and reliability of digital systems on chips (SoCs) is solved by the creation of built-in self-test (BIST) and BISR infrastructure for testing, diagnosis, optimization, and repair due to hardware redundancy and reduction of functional operation performance [2–5].

An illustrative example is the FPGA Family Xilinx® UltraScale™ Architecture configurable logic block—the first ASIC class all programmable architecture—which

V. Hahanov (✉) • T.B. Amer • I. Iemelianov • M. Liubarskyi
Design Automation Department, Kharkov National University of Radio Electronics, Ukraine, Nauka Avenue 14, Kharkov 61166, Ukraine
e-mail: hahanov@icloud.com; tameramer34@yahoo.com; igor@itdelight.com; mlyubarskyy@gmail.com

V. Hahanov, *Cyber Physical Computing for IoT-driven Services*,
DOI 10.1007/978-3-319-54825-8_5

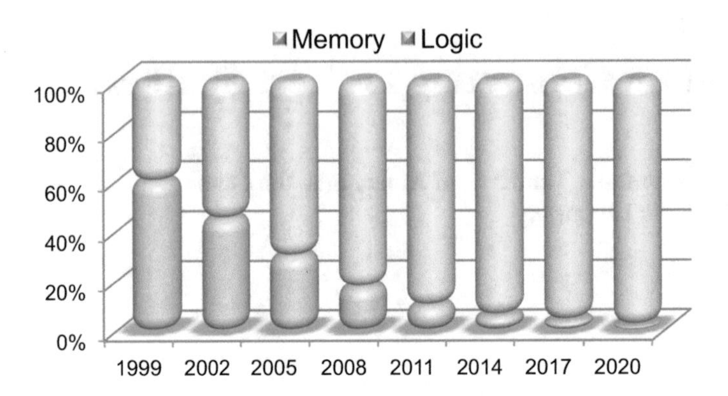

Fig. 5.1 On-chip distribution of memory and logic

belongs to the ASIC class and provides the performance of implemented smart functionalities at hundreds of gigabytes per second. It is manufactured in the form of a 3D-on-3D chip at 20 nm technology based on FinFET transistors and operates as a multiprocessor SoC (MPSoC), providing network performance of 1+ Tb/s of data exchange [www.xilinx.com/support/documentation/user_guides/ug574-ultrascale-clb.pdf]. In the early 2000s, the ratio of the performance of the processor logic and memory was about 100:1. At present, the situation is gradually becoming aligned in the direction of decreasing the difference to 10:1. This means that there is a process of transforming logic into logic in memory in order to improve the reliability of digital SoCs during partial loss of performance. Finally, memory and logic have a common atomic limit of perfection, when the memory bit will be identified with a spin or orbit of an electron. The performance of interatomic transactions based on photon exchange would be commensurate with light speed. Classical transistor logic, which is controlled by the flow or one electron, has the same analogue of perfection—change of an electron state under the influence of a quantum. Thus, the transition of the electronic industry to only-memory computing (memory–address–transaction (MAT)) is inevitable. Hence the conclusion that timely creation of a theory of synthesis and analysis of digital computational units based on MAT computing can bring to the market fundamentally new possibilities for solving the problems of reliability, performance, and energy saving.

The goal is a significant increase in the speed of test synthesis and deductive verification for black box functionality of logical components through the use of compact descriptions in the form of qubit coverage and parallel execution of a minimum number of register logic operations (*shift*, *or*, *not*, *nxor*).

The objectives of this chapter are the following:

1. Testing of a generation method for black box functions based on the use of qubit coverage and parallel execution of register logic operations (*shift*, *or*, *not*, *nxor*).
2. A method for taking the Boolean derivative for test synthesis based on the use of qubit coverage.

3. A test synthesis method based on the use of Boolean derivatives represented by vectors of qubit coverage.
4. A deductive fault simulation method for the functional elements represented by vectors of qubit coverage.
5. A fault simulation processor based on qubit description of functionality, integrating methods developed in the cloud or the infrastructure of SoC test services.
6. Testing and verification of methods for test synthesis and fault simulation by means of parallel execution of register logic operations on qubit descriptions of logic circuits.

The research content is development of test generation methods and evaluation of their quality for functional logic components by means of parallel execution of register logic operations (*shift*, *or*, *not*, *nxor*) on the qubit vector and its derivatives in terms of a qubit simulation processor.

Testing and diagnosis of memory devices is performed by the use of special test generation algorithms: march, running zero and one, and logarithmic division [1–5]. If the memory elements perform the functions of logic circuits, it is necessary to generate tests for their functional verification not related to the physics of storage processes. Therefore, a method for test synthesis of functional circuits represented in the form of a black box, described by qubit vectors, is considered [6, 7].

The market feasibility of the application of quantum computing methods to create computing structures in cyberspace is based on the use of qubit data models, focused on parallel solving of the problems of designing, testing, and discrete optimization [5, 8] through an increase in storage costs. Without going into the details of the physical foundations of quantum mechanics related to nondeterministic interactions of atomic particles [9, 10], we use the notion of the qubit as a binary vector for concurrent and simultaneous definition of the set of all subsets in a discrete cyberspace area based on linear superposition of unitary codes, focused on parallel execution of operations. In the rapidly developing theory of quantum computing the state vectors form the quantum register of n qubits, which constitute a unitary or Hilbert space [11] H, the dimension of which has a power dependence on the number of qubits Dim $H = 2^n$.

The motivation for a new approach to the design of computing systems is explained by the appearance of new cloud services within the framework of the Internet of Things, which involves systems that are specialized and distributed in space, implemented in hardware or software. Any functional components, as well as the structure of the system, can be represented by using a vector form of a truth table, ordered by addresses and implemented in memory. Further, traditional reusable logic functions are not considered. This leads to partially reducing the performance, but given that 94% of SoC is memory [2, 3], the remaining 6% will be implemented in the memory too, which is not critical for most cloud services. In practice, for developing effective computing structures it is necessary to use a theory based on computing components of a high level of abstraction: addressable memory and the transaction.

The feature of data organization in a classical computer is addressing of a bit, byte, or other component. Addressability defines a problem of processing the association of nonaddressable elements of a set that have no order by definition. The solution is a processor in which the image of the universe of n unitary coded primitives uses superposition to form a set of all subsets $|B(A)| = 2^n$ for all possible states [4, 12].

There is an analogy between the vector representation of a set of all subsets and the qubit of a quantum computer. A set of all subsets forming the Cantor alphabet $A^k = \{0,1,X,\varnothing\}, X = \{0,1\}$ can be put in one-to-one correspondence to the quantum qubit for storing information in a quantum computer [9, 10], supporting the superposition $|\psi\rangle = \alpha|0\rangle + \beta|1\rangle$. Here the alphabet primitives are unitarily encoded by the vectors 0 = (10) и 1 = (01). The codes of other symbols are obtained from the primary symbols by using the operation of superposition (10)∨(01) = (11) and intersection (10)∧ (01) = (00) forming a set of all subsets: {(10), (01), (11), (00)} [4, 11, 12].

The correctness of the use of the term "qubit" for digital device models is based on a comparison of linear algebra and Boolean Cantor algebra with the alphabet $A^k = \{0,1,X,\varnothing\}$. Here the first two symbols are the primitives. The third one is defined by superposition: $X = 0 \cup 1$. In a Hilbert space, linear algebra operates by the primitives $\alpha|0\rangle$, $\beta|1\rangle$ of the qubit, and the third symbol is a superposition of two components: $|\psi\rangle = \alpha|0\rangle + \beta|1\rangle$. In linear algebra, there is no symbol corresponding to the empty set; it is a secondary primitive, obtained by using the inverse function of superposition. In the set theory, such an operation is called intersection, giving the empty set of the primitives: $\varnothing = 0 \cap 1$. In a Hilbert space, such an operation is a scalar product or Dirac function $\langle a|b\rangle$ [11] that has a geometric interpretation: $\langle\alpha|\beta\rangle = |\alpha| \times |\beta| \times \cos\angle(\alpha,\beta)$. If the projections a and b of the quantum state vector are orthogonal we obtain the following: $\langle\alpha|\beta\rangle = |\alpha| \times |\beta| \times \cos\angle(\alpha,\beta) = |\alpha| \times |\beta| \times \cos\angle 90° = 0$. The scalar product of the orthogonal vectors is zero, which is an analog of the empty set symbol in the Cantor algebra. A correspondence table of the algebra of sets and linear algebra, represented below, confirms the properties of the isomorphism between the symbols of a set of all subsets and the states of the qubit vector:

Boolean $A^k =$	0	1	$X = 0 \cup 1$	$\varnothing = 0 \cap 1$
Qubit $\vert\psi\rangle =$	$\vert 0\rangle$	$\vert 1\rangle$	$\alpha\vert 0\rangle + \beta\vert 1\rangle$	$\alpha\vert 0\rangle\vert\beta\vert 1\rangle$

Therefore, the data structure "a set of all subsets" can be seen as a deterministic image of the quantum qubit in the algebra of logic, elements of which are unitarily encoded by binary vectors and have the properties of superposition, parallelism, and entanglement. This makes it possible to use the proposed qubit model for improving the performance of digital device analysis on classical computers, and also without modification in quantum computers, which will appear in a few years in the market of electronics.

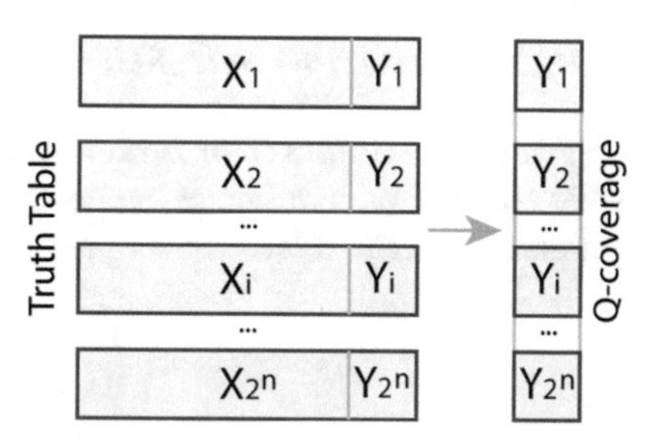

Fig. 5.2 Truth table and qubit coverage

The qubit description of digital functional elements is represented below. The qubit (n qubit) is the vector form of unitary coding of n primitives for a set of all subsets defined as 2^{2^n} by using 2^n binary variables. A superposition in a vector of 2^n states, defined as primitives, is assumed. When setting the binary vector of the logic function, an analog of a qubit is Q coverage or a Q vector [6, 7] as a general vector form of the structure and function description. The format of the circuit qubit component $Q^* = (X,Q,Y)$ includes the input–output variables and the qubit vector Q of the function $Y = Q(X)$. The novelty of the qubit vector lies in replacing truth tables of functional elements with disordered rows by vectors of ordered states of outputs. For instance, if a functional primitive has a binary table, it can be associated with the following qubit or Q coverage:

$$C = \begin{array}{ccc} X_1 & X_2 & X_3 \\ 0 & 0 & 1 \\ 0 & 1 & 1 \\ 1 & 0 & 1 \\ 1 & 1 & 0 \end{array} \rightarrow Q = (1110)$$

Thus, a disruptive idea in the basis of research is to replace a set of input–output relationships of the truth table by a qubit vector of addressable output states (Fig. 5.2).

The compactness and simplicity of the qubit vector form (Q coverage) dictate leveraging of only simple parallel register operations (*not*, *shift*, *or*, *and*, *xor*) to solve all problems of synthesis and analysis of digital devices.

5.2 Qubit Test Generation Method

A test synthesis method using Q coverage or qubit vectors for a functional compact description of the digital device, which is characterized by parallel execution of the logical operations, is proposed.

Q coverage is the vector form of the logic functionality description where each bit has an address concatenated by the binary states of input variables:

$$Q = (Q_1, Q_2, \ldots, Q_i, \ldots, Q_n), Q_i = \{0, 1\}, Q_i = Q(i), i = (X_1X_2, \ldots, X_i, \ldots, X_n).$$

The Q test is the vector form of the test pattern description of digital devices where the vector coordinates create ordered binary sequences, applied to the input variables with the rule:

$$Q_i = Q(i) = \begin{cases} 1 \rightarrow (X_1X_2, \ldots, X_i, \ldots, X_n) = i, \\ 0 \rightarrow (X_1X_2, \ldots, X_i, \ldots, X_n) = \varnothing. \end{cases}$$

In other words, if the coordinate of the vector $Q(i) = 1$, then a test sequence, composed of bits, forming the decimal address i, is applied to the input device. Otherwise, when there is a zero value of the coordinate $Q(i) = 0$, a test sequence is absent.

Naturally, the dimensions of the Q coverage and and generated Q test T for the single fault F are always the same, which makes it possible to perform a parallel analysis of their interaction in the test synthesis of functional elements according to the rule [4], where Q^* is truth table: $T = f(Q^*, F) = Q^* \oplus F$.

Example 1: It is necessary to generate a fault detection test of single stuck-at faults to logic element 2 and, given truth table. To do this, given the activation table of single stuck-at faults of the input variables where the number of cubes equals the number of inputs. Each fault cube has a 1 unit on the input and output variable coordinates, and the other cube coordinates are zero. This means that there is a fault detection axiom: a change of the input variable should cause a change in the output state.

The test synthesis algorithm is based on the truth table [4]:

1. The first step is to execute the coordinate *xor* operation between the truth table and each row of the fault matrix, which generates two tables, containing the rows–candidates in the test, according to the number of input variables:

$$\begin{bmatrix} X_1 & X_2 & Y \\ 0 & 0 & 0 \\ 0 & 1 & 0 \\ 1 & 0 & 0 \\ 1 & 1 & 1 \end{bmatrix} \oplus \begin{bmatrix} F_1 & F_2 & F_Y \\ 1 & 0 & 1 \\ 0 & 1 & 1 \end{bmatrix} = \begin{bmatrix} X_1 & X_2 & \bar{Y}_1 \\ 1 & 0 & 1 \\ 1 & 1 & 1 \\ 0 & 0 & 1 \\ 0 & 1 & 0 \end{bmatrix} \vee \begin{bmatrix} X_1 & X_2 & \bar{Y}_2 \\ 0 & 1 & 1 \\ 0 & 0 & 1 \\ 1 & 1 & 1 \\ 1 & 1 & 0 \end{bmatrix}$$

2. Then the rows of the last two tables are arranged according to the increasing binary codes of the input variables:

$$\begin{matrix} X_1 & X_2 & \bar{Y}_1 \\ 1 & 0 & 1 \\ [1 & 1 & 1 \\ 0 & 0 & 1 \\ 0 & 1 & 0 \end{matrix}] \vee [\begin{matrix} X_1 & X_2 & \bar{Y}_2 \\ 0 & 1 & 1 \\ 0 & 0 & 1 \\ 1 & 1 & 1 \\ 1 & 1 & 0 \end{matrix}] = \begin{bmatrix} X_1 & X_2 & Y'_1 \\ 0 & 0 & 1 \\ 0 & 1 & 0 \\ 1 & 0 & 1 \\ 1 & 1 & 1 \end{bmatrix} = \begin{bmatrix} X_1 & X_2 & Y'_2 \\ 0 & 0 & 1 \\ 0 & 1 & 1 \\ 1 & 0 & 0 \\ 1 & 1 & 1 \end{bmatrix}$$

3. Thereafter, a comparison of the output states (Y'_1, Y'_2) with the output values vector of original truth table Y by the rule of equivalence operations (*nxor*) is executed. If the values of the same coordinates of two vectors are equal, the result of the comparison is 1, otherwise it is 0:

$$\begin{bmatrix} Y \\ 0 \\ 0 \\ 0 \\ 1 \end{bmatrix} \overline{\oplus} \begin{bmatrix} Y'_1 \\ 1 \\ 0 \\ 1 \\ 1 \end{bmatrix} \vee \begin{bmatrix} Y'_2 \\ 1 \\ 1 \\ 0 \\ 1 \end{bmatrix} = \begin{bmatrix} Y^t_1 \\ 0 \\ 1 \\ 0 \\ 1 \end{bmatrix} = \begin{bmatrix} Y^t_2 \\ 0 \\ 0 \\ 1 \\ 1 \end{bmatrix}$$

4. In the final step, the logical *or* operation between the results (Y^t_1, Y^t_2) obtained before is executed, which gives the T vector. The 1 units of the T vector identify the binary input sequences, which should be applied to a digital input device to test it:

$$\begin{bmatrix} Y^t_1 \\ 0 \\ 1 \\ 0 \\ 1 \end{bmatrix} \vee \begin{bmatrix} Y^t_2 \\ 0 \\ 0 \\ 1 \\ 1 \end{bmatrix} = \begin{bmatrix} T \\ 0 \\ 1 \\ 1 \\ 1 \end{bmatrix} \rightarrow \begin{bmatrix} X_1 & X_2 & Y \\ \cdot & \cdot & \cdot \\ 0 & 1 & 0 \\ 1 & 0 & 0 \\ 1 & 1 & 1 \end{bmatrix}$$

In this example, the T vector 0111 contains three 1 units, so its corresponding input sequences are represented by three components: 010, 100, 111.

Example 2: This illustrates the test synthesis algorithm [4], using the truth table of the logic function, with three input variables. A truth table is featured where the last column is the Y qubit vector. Columns $(F_1F_2F_3F_Y)$ form a fault matrix where the essence of each line is one-dimensional activation of input to output. Columns ($Y'_1Y'_2Y'_3$) are obtained as an *xor* operation result of the truth table and fault matrix, which determines the behavior of functionality when applied for each input error, inverted to its good behavior. Columns $(Y^t_1Y^t_2Y^t_3)$ are obtained by vector coordinate comparisons of truth table output states and vectors $(Y'_1Y'_2Y'_3)$, obtained in the previous step. Vector T logically combines sets $(Y'_1Y'_2Y'_3)$ to test for a given functionality:

X_1	X_2	X_3	Y	F_1	F_2	F_3	F_Y	Y_1'	Y_2'	Y_3'	Y_1^t	Y_2^t	Y_3^t	T
0	0	0	1	1	0	0	1	1	1	0	1	1	0	1
0	0	1	1	0	1	0	1	0	0	0	0	0	0	0
0	1	0	0	0	0	1	1	0	0	0	1	1	1	1
0	1	1	1					1	0	1	1	0	1	1
1	0	0	0					0	0	0	1	1	1	1
1	0	1	1					0	1	1	0	1	1	1
1	1	0	1					1	1	1	1	1	1	1
1	1	1	0					0	0	0	1	1	1	1

The last column is the fault detection test $T = (10111111)$ for external input and output variables of the functional element. The test vector sets the input sequences—0001, 0100, 0111, 1000, 1011, 1101, 1110—that should be applied to the inputs of the functional elements to check all stuck-at defects.

The Q method of test synthesis for stuck-at faults of functional primitives, based on the leverage of Q coverage, contains the following items:

1. Negation of Q coverage for all coordinates: $Q_i = \bar{Q}_i, i = \overline{1, 2^n}$.
2. Ordering of inverted Q coverage for each of n input variables $Q_j = S_j(\bar{Q}), j = \overline{1, n}$. This procedure means a qubit vector shift operation where each shifting has its own logical algorithm for every input variable, illustrated by the following scheme (Fig. 5.3), with the three inputs.

Here, the address indices of the second row define the data exchange process between the adjacent coordinates of a qubit vector through a toward shift of the contents of two adjacent coordinates generating a test vector for the first input variable. The third row shows the data exchange between adjacent pairs of qubit vector coordinates through a toward shift of contents generating a test vector for the second input variable. The fourth row specifies the data exchange between adjacent pairs of qubit vector coordinates through a toward shift of contents generating a test vector for the third input variable.

3. T vector generation for each of the n input variables by comparison with the initial qubit coverage of the functional element:

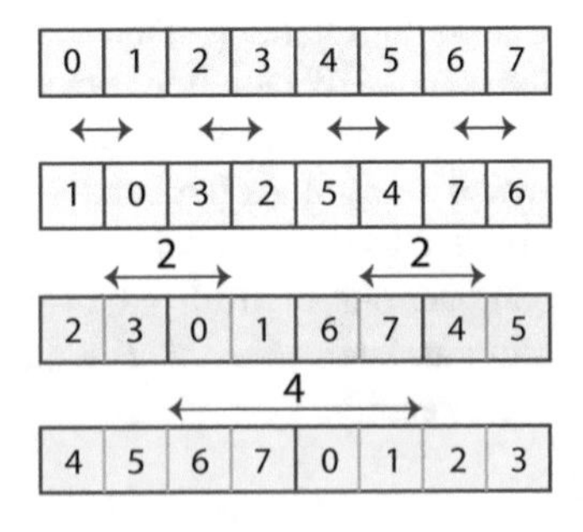

Fig. 5.3 Toward shift operations on registers

$$T_j = Q\overline{\oplus}\bar{Q}_j, j = \overline{1, n}.$$

4. Q test generation for the functionality by logical union of T vectors for each input variable:

$$T = V_{j=1}^{n} T_j.$$

In summary, the algorithm for unconditional test synthesis for functional elements defined by qubit coverage can be represented by the following formula:

$$T = V_{j=1}^{n}\left[Q\overline{\oplus}_{j=1}^{n} S_j(\bar{Q})\right].$$

The computational complexity of the unconditional algorithm for test synthesis based on consistent use of register logic operations (*not*, *shift*, *nxor*, *or*) for a functional element described by a qubit vector is represented by the following expression:

$$q = n + n \times 2^n + n + n = n(2^n + 3).$$

Where n is a number of variables, the four terms consequently mean the computational complexity of the: negation; a logical shift for transposition of coordinate states of a qubit vector relative to each of the n input variables; *nxor* comparison of generated test vectors with the original qubit coverage of the functional element; uniting the test vectors for the input variables.

Abstracting from the concept of the truth table, it is further proposed that a formal unconditional algorithm for qubit test synthesis for functional primitives based on Q coverage be applied to the abovementioned example:

1. Inverting all bits of Q coverage of the functional element with three input variables:

Q	1	1	0	1	0	1	1	0
$\bar{Q}$	0	0	1	0	1	0	0	1

2. Logical shift of bit numbers of the inverted qubit vector in accordance with the sequences of numbers:

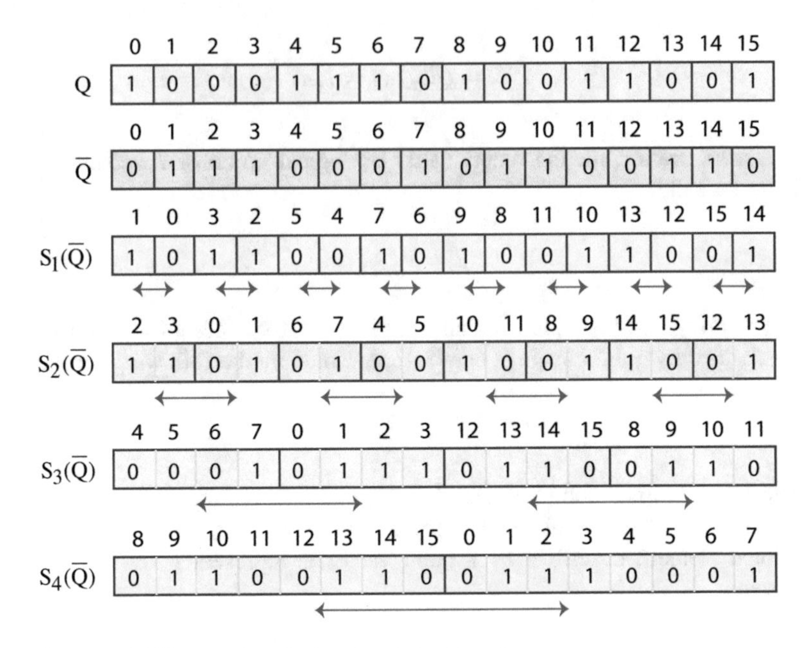

Fig. 5.4 Toward shift operations for four-input functionality

$$Q_1(X_1) = 45670123, Q_2(X_2) = 23016745, Q_3(X_3) = 10325476.$$

A simple logical rule applies: for the first input variable (least significant), parallel data exchange between the adjacent coordinates is performed; for the second input variable, data exchange is carried out between adjacent pairs of coordinates; for the third variable, data exchange is performed between adjacent tetrads of coordinates of the address vector.

A summary of shift operations for the functionality of the four variables is shown in Fig. 5.4. Increasing the number of variables does not change the principle of data shifting: toward shift of two bits, bit pairs, tetrads bits, eights...

The procedure for toward shifting data in the inverse qubit vector is the most time-consuming operation of the algorithm for test synthesis. Therefore, hardware implementation of shift operations provides maximum performance.

The procedural implementation of the number exchange to generate test vectors for checking faults of input variables related to the example under consideration is as follows:

Q	0 1 2 3 4 5 6 7
$\overline{Q_1}$	1 0 3 2 5 4 7 6
$\overline{Q_2}$	2 3 0 1 6 7 4 5
$\overline{Q_3}$	4 5 6 7 0 1 2 3

For the first variable, there is a simple formula for renumbering cells of a qubit vector for test synthesis: $j = j + (-1)^j$. For other variables, such simple relationships have not yet been found.

In view of the imposed rules of the toward shift of qubit vector cells, the implementation of this algorithm step for $Q = 11010110$ is represented by the following table:

$$\begin{array}{llllllll l} Q = & 1 & 1 & 0 & 1 & 0 & 1 & 1 & 0 \\ \bar{Q} = & 0 & 0 & 1 & 0 & 1 & 0 & 0 & 1 \\ \bar{Q}_1 = & 1 & 0 & 0 & 1 & 0 & 0 & 1 & 0 \\ \bar{Q}_2 = & 1 & 1 & 0 & 1 & 0 & 1 & 1 & 0 \\ \bar{Q}_3 = & 0 & 0 & 0 & 1 & 0 & 1 & 1 & 0 \end{array}$$

Thus, each variable makes the partition of the qubit vector into groups of ordered sequences. The first variable (least significant) defines the changes of states, according to the rule: add 1 to the current even address of a cell: $j = j + 1$; subtract 1 from the current odd address of a cell: $j = j - 1$. For the second variable, pairs of cells are considered, which are processed by the rule: add 2 to the current even address of the cell pair: $j = j + 2$; subtract 2 from the current odd address of a cell pair: $j = j - 2$. For the third variable, cell tetrads are considered, which are processed by the rule: add 4 to the current even address of a cell tetrad: $j = j + 4$; subtract 4 from the current odd address of a cell tetrad: $j = j - 4$.

3. Comparing three inverted and reordered Q vectors with initial Q coverage of the functional element by using an equivalence operation:

Q	1 1 0 1 0 1 1 0
$\overline{Q_1}$	0 0 1 0 1 0 0 1
$S_1(\overline{Q})$	1 0 0 1 0 0 1 0
$S_2(\overline{Q})$	1 0 0 0 0 1 1 0
$S_3(\overline{Q})$	0 0 0 1 0 1 1 0
$Q \oplus S_1(\overline{Q})$	1 0 1 1 1 0 1 1
$Q \oplus S_2(\overline{Q})$	1 0 1 0 1 1 1 1
$Q \oplus S_3(\overline{Q})$	0 0 1 1 1 1 1 1

4. Disjunction of the obtained three vectors forms a Q test where the 1 unit coordinates define test patterns for stuck-at faults of inputs and output (Fig. 5.5).

Q	1 1 0 1 0 1 1 0
$\overline{Q}$	0 0 1 0 1 0 0 1
$S_1(\overline{Q})$	1 0 0 1 0 0 1 0
$S_2(\overline{Q})$	1 0 0 0 0 1 1 0
$S_3(\overline{Q})$	0 0 0 1 0 1 1 0
$T_1 = Q \oplus S_1(\overline{Q})$	1 0 1 1 1 0 1 1
$T_2 = Q \oplus S_2(\overline{Q})$	1 0 1 0 1 1 1 1
$T_3 = Q \oplus S_3(\overline{Q})$	0 0 1 1 1 1 1 1
$T = T_1 \vee T_2 \vee T_3$	1 0 1 1 1 1 1 1

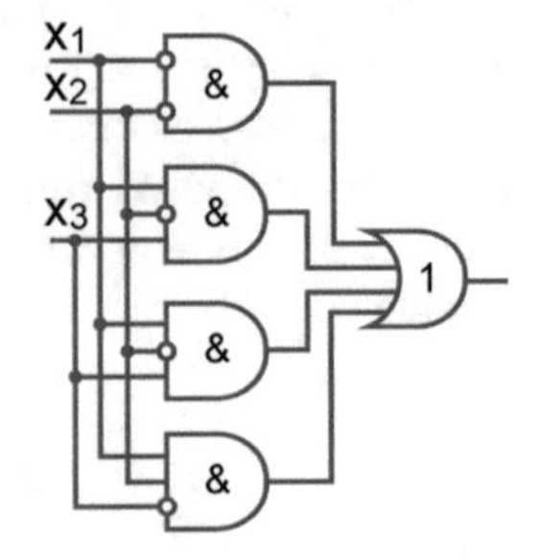

Fig. 5.5 Synthesis of a test table for the circuit

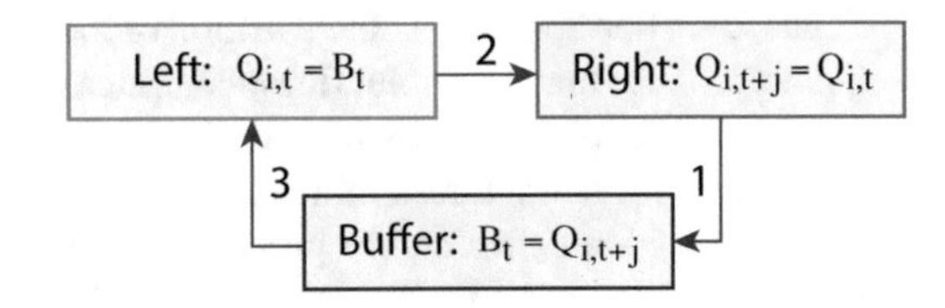

Fig. 5.6 Toward shift of adjacent qubit parts

Given the significance of the toward register shift, a universal algorithm for obtaining transpositions of qubit fragments, depending on the number of input variables, is proposed.

The sequential algorithm for toward data shifting in qubit coverage for Q test generation for logic functionality has three nested loops:

1. Setting the i number of the input variable or step 2^i for toward data shifting in a qubit: $i = \overline{1, n}$.
2. Setting the cycle for processing a qubit with the given increment 2^i (2,4,8,16...), multiple of a power of two: $j = \overline{0, 2^n - 1, 2^i}$. Depending on the number of input variable i, the following index sequences are formed $j = f(i)$: [(0,2,4,6...), (0,4,8,12...), (0,8,16,32...)].
3. Setting the cycle for toward data shifting for a pair of adjacent groups: $t = j, j + 2^{i-1} - 1$. Index values $t = f(i,j)$: processing a qubit of the first variable (0,0; 2,2; 4,4...), second variable (0,1; 4,5; 8,9...), third variable (0,3; 8,11; 16,19...). Execution of shift operations through the use of a buffer register B (Fig. 5.6):

$$\left(B_t = Q_{i,t+j}\right) \rightarrow \left(Q_{i,t+j} = Q_{i,t}\right) \rightarrow \left(Q_{i,t} = B_t\right)$$

Here is the end of the algorithm for toward data shifting in registers.

The computational complexity of the sequential algorithm for processing qubit coverage of the functionality with n input variables is equal to $q = 3n2^n$.

Example 3: It is necessary to generate a test for the functionality defined by the logical equation $Y = \bar{X}_1\bar{X}_2 \vee \bar{X}_1 X_2$. The following table describes the results of the test synthesis algorithm by leveraging a given qubit vector:

Q	0 0 1 1
$\bar{Q}_1$	1 1 0 0
$S_1(\bar{Q})$	0 0 1 1
$S_2(\bar{Q})$	1 1 0 0
$T_1 = Q \overline{\oplus} S_1(\bar{Q})$	1 1 1 1
$T_2 = Q \overline{\oplus} S_2(\bar{Q})$	0 0 0 0
$T = T_1 \vee T_2$	1 1 1 1

The created test contains four input sequences, detected faults of the input variable X_1. However, the test pattern for the X_2 variable does not exist, because $T_2 = 0000$. This means that the variable X_2 cannot be verified, and thus X_2 input is not essential. Indeed, the minimization procedure of the original functions:

$$Y = \bar{X}_1\bar{X}_2 \vee \bar{X}_1 X_2 = \bar{X}_1(\bar{X}_2 \vee X_2) = \bar{X}_1$$

eliminates the redundant X_2 variable. Thus, the test generation method based on qubit coverage provides an additional opportunity to identify the redundancy of variables and minimize Q coverage by reducing the number of variables.

Example 4: It is necessary to generate a test for the functionality defined by the logic equation: $Y = \bar{X}_1 X_2 \vee X_1\bar{X}_2$. The results of the test synthesis algorithm by a given qubit vector based on consistent execution of four register logic operations (*not*, *shift*, *nxor*, *or*) are represented in the following form:

Q	0 1 1 0
$\bar{Q}_1$	1 0 0 1
$S_1(\bar{Q})$	0 1 1 0
$S_2(\bar{Q})$	0 1 1 0
$T_1 = Q \overline{\oplus} S_1(\bar{Q})$	1 1 1 1
$T_2 = Q \overline{\oplus} S_2(\bar{Q})$	1 1 1 1
$T = T_1 \vee T_2$	1 1 1 1

The obtained test includes four input sequences, each of which detects faults for input variables X_1, X_2. Thus, the test generation method based on qubit coverage allows creation of test patterns for functional components but does not provide an opportunity to get a minimal form.

Example 5: It is necessary to generate a test for the functionality defined by the logical equation $Y = \bar{X}_1 X_2 \bar{X}_3 \vee X_1 \bar{X}_2 \bar{X}_3 \vee X_1 X_2 X_3 \vee \bar{X}_1 X_2 X_3$. The results of the test synthesis algorithm by a given qubit vector based on consistent execution of four register logic operations (*not*, *shift*, *nxor*, *or*) are represented in the following form:

Q	0 0 1 1 1 0 0 1
$\overline{Q}$	1 1 0 0 0 1 1 0
$S_1(\overline{Q})$	1 1 0 0 1 0 0 1
$S_2(\overline{Q})$	0 0 1 1 1 0 0 1
$S_3(\overline{Q})$	0 1 1 0 1 1 0 0
$T_1 = Q \overline{\oplus} S_1(\overline{Q})$	0 0 0 0 1 1 1 1
$T_2 = Q \overline{\oplus} S_2(\overline{Q})$	1 1 1 1 1 1 1 1
$T_3 = Q \overline{\oplus} S_3(\overline{Q})$	1 0 1 0 1 0 1 0
$T = T_1 \vee T_2 \vee T_3$	1 1 1 1 1 1 1 1

Fig. 5.7 shows a sequencer for test synthesis of the functional elements defined by qubit coverage. It includes a control unit, which allocates four clocks and creates a test generation cycle consisting of four register operations, executed sequentially: (0) The initial operation, activated by a start signal, provides loading of qubit coverage into the register. (1) Then the clock Clk *N* activates the execution of a *not* operation on the qubit coverage register. (2) The clock signal Clk *S* activates the execution of toward shift operations on the contents of three registers to obtain the following results: S_1 (not$Q$), S_2 (not$Q$), S_3 (not$Q$). (3) Then, the clock signal Clk *C* activates parallel comparison operations of obtained test candidates with initial qubit coverage: *nxor* (notQ, S_1), *nxor* (notQ, S_2), *nxor* (notQ, S_3). (4) The clock

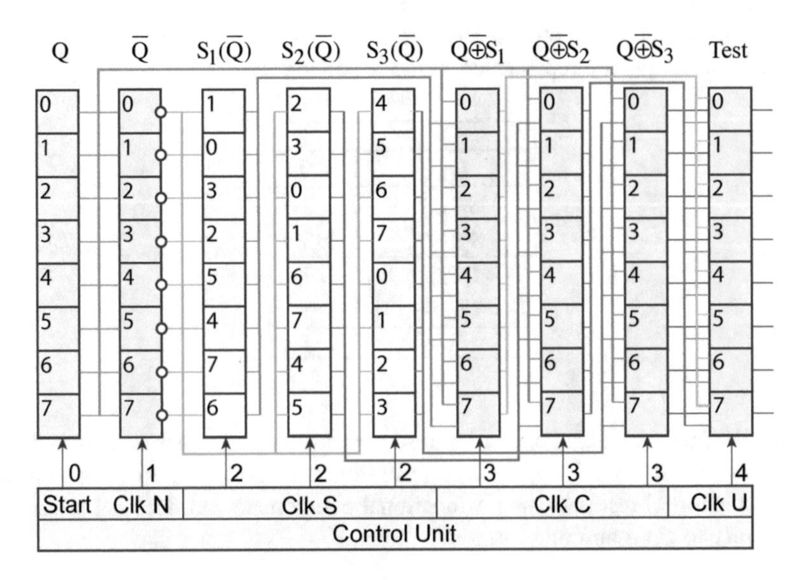

Fig. 5.7 Test synthesis sequencer

signal Clk U activates the execution of the *or* operation on the test candidates—in this case, logical union of the contents of the three registers:

$$T = \mathrm{or}[\mathrm{nxor}(\mathrm{notQ}, S_1), \mathrm{nxor}(\mathrm{notQ}, S_2), \mathrm{nxor}(\mathrm{notQ}, S_3)].$$

If not to save on hardware, the performance of the test generator can be reduced to four automatic cycles $q = 4$ of parallel execution of logic register operations. This allows us to build in the sequencer in the BIST infrastructure of digital SoCs for online testing of functional elements. In this case, the total number of 2^n-bit registers is equal to $N(n) = 1 + 1 + n + n + 1 = (2n + 3)$, and the total volume of register memory is equal to $N(R) = (2n + 3)\ 2^n$ bits where n is the number of input variables of the functional element.

5.3 Taking the Boolean Derivatives for *Q* Test Synthesis

A method for taking the Boolean derivatives on qubit coverage to create activation conditions of input variables for synthesis of qubit tests is considered. An analogy between the two forms of Boolean functions, analytical and vector ones, for taking derivatives is represented. The method is considered in several examples of logical functions:

$$(1)\ f(x) = x_1 \vee x_1\bar{x}_2, (2)\ f(x) = x_1x_2 \vee \bar{x}_1x_3, \quad (3)\ f(x) = \bar{x}_2\bar{x}_3 \vee x_1x_2x_3$$

The objectives to be addressed are (1) taking the derivatives of the first order by using the analytical and qubit form of logical functions; (2) verification of obtained activation conditions by fault-free functionality simulation; and (3) synthesis of activation tests for variables of logical function based on taking derivatives.*

Example 6: It is necessary to define all the derivatives of the first order, described in the qubit form of the logic function $f(x) = x_1 \vee x_1\bar{x}_2$.

Taking the derivative of the analytical expression:

$$\frac{df(x_1,x_2,\ldots,x_i,\ldots,x_n)}{dx_i} = f(x_1,x_2,\ldots,x_i=0,\ldots,x_n) \oplus f(x_1,x_2,\ldots,x_i=1,\ldots,x_n)$$

defines the Boolean derivative of the first order as the *xor* operation result between zero and one residual functions.

Taking into account the given function, the derivative is the following:

$$\frac{df(x_1, x_2)}{dx_1} = f(0, x_2) \oplus f(1, x_2) =$$
$$= (0 \vee 0\bar{x}_2) \oplus (1 \vee 1\bar{x}_2) = 0 \oplus 1 = 1.$$
$$\frac{df(x_1, x_2)}{dx_2} = f(x_1, 0) \oplus f(x_1, 1) =$$
$$= (x_1 \vee x_1 \cdot \bar{0}) \oplus (x_1 \vee x_1 \cdot \bar{1}) =$$
$$= (x_1 \vee x_1 \cdot 1) \oplus (x_1 \vee x_1 \cdot 0) =$$
$$= (x_1 \vee x_1) \oplus (x_1 \vee 0) = x_1 \oplus x_1 = 0.$$

A zero value of the derivative means the conditions' absence for activating the variable x_2, which gives the opportunity to consider it irrelevant and removed from the number of variables forming functionality. The next step offers similar conversion on the qubit coverage of the logic function:

X_1	X_2	Y
0	0	0
0	1	0
1	0	1
1	1	1

$= (0011).$

The derivatives according to the truth table may be taken by sequential setting of ones and zeros in all coordinates of every column corresponding to each variable:

X_1	X_2	Y	Y_1^0	Y_1^1	Y_1'	Y_2^0	Y_2^1	Y_2'
0	0	0	0	1	1	0	0	0
0	1	0	0	1	1	0	0	0
1	0	1	0	1	1	1	1	0
1	1	1	0	1	1	1	1	0

Thus, the derivatives of the first and second variables, stored in the qubit coverage format, are 1111 and 0000. This means that the derivative of the first variable is equal to 1 and that of the second variable is equal to 0, meaning not essential.

However, such a result can be obtained more formally and technologically without considering the input sequences of the truth table, using firstly toward shift operations and then logical *xor* operations on qubit coverage $\{a,b\} = a \oplus b$ where a,b are neighbor qubit subvectors $Q = (a,b)$:

Y	Y_1'	Y_2'
0	1	0
0	1	0
1	1	0
1	1	0

Otherwise, for the first variable, it is necessary to *xor* execute the toward shifted two halves of the first column, and the result is written in both symmetric parts: $\{a, b\} = a \oplus b$, (11,11) = 00 $\oplus$ 11. For the second variable to be considered the two pairs of column coordinates and the overall result is recorded in the each coordinate of pair: $\{a,b\} = a \oplus b$: $(0,0) = 0 \oplus 0 = 0$, $(0,0) = 1 \oplus 1 = 0$. Thus, the *xor* executed result is shared by each pair of adjacent subvectors, whose dimension is determined by the number of the variable, from 0 to $2^n - 1$.

Example 8: It is necessary to define all derivatives of the first order by using the analytical form of the logic function $f(x) = x_1x_2 \vee \bar{x}_1x_3$. The following procedures are executed for the abovementioned function:

$$\begin{aligned}
&\frac{df(x_1,x_2,x_3)}{dx_1} = f(0,x_2,x_3) \oplus f(1,x_2,x_3) = \\
&= \left(0 \cdot x_2 \vee \bar{0} \cdot x_3\right) \oplus \left(1 \cdot x_2 \vee \bar{1} \cdot x_3\right) = \\
&= (0 \vee 1 \cdot x_3) \oplus (x_2 \vee 0 \cdot x_3) = \\
&= x_3 \oplus x_2 = x_2\bar{x}_3 \vee \bar{x}_2x_3.
\end{aligned}$$

$$\begin{aligned}
&\frac{df(x_1,x_2,x_3)}{dx_2} = f(x_1,0,x_2) \oplus f(x_1,1,x_3) = \\
&= \bar{x}_1x_3 \oplus (x_1 \vee x_3) = \\
&\overline{\bar{x}_1x_3}(x_1 \vee x_3) \vee \bar{x}_1x_3\overline{(x_1 \vee x_3)} = \\
&= \left(x_1 \vee \bar{x}_3\right)(x_1 \vee x_3) \vee \bar{x}_1x_3\bar{x}_1\bar{x}_3 = x_1.
\end{aligned}$$

$$\begin{aligned}
&\frac{df(x_1,x_2,x_3)}{dx_3} = f(x_1,x_2,0) \oplus f(x_1,x_2,1) = \\
&= x_1x_2 \oplus \left(\bar{x}_1 \vee x_2\right) = \\
&\overline{\bar{x}_1x_2}\left(\bar{x}_1 \vee x_2\right) \vee x_1x_2\overline{(x_1 \vee x_2)} = \\
&= \left(\bar{x}_1 \vee \bar{x}_2\right)\left(\bar{x}_1 \vee x_2\right) \vee x_1x_2x_1\bar{x}_2 = \bar{x}_1.
\end{aligned}$$

Four activation conditions are obtained for the three variables, which correspond to four logical paths in the circuit structure of the disjunctive form of the function.

Taking the three derivatives of the first order by means of the truth table $f(x) = x_1x_2 \vee \bar{x}_1x_3$ gives the following result:

X_1	X_2	X_3	Y	Y_1^0	Y_1^1	Y_1'	Y_2^0	Y_2^1	Y_2'	Y_3^0	Y_3^1	Y_3'
0	0	0	0	0	0	0	0	0	0	0	1	1
0	0	1	1	1	0	1	1	1	0	0	1	1
0	1	0	0	0	1	1	0	0	0	0	1	1
0	1	1	1	1	1	0	1	1	0	0	1	1
1	0	0	0	0	0	0	0	1	1	0	0	0
1	0	1	0	1	0	1	0	1	1	0	0	0
1	1	0	1	0	1	1	0	1	1	1	1	0
1	1	1	1	1	1	0	0	1	1	1	1	0

If we are to exclude the truth table from consideration, and to leverage Q coverage only, then the process of taking derivatives will be as follows:

Y	Y'_1	Y'_2	Y'_3
0	0	0	1
1	1	0	1
0	1	0	1
1	0	0	1
0	0	1	0
0	1	1	0
1	1	1	0
1	0	1	0

The qubit method of taking the Boolean derivative is as follows:

1. Determine the qubit vector functionality.
2. Run an *xor* operation for adjacent parts of the qubit (bits, pairs, and tetrads) shifted toward each other.
3. The result is written in the two adjacent parts: bits, pairs, and tetrads.

For this the considered example, the implementation of the procedures for taking the derivative for the first variable is of the form $\{a, b\} = a \oplus b$, (0110,0110) $=$ 0101 $\oplus$ 0011. For taking the derivative of the second variable it is necessary to perform sequentially *xor* operations on adjacent pairs of the qubit vector Y and save the overall results in each pair: $\{a, b\} = a \oplus b$: (00,00) $=$ 01 $\oplus$ 01, (11,11) $=$ 00 $\oplus$ 11. For taking the derivative of the third variable it is necessary to perform sequentially *xor* operations on adjacent bits of the qubit vector Y and save the overall results in each adjacent bit: $\{a, b\} = a \oplus b$: (1,1) $= 0 \oplus 1$, (1,1) $= 0 \oplus 1$, (0,0) $= 0 \oplus 0$, (0,0) $= 0 \oplus 0$.

Naturally, the qubit derivative for any input variable, as a vector, is characterized by a relative symmetric equality of subvectors by the construction: the derivative of the first variable is defined by a symmetric equality of two tetrads; the derivative of the second variable is defined by a symmetric equality of each adjacent pair; and the derivative of the third variable is defined by a symmetric equality of each adjacent bit.

Example 9: It is necessary to take all the derivatives of the first order by the truth table of the logical function with three variables:

$$f(x) = \bar{x}_2\bar{x}_3 \vee x_1x_2x_3.$$

The result of taking derivatives by using a truth table [4] of the given functionality, based on logic operation execution on the columns of the input variables, is represented as follows:

$\frac{df}{dx_1} =$

x_1	x_2	x_3	Y	Y_2^0	Y_2^1	$Y_2^{\oplus}$
0	0	0	1	1	1	0
0	0	1	0	0	0	0
0	1	0	0	0	0	0
0	1	1	0	0	1	1
1	0	0	1	1	1	0
1	0	1	0	0	0	0
1	1	0	0	0	0	0
1	1	1	1	0	1	1

$= x_2x_3.$

$\frac{df}{dx_2} =$

x_1	x_2	x_3	Y	Y_2^0	Y_2^1	$Y_2^{\oplus}$
0	0	0	1	1	0	1
0	0	1	0	0	0	0
0	1	0	0	1	0	1
0	1	1	0	0	0	0
1	0	0	1	1	0	1
1	0	1	0	0	1	1
1	1	0	0	1	0	1
1	1	1	1	0	1	1

$= \bar{x}_1\bar{x}_3 \vee x_1\bar{x}_3 \vee x_1x_3 = \bar{x}_3 \vee x_1x_3.$

$\frac{df}{dx_3} =$

x_1	x_2	x_3	Y	Y_3^0	Y_3^1	$Y_3^{\oplus}$
0	0	0	1	1	0	1
0	0	1	0	1	0	1
0	1	0	0	0	0	0
0	1	1	0	0	0	0
1	0	0	1	1	0	1
1	0	1	0	1	0	1
1	1	0	0	0	1	1
1	1	1	1	0	1	1

$= \bar{x}_1\bar{x}_2 \vee x_1\bar{x}_2 \vee x_1x_2 = \bar{x}_2 \vee x_1x_2.$

As an alternative, the results of taking the derivatives of the first order for three variables by using cubic coverage or a truth table of the functionality is represented below:

X_1	X_2	X_3	Y	Y_1^0	Y_1^1	Y_1'	Y_2^0	Y_2^1	Y_2'	Y_3^0	Y_3^1	Y_3'
0	0	0	1	1	1	0	1	0	1	1	0	1
0	0	1	0	0	0	0	0	0	0	1	0	1
0	1	0	0	0	0	0	1	0	1	0	0	0
0	1	1	0	0	1	1	0	0	0	0	0	0
1	0	0	1	1	1	0	1	0	1	1	0	1
1	0	1	0	0	0	0	0	1	1	1	0	1
1	1	0	0	0	0	0	1	0	1	0	1	1
1	1	1	1	0	1	1	0	1	1	0	1	1

Further, the process of taking derivatives on qubit coverage without consideration of the truth table is shown at a formal level:

Y	Y_1'	Y_2'	Y_3'
1	0	1	1
0	0	0	1
0	0	1	0
0	1	0	0
1	0	1	1
0	0	1	1
0	0	1	1
1	1	1	1

Naturally, the derivatives of the functions, represented by qubit coverage, are the functions defined by vectors of the same dimension. They can be written in analytical form (DNF) on the 1-unit values of the variables, which form cell addresses of a qubit vector:

$$Y_1' = 011 \vee 111 = \bar{X}_1X_2X_3 \vee X_1X_2X_3 = X_2X_3(\bar{X}_1 \vee X_1) = X_2X_3.$$
$$Y_2' = 000 \vee 010 \vee 100 \vee 101 \vee 110 \vee 111 = \bar{X}_1\bar{X}_3 \vee X_1.$$
$$Y_3' = 000 \vee 001 \vee 100 \vee 101 \vee 110 \vee 111 = \bar{X}_1\bar{X}_2 \vee X_1.$$

Minimization of Boolean functions, corresponding to derivatives, leads to analytical expressions where there are no variables on which the derivative is taken. Thus, all the results of taking the derivatives by using three forms (analytical, tabular, and vector) of the function definition are identical. The most technological technique for taking a derivative is a qubit coverage method. It possesses lower computational complexity due to its compact functionality form. An analytical form suggests a significant increase in the complexity of the algorithms associated with the application of Boolean algebra and function minimization principles, which limits its leverage to solve practical problems.

For comparison, a method for obtaining test $T = [T_{ij}], i = \overline{1,k}; j = \overline{1,n}$ for combinational functionality, defined by qubit coverage or a truth table, is described below [4]:

$$1.\ f'(x_i) = f(x_1, x_2, \ldots, x_i = 0, \ldots, x_n) \oplus f(x_1, x_{2,\ldots,} x_i = 1, \ldots, x_n);$$
$$2.\ T = \bigcup_{i=1}^{n} [f'(x_i)^*(x_i = 0) \vee (x_i = 1)];$$
$$3.\ T_{ij} = T_{i-1,j} \leftarrow T_{ij} = X; T_{1j} = 1 \leftarrow T_{1j} = X;$$
$$4.\ T = T \backslash T_i \leftarrow T_i = T_{i-r}, r = \overline{1, i-1}, i = \overline{2, n}.$$

(1) taking derivatives by all n variables of the functionality by using cubic coverage (a compact form of a truth table, based on the multivalued alphabet for coordinate description); (2) union of all activation conditions (vectors) in a table where each vector is assigned with a change of the variable on which the derivative was taken by concatenating (*)—this procedure means doubling the number of test patterns in relation to the total number (k) of activation conditions; and

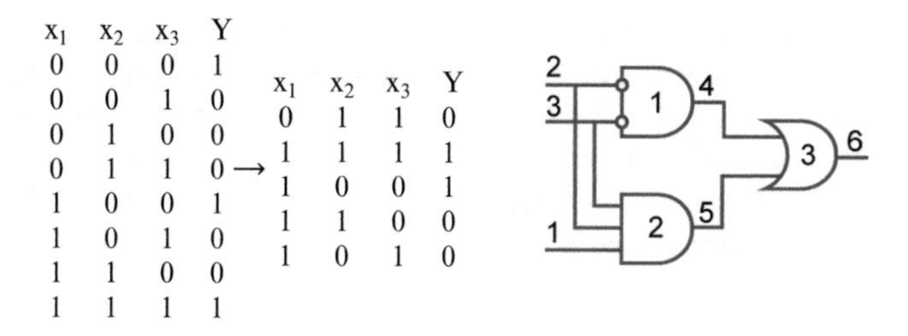

x1	x2	x3	Y
0	0	0	1
0	0	1	0
0	1	0	0
0	1	1	0
1	0	0	1
1	0	1	0
1	1	0	0
1	1	1	1

→

x1	x2	x3	Y
0	1	1	0
1	1	1	1
1	0	0	1
1	1	0	0
1	0	1	0

Fig. 5.8 Test generation for the circuit structure of a Boolean function

(3) minimization of the test vectors by removing repetitive input sequences. Fig. 5.8 shows the tables of test obtaining for the functionality $f(x) = \bar{x}_2\bar{x}_3 \vee x_1x_2x_3$ in accordance with algorithm steps 2–3.

As an alternative to the above technique, a technologically simple test synthesis method is further proposed, based on taking the derivatives of the qubit coverage of functional elements, without considering the states of the input variables.

1. Initial definition of logical functionality by qubit coverage.
2. Execution of toward shift operations of qubit vector parts and then *xor* operations to get derivative vectors for each input variable.
3. Logical union of derivative vectors that form a qubit test of the same size.
4. If necessary to get the minimum test, the coverage problem is solved (already in the matrix of qubit derivatives) by finding the minimum number of pairs of 1-unit coordinates (bold red color) of qubit derivatives of all variables where a pair of 1 units should detect the stuck-at faults of each input. The procedure for choosing a pair of 1 units in the qubit derivative is determined by the presence of two components, one from each of the even and odd parts of a vector.

The qubit coverage (10001001) corresponds to the logic function of three variables $f(x) = \bar{x}_2\bar{x}_3 \vee x_1x_2x_3$. The following table shows the results of test synthesis for the mentioned functionality:

X_1	X_2	X_3	Y	Y_1'	Y_2'	Y_3'	T
0	0	0	1	0	1	1	1
0	0	1	0	0	0	1	1
0	1	0	0	0	1	0	1
0	1	1	0	1	0	0	1
1	0	0	1	0	1	1	0
1	0	1	0	0	1	1	0
1	1	0	0	0	1	1	0
1	1	1	1	1	1	1	1

X_1	X_2	X_3	Y
0	1	1	0
1	1	1	1
0	1	0	0
0	0	0	1
0	0	1	0

According to step 3, the synthesis algorithm provides the minimum test for detecting the input variables, which contains five sequences, represented in the *T* column, as well as being duplicated explicitly in the right table.

An interesting fact is that the result of the execution procedure for taking the derivative of the qubit vector already contains the activation test of each variable. The integrated test detects all the stuck-at faults of the input variables and can also be used to diagnose the faults, because all derivative vectors for essential inputs will be different. In fact, taking the derivatives on all variables of the qubit coverage forms a Q test, no more and no less.

5.4 Deductive Qubit Fault Simulation Method for Digital Structures

The qubit method is applied to evaluate the test quality relative to stuck-at faults of primary inputs and outputs of logic functionality. There is a well-developed theory of deductive analysis [4] focused on parallel processing of fault lists for the RTL level of digital system description. Basic concepts of the theory of fault simulation are represented by means of fault list transportation through the functional logic elements [13–19]. The deductive functions of the concurrent fault simulation on an exhaustive test for the basic functional elements *and*, *or*, *not* are represented below. Forming the deductive converter for the *and* function is shown in the expression:

$$\begin{aligned}
&L[T=(00,01,10,11),F=(X_1\wedge X_2)]=\\
&=L\{(\bar{x}_1\bar{x}_2\vee\bar{x}_1x_2\vee x_1\bar{x}_2\vee x_1x_2)\wedge[(X_1\oplus T_{t1}\wedge X_2\oplus T_{t2})\oplus T_{t3})]\}=\\
&=(\bar{x}_1\bar{x}_2)\{[X_1\oplus 0)\wedge(X_2\oplus 0)]\oplus 0\}\vee(\bar{x}_1x_2)\{[X_1\oplus 0)\wedge(X_2\oplus 1)]\oplus 0\}\vee\\
&\vee(x_1\bar{x}_2)\{[X_1\oplus 1)\wedge(X_2\oplus 0)]\oplus 0\}\vee(x_1x_2)\{[X_1\oplus 1)\wedge(X_2\oplus 1)]\oplus 1\}=\\
&=(\bar{x}_1\bar{x}_2)(X_1\wedge X_2)\vee(\bar{x}_1x_2)(X_1\wedge\bar{X}_2)\vee(x_1\bar{x}_2)(\bar{X}_1\wedge X_2)\vee(x_1x_2)(X_1\vee X_2).
\end{aligned}$$

Similar operations are performed for the *or* function:

$$\begin{aligned}
&L[T=(00,01,10,11),F=(X_1\vee X_2)]=\\
&=L\{(\bar{x}_1\bar{x}_2\vee\bar{x}_1x_2\vee x_1\bar{x}_2\vee x_1x_2)\wedge[(X_1\oplus T_{t1}\vee X_2\oplus T_{t2})\oplus T_{t3})]\}=\\
&=(\bar{x}_1\bar{x}_2)\{[X_1\oplus 0)\vee(X_2\oplus 0)]\oplus 0\}\vee(\bar{x}_1x_2)\{[X_1\oplus 0)\vee(X_2\oplus 1)]\oplus 0\}\vee\\
&\vee(x_1\bar{x}_2)\{[X_1\oplus 1)\vee(X_2\oplus 0)]\oplus 0\}\vee(x_1x_2)\{[X_1\oplus 1)\vee(X_2\oplus 1)]\oplus 1\}=\\
&=(\bar{x}_1\bar{x}_2)(X_1\vee X_2)\vee(\bar{x}_1x_2)(X_1\wedge\bar{X}_2)\vee(x_1\bar{x}_2)(X_1\wedge\bar{X}_2)\vee(x_1x_2)(X_1\wedge X_2).
\end{aligned}$$

Here $T_t=(T_{t1},T_{t2},T_{t3}),\ (t=\overline{1,4})$ is a test vector with three coordinates where the last one determines the state of the output of the two-input element *and* (*or*); L is an output fault list; X is a fault list of a primitive input; and $x=\{0,1\}$ is the binary value of the input. In the next conversion, a test vector $T_t=(T_{t1},T_{t2}),\ (t=\overline{1,2})$ is used. It is based on only two coordinates; the second one is the output state of the *not* operation element:

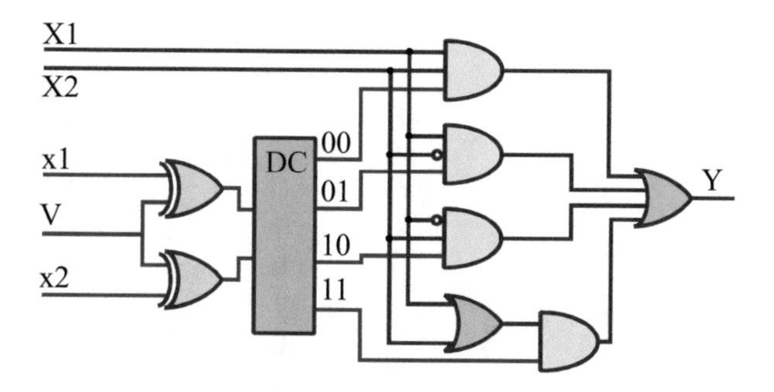

Fig. 5.9 Deductive parallel fault simulator

$$L[T=(0,1),F=\bar{X}_1]=L\{(\bar{x}_1 \vee x_1)[(\overline{X_1 \oplus T_{t1}}) \oplus \underline{T}_{t2}]\}=$$
$$=\bar{x}_1[\overline{X_1 \oplus 0}) \oplus 1] \vee x_1[\overline{X_1 \oplus 1}) \oplus 0]=\bar{x}_1\bar{X}_1 \vee x_1\bar{X}_1=\bar{x}_1 X_1 \vee x_1 X_1.$$

The last expression shows the invariance of the *not* operation to the input set for fault transportation. It is transformed into a repeater. Therefore, this function does not appear at the outputs of the deductive elements. Integrated hardware implementation of deductive functions for the two-input element *and*, *or* applied in exhaustive tests is represented by a common functional circuit (Fig. 5.9) for deductive parallel fault simulation.

Boolean ($x1$,$x2$) and register ($X1$,$X2$) inputs for coding faults, the variable for choosing the type of fault-free function (*and*, *or*), and output register variable Y are represented in the simulator. The states of the binary inputs $x1$,$x2$ and the variable V for choosing an element define one of the four deductive functions for getting the vector Y of testable faults.

Technological implementation of deductive simulation based on the qubit form of function description is proposed below, which is different from the above-described method by parallel execution of logical operations on Q vectors, and also the ability to use the method for any digital structures.

A set of qubit derivatives of all the input variables, taken on the qubit coverage, is a qubit matrix for implementing the deductive fault simulation method. The matrix row creates conditions for the transportation of the fault lists from the external inputs to the outputs according to the rule; the 1-unit values create a union of input lists and zero signals indicate the input lists, which must be subtracted from the result of the union. The presence of all zero signals in a qubit matrix row creates conditions for the intersection of the lists.

The next example shows the construction of deductive formulae of fault list transportation from input variables to the output of the functionality, leveraging the input test sequences, and the qubit matrix of derivatives:

x_1	x_2	x_3	Y	X_1	X_2	X_3
0	0	0	1	0	1	1
0	0	1	0	0	0	1
0	1	0	0	0	1	0
0	1	1	0	1	0	0
1	0	0	1	0	1	1
1	0	1	0	0	1	1
1	1	0	0	0	1	1
1	1	1	1	1	1	1

To increase the speed of the fault simulation, it is necessary to exclude analytical expressions in order to operate only with qubit vectors. In this case, a restriction on the presence of only single stuck-at faults in the circuit should be borne in mind, which cannot simultaneously occur at several inputs of the functional module. In this case, the compact and expanded formulae of deductive qubit fault simulation for the logic black box represented in the form of qubit vectors have the following forms:

$$L = \bigvee_{i=1}^{2^n} \{x_i \wedge [(X_i^1) \wedge \bar{L}(X_i^0)]\}.$$

$$L = \bigvee_{i=1}^{2^n} \left\{ (x_{i1}, x_{i2}, \ldots, x_{i2}, \ldots x_{in}) \wedge \left[\bigvee_{X_{ij}=1} L(X_{ij}) \bigwedge_{X_{ij}=0} \bar{L}(X_{ij}) \right] \right\}.$$

Here, L is the list of output faults, $L(X)$ is the list of fault vectors at the inputs, x is the matrix of the input test vectors, X is the matrix of the qubit derivatives of the qubit coverage, and n is the number of input variables.

An algorithm for constructing a deductive formula for a given functionality includes the following steps:

1. Setting qubit coverage of the functionality.
2. Taking the qubit derivatives on the input variables to obtain the corresponding matrix.
3. Formation of the analytical and matrix vector form for defining output fault lists by logical multiplication of matrices of input test patterns and the derivative matrix.

Creation of the analytical form for deductive fault simulation of logic functionality based on the derivative matrix is represented below:

$$\begin{aligned}
T &= (000, 001, 010, 011, 100, 101, 110, 111). \\
Q &= (011, 001, 010, 100, 011, 011, 011, 111)]. \\
L &= (000 \wedge 011) \vee (001 \wedge 001) \vee (010 \wedge 010) \vee (011 \wedge 100) \vee \\
&\vee (100 \wedge 011) \vee (101 \wedge 011) \vee (110 \wedge 011) \vee (111 \wedge 111). \\
L &= \left(\bar{x}_1\bar{x}_2\bar{x}_3 \wedge \bar{X}_1\bar{X}_2X_3\right) \vee \left(\bar{x}_1\bar{x}_2x_3 \wedge \left(\bar{X}_1 \vee \bar{X}_2\right)X_3\right) \vee \left(\bar{x}_1x_2\bar{x}_3 \wedge \left(\bar{X}_1 \vee \bar{X}_3\right)X_2\right) \vee \\
&\vee \left(\bar{x}_1x_2x_3 \wedge X_1\left(\bar{X}_2 \vee \bar{X}_3\right)\right) \vee \left(x_1\bar{x}_2\bar{x}_3 \wedge \bar{X}_1X_2X_3\right) \vee \left(x_1\bar{x}_2x_3 \wedge \bar{X}_1X_2X_3\right) \vee \\
&\vee \left(x_1x_2\bar{x}_3 \wedge \bar{X}_1X_2X_3\right) \vee \left(x_1x_2x_3 \wedge X_1X_2X_3\right).
\end{aligned}$$

For technological complexity a comparison of the proposed qubit method, the procedure for obtaining formulae of the deductive fault simulation method described in [4], based on the analytical expression of the functionalities, is shown below:

$$\begin{aligned}
L = {} & \bar{x}_1\bar{x}_2\bar{x}_3\left\{\left[\left(\bar{X}_2 \oplus 0\right)\left(\bar{X}_3 \oplus 0\right) \vee (X_1 \oplus 0)(X_2 \oplus 0)(X_3 \oplus 0)\right] \oplus 1\right\} \\
& \vee \bar{x}_1\bar{x}_2x_3\left\{\left[\left(\bar{X}_2 \oplus 0\right)\left(\bar{X}_3 \oplus 1\right) \vee (X_1 \oplus 0)(X_2 \oplus 0)(X_3 \oplus 1)\right] \oplus 0\right\} \\
& \vee \bar{x}_1x_2\bar{x}_3\left\{\left[\left(\bar{X}_2 \oplus 1\right)\left(\bar{X}_3 \oplus 0\right) \vee (X_1 \oplus 0)(X_2 \oplus 1)(X_3 \oplus 0)\right] \oplus 0\right\} \\
& \vee \bar{x}_1x_2x_3\left\{\left[\left(\bar{X}_2 \oplus 1\right)\left(\bar{X}_3 \oplus 1\right) \vee (X_1 \oplus 0)(X_2 \oplus 1)(X_3 \oplus 1)\right] \oplus 0\right\} \\
& \vee x_1\bar{x}_2\bar{x}_3\left\{\left[\left(\bar{X}_2 \oplus 0\right)\left(\bar{X}_3 \oplus 0\right) \vee (X_1 \oplus 1)(X_2 \oplus 0)(X_3 \oplus 0)\right] \oplus 1\right\} \\
& \vee x_1\bar{x}_2x_3\left\{\left[\left(\bar{X}_2 \oplus 0\right)\left(\bar{X}_3 \oplus 1\right) \vee (X_1 \oplus 1)(X_2 \oplus 0)(X_3 \oplus 1)\right] \oplus 0\right\} \\
& \vee x_1x_2\bar{x}_3\left\{\left[\left(\bar{X}_2 \oplus 1\right)\left(\bar{X}_3 \oplus 0\right) \vee (X_1 \oplus 1)(X_2 \oplus 1)(X_3 \oplus 0)\right] \oplus 0\right\} \\
& \vee x_1x_2x_3\left\{\left[\left(\bar{X}_2 \oplus 1\right)\left(\bar{X}_3 \oplus 1\right) \vee (X_1 \oplus 1)(X_2 \oplus 1)(X_3 \oplus 1)\right] \oplus 1\right\}.
\end{aligned}$$

$$\begin{aligned}
L = {} & \bar{x}_1\bar{x}_2\bar{x}_3\left(\overline{\bar{X}_2\bar{X}_3 \vee X_1X_2X_3}\right) \vee \bar{x}_1\bar{x}_2x_3\left(X_2X_3 \vee X_1X_2\bar{X}_3\right) \vee \bar{x}_1x_2\bar{x}_3\left(X_2\bar{X}_3 \vee X_1\bar{X}_2\bar{X}_3\right) \\
& \vee \bar{x}_1x_2x_3\left(X_2X_3 \vee X_1\bar{X}_2\bar{X}_3\right) \vee \bar{x}_1\bar{x}_2x_3\left(\overline{\bar{X}_2\bar{X}_3 \vee \bar{X}_1X_2X_3}\right) \\
& \times \vee x_1\bar{x}_2x_3\left(\bar{X}_2X_3 \vee \bar{X}_1X_2\bar{X}_3\right) \vee x_1x_2\bar{x}_3\left(X_2\bar{X}_3 \vee \bar{X}_1\bar{X}_2X_3\right) \\
& \vee x_1x_2x_3\left(\overline{X_2X_3 \vee \bar{X}_1\bar{X}_2\bar{X}_3}\right).
\end{aligned}$$

$$\begin{aligned}
L = {} & \bar{x}_1\bar{x}_2\bar{x}_3\left(\overline{\bar{X}_2\bar{X}_3} \vee \overline{X_1X_2X_3}\right) \vee \bar{x}_1\bar{x}_2x_3\left(\bar{X}_2X_3 \vee X_1X_2\bar{X}_3\right) \vee \bar{x}_1x_2\bar{x}_3\left(X_2\bar{X}_3 \vee X_1\bar{X}_2\bar{X}_3\right) \\
& \vee \bar{x}_1x_2x_3\left(X_2X_3 \vee X_1\bar{X}_2\bar{X}_3\right) \vee x_1\bar{x}_2\bar{x}_3\left(\overline{\bar{X}_2\bar{X}_3} \vee \overline{\bar{X}_1X_2X_3}\right) \\
& \times \vee x_1\bar{x}_2x_3\left(\bar{X}_2X_3 \vee \bar{X}_1X_2\bar{X}_3\right) \vee x_1x_2\bar{x}_3\left(X_2\bar{X}_3 \vee \bar{X}_1\bar{X}_2X_3\right) \\
& \vee x_1x_2x_3\left(\overline{X_2X_3} \vee \overline{\bar{X}_1\bar{X}_2\bar{X}_3}\right).
\end{aligned}$$

$$\begin{aligned}
L = {} & \bar{x}_1\bar{x}_2\bar{x}_3\left(X_2 \vee X_3 \vee \left(\bar{X}_1 \vee \bar{X}_2 \vee \bar{X}_3\right) \vee \bar{x}_1\bar{x}_2x_3\left(\bar{X}_2X_3 \vee X_1X_2\bar{X}_3\right) \vee \bar{x}_1x_2\bar{x}_3\left(X_2\bar{X}_3\right.\right. \\
& \left.\vee X_1\bar{X}_2X_3\right) \vee \bar{x}_1x_2x_3\left(X_2X_3 \vee X_1\bar{X}_2\bar{X}_3\right) \vee x_1\bar{x}_2\bar{x}_3(X_2 \vee X_3) \wedge \left(X_1\right. \\
& \left.\left.\vee \bar{X}_2 \vee \bar{X}_3\right)\right] \vee x_1\bar{x}_2x_3\left(\bar{X}_2X_3 \vee \bar{X}_1X_2\bar{X}_3\right) \vee x_1x_2\bar{x}_3\left(X_2\bar{X}_3 \vee \bar{X}_1\bar{X}_2X_3\right) \\
& \vee x_1x_2x_3\left(\bar{X}_2 \vee \bar{X}_3\right) \wedge \left(X_1 \vee X_2 \vee X_3\right)].
\end{aligned}$$

$$\begin{aligned}
L = {} & \bar{x}_1\bar{x}_2\bar{x}_3\left(\bar{X}_1X_2 \vee \bar{X}_1X_3 \vee \bar{X}_2\bar{X}_3 \vee X_2\bar{X}_3\right) \vee \bar{x}_1\bar{x}_2x_3\left(\bar{X}_2X_3 \vee X_1X_2\bar{X}_3\right) \\
& \qquad \vee X_1\bar{X}_2X_3\right) \vee \bar{x}_1x_2x_3\left(X_2X_3 \vee X_1\bar{X}_2\bar{X}_3\right) \vee x_1\bar{x}_2\bar{x}_3(X_1X_2 \vee X_1X_3) \\
& \times \vee \bar{x}_1x_2\bar{x}_3\left(X_2\bar{X}_3 \vee \bar{X}_2X_3 \vee X_2\bar{X}_3\right) \vee x_1\bar{x}_2x_3\left(\bar{X}_2X_3 \vee \bar{X}_1X_2\bar{X}_3\right) \vee x_1x_2\bar{x}_3\left(X_2\bar{X}_3\right. \\
& \qquad \left.\vee \bar{X}_1\bar{X}_2X_3\right) \vee x_1x_2x_3\left(X_1\bar{X}_2 \vee X_1\bar{X}_3 \vee X_2\bar{X}_3 \vee \bar{X}_2X_3\right).
\end{aligned}$$

Creation of vector forms for deductive qubit fault simulation of the basic logical elements *or*, *and*, *xor* is represented in the following table for ease of understanding and programming of software/hardware:

x_1	x_2	$Y^{\vee}$	$X_1^{\vee}$	$X_2^{\vee}$	$Y^{\wedge}$	$X_1^{\wedge}$	$X_2^{\wedge}$	$Y^{\oplus}$	$X_1^{\oplus}$	$X_2^{\oplus}$
0	0	0	1	1	0	0	0	0	1	1
0	1	1	0	1	0	1	0	1	1	1
1	0	1	1	0	0	0	1	1	1	1
1	1	1	0	0	1	1	1	0	1	1

Here, two vector columns of a truth table for two input variables (x_1, x_2), the qubit coverage of the function or $Y^{\vee}$, and two derivatives $X_1^{\vee}, X_2^{\vee}$ for each input variable are defined. The next columns show the qubit vector of the *and* function and two columns of derivatives; and the qubit vector of the *xor* function and two columns of input derivatives. The formulae of the deductive simulation are written according to the table rows:

$$L^{\vee} = \bar{x}_1\bar{x}_2(X_1 \vee X_2) \vee \bar{x}_1 x_2(\bar{X}_1 \wedge X_2) \vee x_1\bar{x}_2(X_1 \wedge \bar{X}_2) \vee x_1 x_2(X_1 \wedge X_2);$$
$$L^{\wedge} = \bar{x}_1\bar{x}_2(X_1 \wedge X_2) \vee \bar{x}_1 x_2(X_1 \wedge \bar{X}_2) \vee x_1\bar{x}_2(\bar{X}_1 \wedge X_2) \vee x_1 x_2(X_1 \vee X_2);$$
$$L^{\oplus} = \bar{x}_1\bar{x}_2(X_1 \vee X_2) \vee \bar{x}_1 x_2(X_1 \wedge X_2) \vee x_1\bar{x}_2(X_1 \vee X_2) \vee x_1 x_2(X_1 \vee X_2) = (X_1 \vee X_2).$$

Here are the input variables: x_i connected with the $\wedge$ operation and derivatives with the $\vee$ operation. The transportation variables of the input fault lists, signed by X_i, are connected to each other by the $\vee$ operation in accordance with the coordinate states of the derivative columns.

Thus, the proposed qubit fault simulation method based on qubit coverage of the functionality and derivative has no analogs for availability of understanding, implementation, and performance. To obtain the deductive formulae for the fault transportation of any functionality, it is necessary and sufficient to get the qubit derivatives of all input variables by means of shift and *xor* operations. The computational complexity of these operations depends on the accessible hardware volume and at the limit can be reduced to a linear dependence on the number of input variables.

A qubit simulation processor of digital devices is shown in Fig. 5.10. It includes the following structures: a fault-free interpretative qubit simulator, a deductive qubit fault simulator designed to evaluate the test quality and build a fault detection table, and also a structure for testing and diagnosing faults in the stages of design and operation. The main difference from the existing solutions is leveraging qubit coverage, which gives an opportunity to significantly increase the speed of simulation by performing parallel register operations.

Here are the following components: a control unit designed for synchronization of units in qubit operation for fault-free simulation, fault simulation, test generation, and fault diagnosis based on online testing; a fault matrix (FM) of input faults of considered functionality; a test bench (TB), which is an ordered set of input test sequences where the current input pattern is identified as x_i; and a circuit description

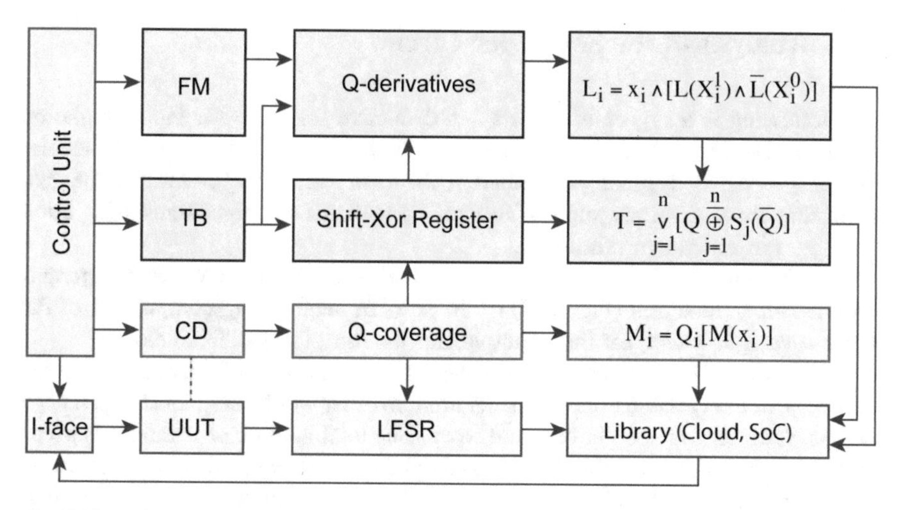

Fig. 5.10 Qubit simulation processor for digital devices

(CD) of a digital device where the functional elements are represented by a qubit vector (*Q* coverage). Their processing is carried out by the fault-free simulation unit $M_i = Q_i[M(x_i)]$, which executes addressable transactions between qubit coverage and the simulation vector *M*. The results of fault-free simulation of input test patterns form the matrix GM (good simulation matrix), written into the library. A shift *xor* register unit generates a derivative matrix in the qubit form (*Q* derivatives), using shift and *xor* register operations. The *L* unit creates an output fault list by using the following deductive formula, operating test vector, input fault lists *L*(*X*), and derivative *X* matrix rows:

$$L_i = x_i \wedge \left[L\left(X_i^1\right) \wedge \bar{L}\left(X_i^0\right)\right]$$

The results of the deductive qubit analysis form an output fault list corresponding to the input sequences, which are combined to the DM (fault detected matrix) and written in the library. The *T* module generates a *Q* test and evaluates its quality by using the metric of stuck-at faults of external inputs and outputs of the functional elements, which are written in the library.

Tests and fault detection matrices create library = {signature, *Q* coverage, *Q* test, quality, DM, GM}, which can be reused as a cloud or built-in SoC service for testing and/or diagnosis of the UUT (unit under test) functionalities through the use of an interface (I face) supporting the standard IEEE 1500 SECT IP (Internet protocol). The search of test services in the library is carried out by a qubit vector previously compressed in a 16-bit binary code signature (Sign), based on the linear feedback shift register (LFSR) that structures the library for quick extraction of the data and online testing. The LFSR unit performs two compression functions in a 16-bit code signature of the following bit vectors: (1) *Q* coverage and (2) UUT responses to the input test patterns for the next diagnosis of the location, cause, and type of fault.

5.5 Analysis of the Schneider Circuit

The following is a synthesis of tests and deductive formulae for fault simulation based on Boolean derivatives with respect to input variables for the Schneider circuit (Fig. 5.11), originally described by the truth table and Q coverage. The goal is to show the technology and performance of methods and algorithms using qubit coverages of functional primitives.

The formula for determining the derivative Q' to the input variables operates with the *xor* interaction (Fig. 5.12) of the pairs of neighboring components of the qubit coverage, which, for this example, has the form (1000000000000001).

The power of each component, defined as the i number of bits, depends on the number k of the variable under consideration. In other words, neighboring pairs can be bits, couples, fours, eight bits, and so on. The total number of variables is given by the number n. An increase in the number of the variable leads to an increase in the width of the neighboring components and to a corresponding decrease in their total number necessary to cover the qubit vector.

To obtain a qubit derivative to the first variable of the Schneider scheme, it is necessary to *xor* operate adjacent bits in each pair of the qubit vector Q and write the result to each bit of the pair according to the rule $\{a, b\} = a \oplus b$: $\{1,1\} = 1 \oplus 0$, $(0,0) = 0 \oplus 0$, $(0,0) = 0 \oplus 0$, $(0,0) = 0 \oplus 0$, $(0,0) = 0 \oplus 0$, $(0,0) = 0 \oplus 0$, $(0,0) = 0 \oplus 0$, $(1,1) = 0 \oplus 1$.

To obtain a qubit derivative to the second variable, it is necessary to *xor* operate the neighboring pairs of the qubit vector Q and write the result to each pair $\{a, b\} = a \oplus b$: $\{10,10\} = 10 \oplus 00$, $(00,00) = 00 \oplus 00$, $(00,00) = 00 \oplus 00$, $(01.01) = 00 \oplus 01$.

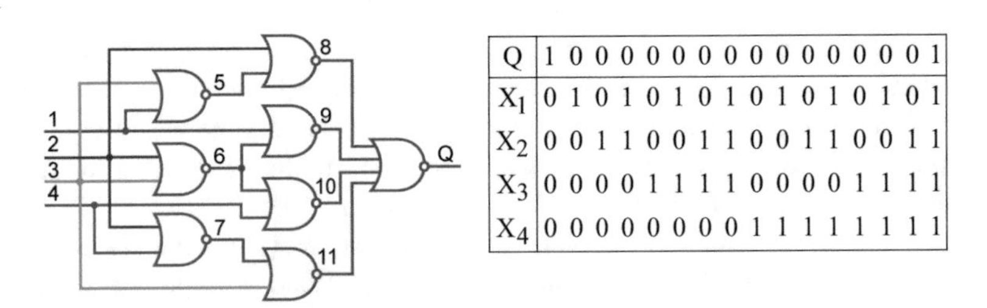

Q	1	0	0	0	0	0	0	0	0	0	0	0	0	0	0	1
X_1	0	1	0	1	0	1	0	1	0	1	0	1	0	1	0	1
X_2	0	0	1	1	0	0	1	1	0	0	1	1	0	0	1	1
X_3	0	0	0	0	1	1	1	1	0	0	0	0	1	1	1	1
X_4	0	0	0	0	0	0	0	0	1	1	1	1	1	1	1	1

Fig. 5.11 Schneider circuit and truth table

Fig. 5.12 Taking the Q derivative

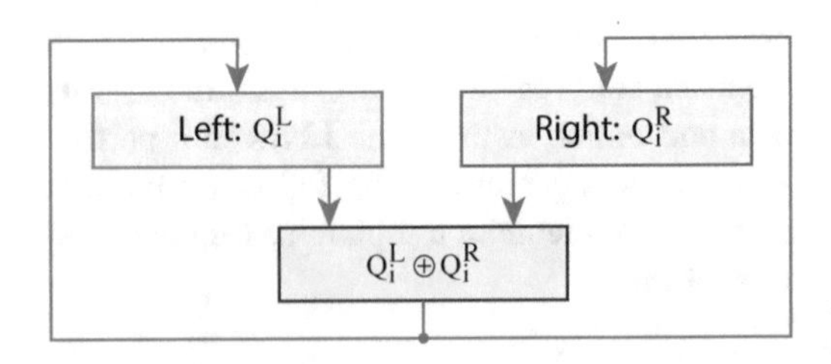

To obtain a qubit derivative to the third variable, it is necessary to *xor* operate the pairs of adjacent tetrads of the qubit vector Q and write the result in both tetrads $\{a, b\} = a \oplus b$: {1000,1000} = 1000 ⊕ 0000, (0001,0001) = 0000 ⊕ 0001.

To obtain a qubit derivative to the fourth variable, it is necessary to *xor* operate the neighboring eight bits of the qubit vector Q and write the result to both ones $\{a, b\} = a \oplus b$: {10000001,10000001} = 10000000 ⊕ 00000001. The following table contains a Q coverage and four vectors that are derivatives of four variables:

Q'(X)	1 0 0 0 0 0 0 0 0 0 0 0 0 0 0 1
X_1	1 1 0 0 0 0 0 0 0 0 0 0 0 0 1 1
X_2	1 0 1 0 0 0 0 0 0 0 0 0 0 1 0 1
X_3	1 0 0 0 1 0 0 0 0 0 0 1 0 0 0 1
X_4	1 0 0 0 0 0 0 1 1 0 0 0 0 0 0 1

The analytical form of the qubit derivatives, written according to the 1 unit of the qubit coverage in the format of the input variables, has the following form:

$$Q'(X_1) = \bar{X}_1\bar{X}_2\bar{X}_3\bar{X}_4 \vee X_1\bar{X}_2\bar{X}_3\bar{X}_4 \vee \bar{X}_1X_2X_3X_4 \vee X_1X_2X_3X_4 = \\ = \bar{X}_2\bar{X}_3\bar{X}_4 \vee X_2X_3X_4.$$

$$Q'(X_2) = \bar{X}_1\bar{X}_2\bar{X}_3\bar{X}_4 \vee \bar{X}_1X_2\bar{X}_3\bar{X}_4 \vee X_1\bar{X}_2X_3X_4 \vee X_1X_2X_3X_4 = \\ = \bar{X}_1\bar{X}_3\bar{X}_4 \vee X_1X_3X_4.$$

$$Q'(X_3) = \bar{X}_1\bar{X}_2\bar{X}_3\bar{X}_4 \vee \bar{X}_1\bar{X}_2X_3\bar{X}_4 \vee X_1X_2\bar{X}_3X_4 \vee X_1X_2X_3X_4 = \\ = \bar{X}_1\bar{X}_2\bar{X}_4 \vee X_1X_2X_4.$$

$$Q'(X_4) = \bar{X}_1\bar{X}_2\bar{X}_3\bar{X}_4 \vee \bar{X}_1\bar{X}_2\bar{X}_3X_4 \vee X_1X_2X_3\bar{X}_4 \vee X_1X_2X_3X_4 = \\ = \bar{X}_1\bar{X}_2\bar{X}_3 \vee X_1X_2X_3.$$

The procedure for the test synthesis of the Schneider scheme is based on derivatives. Each function obtained above is the condition for activating an input variable, over which the derivative is taken. This means that two test vectors with changes in input signals 01 or 10 detect stuck-at faults on the logical path specified by the activation conditions. With reference to Schneider's circuit, such a test contains 16 input sets. A minimal test in a variable format $(X_1X_2X_3X_4)$ that does not contain duplicate input sets has only 10 vectors written in the right column:

Activation	Test	Min
$\overline{X}_2\overline{X}_3\overline{X}_4 \vee X_2X_3X_4$	0000	0000
	1000	1000
	0111	0111
	1111	1111
$\overline{X}_1\overline{X}_3\overline{X}_4 \vee X_1X_3X_4$	0000	
	0100	0100
	1011	1011
	1111	
$\overline{X}_1\overline{X}_2\overline{X}_4 \vee X_1X_2X_4$	0000	
	0010	0010
	1101	1101
	1111	
$\overline{X}_1\overline{X}_2\overline{X}_3 \vee X_1X_2X_3$	0000	
	0001	0001
	1110	1110
	1111	

The result of the test synthesis obtained by taking the derivatives in the Q vector format is the following:

$$T(A) = \boxed{1\ 1\ 1\ 0\ 1\ 0\ 0\ 1\ 1\ 0\ 0\ 1\ 0\ 1\ 1\ 1}.$$

The synthesis of Q tests based on the toward shifting of the vector's components of the qubit coverage leverages the following formula:

$$T(S) = \bigvee_{j=1}^{n} \left[Q \overline{\oplus}_{j=1}^{n} S_j(\bar{Q}) \right].$$

The result of leveraging a test synthesis formula that includes four logical register operations is shown below:

Q – Test Synthesis	1 0 0 0 0 0 0 0 0 0 0 0 0 0 0 1
$\overline{Q}$	0 1 1 1 1 1 1 1 1 1 1 1 1 1 1 0
$S_1(\overline{Q})$	1 0 1 1 1 1 1 1 1 1 1 1 1 1 0 1
$S_2(\overline{Q})$	1 1 0 1 1 1 1 1 1 1 1 1 1 0 1 1
$S_3(\overline{Q})$	1 1 1 1 0 1 1 1 1 1 1 0 1 1 1 1
$S_4(\overline{Q})$	1 1 1 1 1 1 1 0 0 1 1 1 1 1 1 1
$T_1 = Q \overline{\oplus} S_1(\overline{Q})$	1 1 0 0 0 0 0 0 0 0 0 0 0 0 1 1
$T_2 = Q \overline{\oplus} S_2(\overline{Q})$	1 0 1 0 0 0 0 0 0 0 0 0 0 1 0 1
$T_3 = Q \overline{\oplus} S_3(\overline{Q})$	1 0 0 0 1 0 0 0 0 0 0 1 0 0 0 1
$T_4 = Q \overline{\oplus} S_4(\overline{Q})$	1 0 0 0 0 0 0 1 1 0 0 0 0 0 0 1
$T(S) = T_1 \vee T_2 \vee T_3 \vee T_4$	1 1 1 0 1 0 0 1 1 0 0 1 0 1 1 1

Thus, both methods of test synthesis (based on toward shifting and input activation) give the same result, containing ten input vectors that detect all stuck-at faults on the lines of all logical paths in the circuit:

$$T(S) = T(A) = (1110100110010111).$$

However, the test can be obtained more simply by using a parallel logical vector operation of disjunction over the coordinates of the qubit derivatives:

$$T(Q') = \bigvee_{i=1}^{n} Q'(X_i).$$

The result of the register disjunction operation is presented in the bottom line of the following table of the qubit derivatives:

$Q'(X)$	1 0 0 0 0 0 0 0 0 0 0 0 0 0 0 1
$Q'(X_1)$	1 1 0 0 0 0 0 0 0 0 0 0 0 0 1 1
$Q'(X_2)$	1 0 1 0 0 0 0 0 0 0 0 0 0 1 0 1
$Q'(X_3)$	1 0 0 0 1 0 0 0 0 0 0 1 0 0 0 1
$Q'(X_4)$	1 0 0 0 0 0 0 1 1 0 0 0 0 0 0 1
$T(Q') = \bigvee_{i=1}^{n} Q'(X_i)$	1 1 1 0 1 0 0 1 1 0 0 1 0 1 1 1

The presented three methods of the test synthesis based on the analysis of functional qubit coverage, with different computational complexity, give the same result:

$$T(Q') = T(S) = T(A) = (1110100110010111).$$

Further minimization of the test due to the fault simulation gives only six vectors represented by the following fault-free and fault simulation tables:

1	2	3	4	5	6	7	8	9	10	11	12
0	0	0	0	1	1	1	0	0	0	0	1
0	0	0	1	0	0	1	0	0	1	0	0
0	0	1	0	0	0	1	1	1	1	0	0
0	1	0	0	1	0	0	0	1	1	1	0
1	0	0	0	0	1	1	1	0	0	0	0
1	1	1	1	0	0	0	0	0	0	0	1

1	2	3	4	5	6	7	8	9	10	11	12	FD	FC
1	1	1	1	0	0	0	1	1	1	1	0	37	37
.	.	.	0	.	.	0	.	.	0	.	1	16	54
.	.	0	.	.	.	0	.	.	.	.	1	16	66
.	0	.	.	0	.	.	1	.	.	.	1	16	79
0	.	.	.	.	0	.	.	1	.	.	1	16	87
0	0	0	0	1	1	1	0	0	0	0	0	50	100

Qubit Deductive Method of Fault Simulation: The analytic interpretation of the $Q'(X)$ table of qubit derivatives, rotated 90° to write deductive formulae and further fault simulation, is also easy to understand. In practice, the procedure for forming the output list (vector) of transported stuck-at faults $L(X)$ consists of disjunction of the lists of input faults written down by 1 unit of the derivative table row logically multiplied by the negation of conjunction written according to the zero coordinates of the derivative table row:

$$\begin{aligned} L^S(X) &= \bigvee_{i=1}^{2^n} \left\{ \left[\bigvee_{X_{ij}=1} L(X_{ij}) \right] \& \left[\overline{\bigvee_{X_{ij}=0} L(X_{ij})} \right] \right\} \\ &= \bigvee_{i=1}^{2^n} \left\{ \left[\bigvee_{X_{ij}=1} L(X_{ij}) \right] \& \left[\bigvee_{X_{ij}=0} \bar{L}(X_{ij}) \right] \right\}. \end{aligned}$$

Following this formula, a further expression table is synthesized for deductive fault simulation with the number of terms equal to the length of the qubit coverage:

x_1	x_2	x_3	x_4	Y
0	0	0	0	1
0	0	0	1	0
0	0	1	0	0
0	0	1	1	0
0	1	0	0	0
0	1	0	1	0
0	1	1	0	0
0	1	0	1	0
1	0	0	0	0
1	0	0	1	0
1	0	1	0	0
1	0	1	1	0
1	1	0	0	0
1	1	0	1	0
1	1	1	0	0
1	1	1	1	0

X_1	X_2	X_3	X_4	$L^S(X)$
1	1	1	1	$X_1 \vee X_2 \vee X_3 \vee X_4$
0	0	0	1	$X_4(\overline{X_1X_2X_3})$
0	0	1	0	$X_3(\overline{X_1X_2X_4})$
0	0	0	0	$X_1X_2X_3X_4$
0	1	0	0	$X_2(\overline{X_1X_3X_4})$
0	0	0	0	$X_1X_2X_3X_4$
0	0	0	0	$X_1X_2X_3X_4$
1	0	0	0	$X_1(\overline{X_2X_3X_4})$
1	0	0	0	$X_1(\overline{X_2X_3X_4})$
0	0	0	0	$X_1X_2X_3X_4$
0	0	0	0	$X_1X_2X_3X_4$
0	1	0	0	$X_2(\overline{X_1X_3X_4})$
0	0	0	0	$X_1X_2X_3X_4$
0	0	1	0	$X_3(\overline{X_1X_2X_4})$
0	0	0	1	$X_4(\overline{X_1X_2X_3})$
1	1	1	1	$X_1 \vee X_2 \vee X_3 \vee X_4$

Here, the conjunctive terms related to 1 unit of the derivative column (right table) represent approximate conditions for transporting lists of input stuck-at faults by the input test vector specified by the corresponding line addresses (0000, 0001, ...). For instance, the first and last table lines combine all lists of input faults. The second line of the table indicates the transportation to the function output—all defects from the first input minus all those that will simultaneously be present at inputs 2,3,4. If there are all zero coordinates in the line of the table, a logical intersection of the single fault lists is performed for all inputs that will be transported to the function output. The exact deductive simulation formulae have the larger dimension where for each input test vector it is necessary to build a DNF, which in size has an upper limit equal to the truth table. Fig. 5.13 contains the list of input defect vectors that need to be transported through the Schneider scheme as a primitive, on six test vectors: 0000, 0001, 0010, 0100, 1000, 1111. One result is shown as the vector of output faults verified on the input vector 0000, obtained by using the first column of the derivative matrix $Q'(X)$ to simulate single defects. Naturally, each fault from the input list of the test vector is inverse to the state of the considered input variable.

The remaining results of input fault simulation on six input vectors (test columns), leveraging the derivative columns (derivatives) for the Schneider circuit, are presented below:

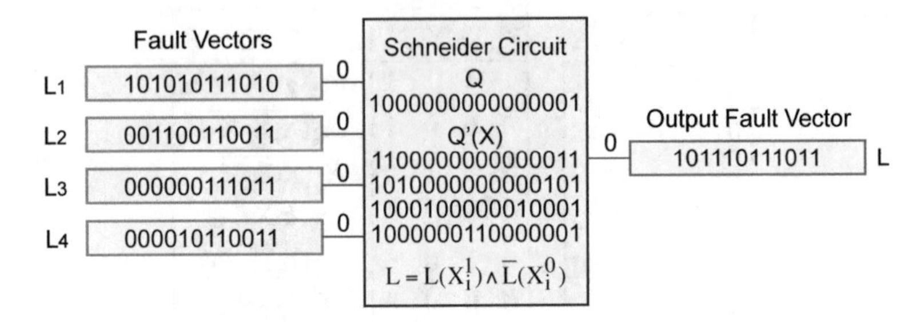

Fig. 5.13 Transportation of fault detection vectors

Input Faults	Test Column	Derivatives		Output Faults
				111111111111
101010111010	0 1 0 0 0 1	1 1 0 0 0 1		000010000000
001100110011	0 0 1 0 0 1	1 0 1 0 0 1	=	000000000000
000000111011	0 0 0 1 0 1	1 0 0 1 0 1		000000000000
111101001100	0 0 0 0 1 1	1 0 0 0 1 1		010001000100
				111111111111

The back side of the high performance of the qubit deductive simulation is a little decrease in accuracy. This can theoretically affect the results of the simulation. For comparison, tables for building exact deductive analysis formulae for seven test input sequences of the Schneider circuit are further proposed. The left column is a truth table; below are the input test vector and also two equivalent formulae for deductive simulation of single faults. In the formation of terms of deductive analysis, only those truth table lines that specify unit values of the function Y are used. If the input test pattern reverses the function in 1, then all deductive terms have to be negated.

X_1	X_2	X_3	X_4	Y	L_i	$L_i \oplus Y(x)$	DNF	$L^S(x,X)$
0	0	0	0	1	1111	$\overline{1111}$	$(\overline{X_1X_2X_3X_4})$	$\bar{X}_1 \vee \bar{X}_2 \vee \bar{X}_3 \vee \bar{X}_4$
0	0	0	1	0	.	.	.	.
0	0	1	0	0	.	.	.	.
0	0	1	1	0	.	.	.	.
0	1	0	0	0	.	.	.	.
0	1	0	1	0	.	.	.	.
0	1	1	0	0	.	.	.	.
0	1	1	1	0	.	.	.	.
1	0	0	0	0	.	.	.	.
1	0	0	1	0	.	.	.	.
1	0	1	0	0	.	.	.	.
1	0	1	1	0	.	.	.	.
1	1	0	0	0	.	.	.	.
1	1	0	1	0	.	.	.	.
1	1	1	0	0	.	.	.	.
1	1	1	1	1	0000	$\overline{0000}$	$\overline{\bar{X}_1\bar{X}_2\bar{X}_3\bar{X}_4}$	$X_1 \vee X_2 \vee X_3 \vee X_4$

$$x = 1111; L_i = x(1111) \oplus X_i^1; Y(x) = 1; L_i = L_i \oplus Y(x) = \overline{L_i};$$
$$L = (X_1 \vee X_2 \vee X_3 \vee X_4)\left(\bar{X}_1 \vee \bar{X}_2 \vee \bar{X}_3 \vee \bar{X}_4\right);$$
$$L = (X_1 \vee X_2 \vee X_3 \vee X_4)\left(L = \left(\overline{X_1X_2X_3X_4}\right).\right.$$

The shortcut tables are represented below. Nonessential rows for obtaining results in the form of analytical expressions for deductive simulation of single faults of functional components are excluded.

X_1	X_2	X_3	X_4	Y	L_i	$L_i \oplus Y(x)$	DNF	$L^S(x,X)$
0	0	0	0	1	0000	$\overline{0000}$	$\overline{\bar{X}_1\bar{X}_2\bar{X}_3\bar{X}_4}$	$X_1 \vee X_2 \vee X_3 \vee X_4$
0	0	0	1	0	.	.	.	.
.	.	.	.	.	.	.	.	.
1	1	1	1	1	1111	$\overline{1111}$	$\overline{X_1X_2X_3X_4}$	$\bar{X}_1 \vee \bar{X}_2 \vee \bar{X}_3 \vee \bar{X}_4$

$$x = 0000; L_i = x(0000) \oplus X_i^1; Y(x) = 1; L_i = L_i \oplus Y(x) = \overline{L_i};$$
$$L = (X_1 \vee X_2 \vee X_3 \vee X_4)\left(\bar{X}_1 \vee \bar{X}_2 \vee \bar{X}_3 \vee \bar{X}_4\right);$$
$$L = (X_1 \vee X_2 \vee X_3 \vee X_4)\left(L = \left(\overline{X_1X_2X_3X_4}\right).\right.$$

X_1	X_2	X_3	X_4	Y	L_i	$L_i \oplus Y(x)$	DNF	$L^S(x,X)$
0	0	0	0	1	0001	0001	$\bar{X}_1\bar{X}_2\bar{X}_3X_4$	$(\overline{X_1 \vee X_2 \vee X_3})X_4$
0	0	0	1	0	.	.	.	.
.	.	.	.	.	.	.	.	.
1	1	1	1	1	1110	1110	$X_1X_2X_3\bar{X}_4$	$(X_1X_2X_3)\bar{X}_4$

$$x = 0001; L_i = x(0001) \oplus X_i^1; Y(x) = 0; L_i = L_i \oplus Y(x) = L_i;$$
$$L = \bar{X}_1\bar{X}_2\bar{X}_3X_4 \vee X_1X_2X_3\bar{X}_4;$$
$$L = \left(\overline{X_1X_2X_3}\right)X_4 \vee (X_1X_2X_3)\bar{X}_4.$$

X_1	X_2	X_3	X_4	Y	L_i	$Y(x) = 0 \rightarrow L_i$	DNF	$L^S(x,X)$
0	0	0	0	1	1100	1100	$X_1X_2\bar{X}_3\bar{X}_4$	$X_1X_2(\overline{X_3 \vee X_4})$
0	0	0	1	0	.	.	.	.
.	.	.	.	.	.	.	.	.
.	.	1	1	.	0011	0011	$\bar{X}_1\bar{X}_2X_3X_4$	$(\overline{X_1 \vee X_2})X_3X_4$

$$x = 1100; L_i = x(1100) \oplus X_i^1; L_i = L_i \oplus Y(x) = L_i;$$
$$L = X_1X_2\bar{X}_3\bar{X}_4 \vee \bar{X}_1\bar{X}_2X_3X_4;$$
$$L = X_1X_2\left(\overline{X_3 \vee X_4}\right) \vee \left(\overline{X_1 \vee X_2}\right)X_3X_4.$$

X_1	X_2	X_3	X_4	Y	L_i	$L_i \oplus Y(x)$	DNF	$L^S(x,X)$
0	0	0	0	1	0010	0010	$\bar{X}_1\bar{X}_2X_3\bar{X}_4$	$(\overline{X_1 \vee X_2 \vee X_4})X_3$
0	0	0	1	0	.	.	.	.
.	.	.	.	.	.	.	.	.
1	1	1	1	1	1101	1101	$X_1X_2\bar{X}_3X_4$	$(X_1X_2X_4)\bar{X}_3$

$$x = 0010; L_i = x(0010) \oplus X_i^1; Y(x) = 0; L_i = L_i \oplus Y(x) = L_i;$$
$$L = \bar{X}_1\bar{X}_2X_3\bar{X}_4 \vee X_1X_2\bar{X}_3X_4;$$
$$L = \left(\overline{X_1 \vee X_2 \vee X_4}\right)X_3 \vee (X_1X_2X_4)\bar{X}_3.$$

X_1	X_2	X_3	X_4	Y	L_i	$L_i \oplus Y(x)$	DNF	$L^S(x,X)$
0	0	0	0	1	0100	0100	$\bar{X}_1X_2\bar{X}_3\bar{X}_4$	$(\overline{X_1 \vee X_3 \vee X_4})X_2$
0	0	0	1	0	.	.	.	.
.	.	.	.	.	.	.	.	.
1	1	1	1	1	1011	1011	$X_1\bar{X}_2X_3X_4$	$(X_1X_3X_4)\bar{X}_2$

$$x = 0100; L_i = x(0100) \oplus X_i^1; Y(x) = 0; L_i = L_i \oplus Y(x) = L_i;$$
$$L = \bar{X}_1X_2\bar{X}_3\bar{X}_4 \vee X_1\bar{X}_2X_3X_4;$$
$$L = \left(\overline{X_1 \vee X_3 \vee X_4}\right)X_2 \vee (X_1X_3X_4)\bar{X}_2.$$

X_1	X_2	X_3	X_4	Y	L_i	$L_i \oplus Y(x)$	DNF	$L^S(x, X)$
0	0	0	0	1	1000	1000	$X_1\bar{X}_2\bar{X}_3\bar{X}_4$	$(\overline{X_2 \vee X_3 \vee X_4})X_1$
0	0	0	1	0	.	.	.	.
.	.	.	.	.	.	.	.	.
1	1	1	1	1	0111	0111	$\bar{X}_1X_2X_3X_4$	$(X_2X_3X_4)\bar{X}_1$

$$x = 1000; L_i = x(1000) \oplus X_i^1; Y(x) = 0; L_i = L_i \oplus Y(x) = L_i;$$
$$L = X_1\bar{X}_2\bar{X}_3\bar{X}_4 \vee \bar{X}_1X_2X_3X_4;$$
$$L = \left(\overline{X_2 \vee X_3 \vee X_4}\right)X_1 \vee (X_2X_3X_4)\bar{X}_1.$$

The leverage of accurate deductive simulation formulae for four fault lists on six input vectors for the Schneider circuit gives the same result that was obtained by analyzing faults based on Q vectors of Boolean derivatives:

Input Faults	Deductive formulas	Output Faults
	$L = 0000 \wedge (X_1 \vee X_2 \vee X_3 \vee X_4)(\overline{X_1X_2X_3X_4})$	111111111111
101010111010	$L = 0001 \wedge (\overline{X_1 \vee X_2 \vee X_3})X_4 \vee (X_1X_2X_3)\bar{X}_4$	000010000000
001100110011	$L = 0010 \wedge (\overline{X_1 \vee X_2 \vee X_4})X_3 \vee (X_1X_2X_4)\bar{X}_3$	000000000000
000000111011	$L = 0100 \wedge (\overline{X_1 \vee X_3 \vee X_4})X_2 \vee (X_1X_3X_4)\bar{X}_2$	000000000000
111101001100	$L = 1000 \wedge (\overline{X_2 \vee X_3 \vee X_4})X_1 \vee (X_2X_3X_4)\bar{X}_1$	010001000100
	$L = 1111 \wedge (X_1 \vee X_2 \vee X_3 \vee X_4)(\overline{X_1X_2X_3X_4})$	111111111111

Thus, to increase the performance of deductive fault analysis, we can sacrifice a few percent of accuracy, which does not significantly affect the design of large-scale digital systems. The inaccuracy of the qubit deductive method is associated with the identification of faults, which is an effect of multiple path fault activation on the reconvergent fan-out structure. The benefit from leveraging a practically focused method is a significant reduction in the memory costs for storing data structures, and also high performance of the deductive analysis algorithm based on the use of qubit derivatives.

The qubit description form of digital systems by using the metric (compactness, performance, and quality), surpasses all existing methods for specifying computing devices. The qubit coverage of functionality is the most technological tool for solving problems of analysis, synthesis, testing, and modeling of digital components.

Taking into account that large-size qubit coverage, as a rule, is not defined for all coordinates, it makes sense to create a compact qubit vector by introducing redundancy in the form of a decoder vector of actual and model addresses. For example, a qubit coverage with indefinite values (10xxxx0xxxxxxxx1) can be written compactly by using the significant states of the bits as (1010), saving the original bit addresses in the additional vector (0,1,6,15). In addition, based on the qubit theory of design and test, it is necessary to move away from fault models for reusable logic lines by creating new methods for simulating qubit faults.

5.6 Conclusion

Several formulations of the scientific novelty and practical value of the described research are represented below:

1. A method and sequencer for qubit test synthesis of functional logic components has been developed. It is characterized by parallel execution of register logic operations (*shift*, *or*, *not*, *nxor*) on the qubit vector and its derivatives, which makes it possible to significantly reduce the test generation and online device testing time.
2. A method for taking derivatives to generate tests of the functional components has been designed. It is characterized by parallel implementation of register logic operations (*shift*, *or*, *not*, *nxor*) on the qubit coverage, which allows a significant reduction in the time needed for input test vector synthesis and device testing through hardware redundancy.
3. A deductive qubit fault simulation method for functional components has been developed. It is characterized by parallel execution of register logic operations (*shift*, *or*, *not*, *nxor*) on the qubit coverage and its derivatives, which makes it possible to significantly reduce the time needed for online test verification of digital devices.
4. A qubit simulation processor of digital devices implemented in an SoC or cloud service has been developed for fault-free and fault simulation based on the qubit coverage description of functional elements. It differs from the known solutions by applying a minimum set of register logic operations and high performance of the abovementioned procedures.
5. The practical value of the research is the ability for cloud implementation of high-speed qubit methods for test synthesis and fault simulations of functional logic components based on the parallel execution of register logic operations (*shift*, *or*, *not*, *nxor*) on qubit coverage and its derivatives. This makes it possible to generate input test patterns and evaluate their quality online. In addition, a cloud microservice of test synthesis and fault simulation of functional logic components can be claimed for educational and scientific investigations in the processes of synthesis and analysis of digital architectures.

6. The proposed qubit test synthesis method for the functionality can be used as an embedded BIST service of SoCs based on the boundary scan standard IEEE 1500 SECT or as an online cloud service for hardware testing via IP protocol.
7. Further research in this area will be focused on creation of the quantum (qubit) digital systems design and test theory; development of cloud services for qubit synthesis and analysis of digital systems; and creation of software/hardware test generators, fault simulators, fault-free simulators, diagnosis algorithms, and library solutions embedded in the SoC infrastructure and/or cloud services, using a qubit description of logic black box functionality.

References

1. Ryabtsev, V. G., & Muamar, D. N. (2010). Method and algorithm visualization tools of memory device diagnosis tests. *Electronic Modeling Journal, 32*(3), 43–52.
2. Zorian, Y., & Shoukourian, S. (2013). *Test solutions for nanoscale systems-on-chip: Algorithms, methods and test infrastructure*. In: Ninth International Conference on Computer Science and Information Technologies Revised Selected Papers, Yerevan, 2013.
3. Tshagharyan, G., Harutyunyan, G., Shoukourian, S., Zorian, Y. (2015). *Overview study on fault modeling and test methodology development for FinFET-based memories*. In: 2015 I.E. East-West Design & Test Symposium (EWDTS), Batumi, 2015.
4. Hahanov, V. I., Hahanova, I. V., Litvinova, E. I., & Guz, O. A. (2010). *Design and verification of digital system-on-chip*. Novoye Slovo: Kharkov.
5. Abramovici, M., Breuer, M. A., & Friedman, A. D. (1998). *Digital systems testing and testable design*. Computer Science Press: New York.
6. Hahanov, V. I., Gharibi, W., Litvinova, E. I., & Shkil, A. S. (2015). Qubit data structures of digital devices. *Electronic Modeling Journal, 37*(1), 76–99.
7. Hahanov, V. I., Amer, T. B., Chumachenko, S. V., & Litvinova, E. I. (2015). Qubit technologies for analysis and diagnosis of digital devices. *Electronic Modeling Journal, 37*(3), 17–40.
8. Badulin, S. S., Barnaulov, Y. M., et al. (1981). *Computer-aided design of digital devices*. Moscow: Radio and Communications.
9. Nielsen, M. A., & Chuang, I. L. (2010). *Quantum computation and quantum information*. Cambridge University Press: Cambridge.
10. Nfrfhara, M. (2010). *Quantum computing. An overview*. Higashi-Osaka: Kinki University.
11. Kurosh, F. G. (1968). *The course of higher algebra*. Moscow: Science.
12. Bondarenko, M. F., Hahanov, V. I., & Litvinova, E. I. (2012). The structure of the logical association multiprocessor. *Automation and Remote Control Journal, 10*, 71–92.
13. Molnar, L., & Gontean, A. (2016). *Fault simulation methods*. In: 2016 12th IEEE International Symposium on Electronics and Telecommunications (ISETC), Timisoara, Romania, 2016.
14. Hadjitheophanous, S., Neophytou, S.N., Michael, M.K. (2016). *Scalable parallel fault simulation for shared-memory multiprocessor systems*. In: 2016 I.E. 34th VLSI Test Symposium (VTS), Las Vegas, NV, 2016.
15. Pomeranz, I., & Reddy Sudhakar, M. (1995). Aliasing computation using fault simulation with fault dropping. *IEEE Transactions on Computers Journal, 44*(1), 139–144.
16. Ubar, R., Kõusaar, J., Gorev, M., Devadze, S. (2015). *Combinational fault simulation in sequential circuits*. In: 2015 I.E. International Symposium on Circuits and Systems (ISCAS), Lisbon, 2015.

17. Gorev, M., Ubar, R., Devadze, S. (2015). *Fault simulation with parallel exact critical path tracing in multiple core environment*. In: 2015 Design, Automation & Test in Europe Conference & Exhibition (DATE), Grenoble, 2015.
18. Pomeranz, I. (2014). *Fault simulation with test switching for static test compaction*. In: 2014 I. E. 32nd VLSI Test Symposium (VTS), Napa, CA, 2014.
19. Mirkhani, S., Abraham, J.A. (2014). *EAGLE: A regression model for fault coverage estimation using a simulation based metric*. In: 2014 International Test Conference, Seattle, WA, 2014.

Chapter 6
QuaSim Cloud Service for Quantum Circuit Simulation

Ivan Hahanov, Tamer Bani Amer, Igor Iemelianov, Mykhailo Liubarskyi, and Vladimir Hahanov

6.1 The State of the Art

The aim of this chapter is significant improvement in the dependability, quality, and reliability of computing systems, through circuit element addressability, which allows online repair and increases the performance of the testing, modeling, and simulation of complicated digital units by dimensional reducion of the truth table models of the logical elements and creation of addressable components of data structures.

The research has the following objectives: (1) development of an quantum automaton model [1–5] based on qubit vectors to define the functionalities; (2) synthesis of qubit (quantum) models for digital combinational and sequential primitives, including the logic, flip-flops, counters, registers, coders, and multiplexors; (3) synthesis and analysis of digital circuit qubit models; and (4) modeling and simulation of digital devices through leverage of primitive qubit coverages.

The motivation and state of the art are defined as follows. (1) The digital system-on-a-chip (SoC) sustainable trend decreases the logic from 6% to 0% and increases the memory from 94% to 100%. It is well known that in-SoC logic gives 50% problems in verification, testing, simulation, diagnosis, and repair [6, 7]. (2) Leveraging of memory-only elements in computing creates a regular structure, which provides technological advantages in design, manufacture, and service processes, including simulation, verification, testing, diagnosis, and online repair by using memory spare fields [6, 7]. (3) Memory-used logic creates regular data structures for high-performance fault-free and fault simulation based on transaction

I. Hahanov (✉) • T.B. Amer • I. Iemelianov • M. Liubarskyi • V. Hahanov
Design Automation Department, Kharkov National University of Radio Electronics, Ukraine, Nauka Avenue 14, Kharkov 61166, Ukraine
e-mail: ivanhahanov@icloud.com; tameramer34@yahoo.com; igor@itdelight.com; mlyubarskyy@gmail.com; hahanov@icloud.com

V. Hahanov, *Cyber Physical Computing for IoT-driven Services*,
DOI 10.1007/978-3-319-54825-8_6

operations (read–write) between addressable memory cells [3–5, 8, 11, 12]. (4) Power consumption when changing the logic on the memory elements is increased by a few percent, which is, in fact, considered the "price to pay" for the abovementioned advantages associated with the increase in the yield and the dependability of computing systems, reducing the time to market and enabling online human-free remote repair. However, energy-efficient solutions in memory-driven computing processes [9, 10, 13, 14] suggest that the energy consumption would not be increased.

6.2 Model of a Quantum Processor for Binary Simulation

The quantum processor may be of any finite dimension: vector, matrix, and cube. For a structure comprising two measurements, it contains a matrix of columns or Q vectors, which form the corresponding cells of simulation vector M (Fig. 6.1, left). The vector M, together with the X vector of input primitive variables, creates a data structure of relationships between the column elements. The address of Q coverage cell Q_i, defining the M_i state, is determined by the concatenation of M vector cells, found at the addresses specified by input variable vector X_i. Each vector Q_i, as well as the vector of input line numbers X_i, has an address connection with the M_i cell of the simulation vector M. The quantum processor as the scalable architecture has the following structure:

The structural–analytical memory–address–transaction model W contains the following components. (1) The addressable Q set of qubit elements creates the system functionality. (2) The computing state vector M connects all primitive components into a system leveraging the identificators of essential variables: input, output, and internal ones. (3) Vector X of the input variables for each circuit

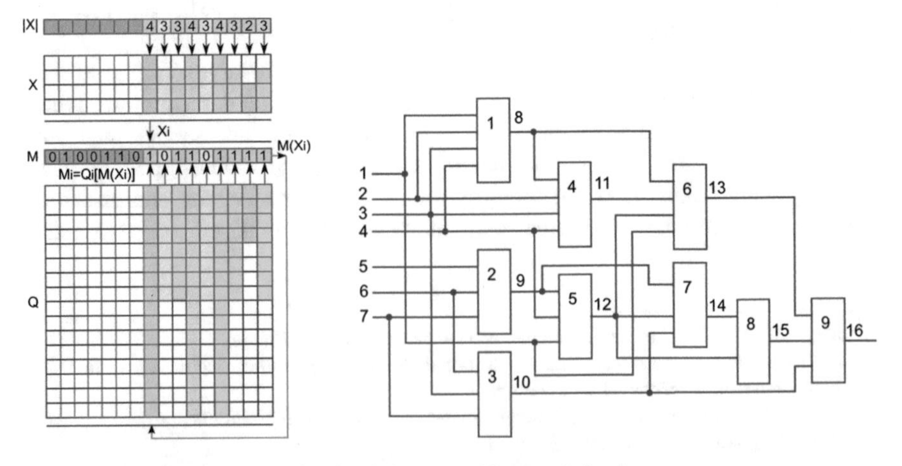

Fig. 6.1 Qubit data structures for simulating a combinational circuit

primitive deliver addressable access to Q vector cells of every primitive (Fig. 6.1). The vector |X| of the input variable number of every primitive creates an address space length of all inputs and Q coverages. The primitive inputs are represented by a table of variables, which define numbers of lines for defining the address of the Q coverage bit. (4) The memory–address–transaction computing leverages an equation that specifies the quantum processor based on read–write operations between Q vectors and simulation vector M.

The circuit of a digital device, corresponding to the data structures described above, is represented in Fig. 6.1: M is a simulation vector, X is a matrix of inputs, and Q is a matrix of coverage. It contains nine primitives; each of them has Q coverage in the form of a quantum vector that carries out some functionality. A special feature of quantum data structures representing a model of the digital circuit is complete addressability of all device components without wire connections.

The following are the axioms of a quantum (memory-only-based) processor. (1) There is nothing in a quantum processor besides addressable memory. (2) The computational process is represented by the sole universal transaction between addressable memory components $M_i = Q_i[M(X_i)]$. (3) The transaction is a universal procedure of reading and writing data to the nonempty set of addressable memory elements. (4) All memory components are repairable online, due to their wireless but addressable connection. (5) Reusable logic elements and also sequential components are implemented in the memory elements. (6) The union of all components in a computing system is performed by digital connection identifiers of input–output variables of the circuit components forming the simulation vector, which stores the states of all the essential lines of the digital system. (7) All components of the quantum model of the digital system $W = \langle Q,M,X \rangle$, including functional modules, the simulation vector, and the vector of input variable addresses, are programmable online, so they are repairable online. (8) A primitive of a digital system has the format $W = \langle Q,Y,X \rangle$ because a separate element does not have the connections and vector M creating the system of the individual components.

6.3 Simulation Method Based on Logic Qubit Coverages

The Q method of fault-free simulation leverages data structures of memory-based logic primitives for addressable transaction analysis of digital system components. Such data structures are associated with the memory [5, 6] cells (FPGA look-up table), which store functions in the form of Q coverages, where each vector bit has an address, which is defined by the input sequence. Q coverage–driven software implementation of the fault-free simulation method is becoming performance competitive in the world market of design and testing of digital SoCs due to the total addressability of data structures.

The leverage of Q coverages allows us to directly define the states of output and internal lines of the digital circuit by using the universal computing equation $M_i = Q_i[M(X_i)]$. The simulation vector M represents the register form of the circuit data structures, where no-input lines are directly connected to the outputs of functional primitives. The applied signals of the input variables specify an address of the Q vector bit defining the state of considering no-input line. It means the one-output Q coverage describes the functionality associated with only one no-input M coordinate. If the qubit functionality has several outputs, then a table represents the Q coverage, where the row's set is equal to the number of outputs. The benefit of a multioutput primitive is parallel computation of the output's states through the one request to the table at the current address. It is considered as the valuable positive argument for qubit synthesis of the digital device's fragments or the whole circuit, which can be simultaneously calculated on one time clock, allowing the simulation performance to be increased by several times. So, the qubit model of digital structure simulation $M_i = Q_i[M(X_i)]$ is simplified up to calculation of two addresses, by excluding the address of primitive output for writing the output states in the coordinates of the M vector.

The fault-free quantum simulation method of qubit-described digital systems is reduced to filling the binary states of the M vector (Fig. 6.2). (0) The initial parameters and conditions for simulation are defined. (1) The next binary test pattern on the input coordinates of the simulation vector M is specified. (2) Number i is assigned as the next processed primitive by incrementing $i = i + 1$. (3) The bit states of the M vector corresponding to the amount of the input variable vector X_i are concatenated; the needed bit from the function qubit coverage Q_i is read by leveraging the binary vector address obtained with the help of M vector bit concatenation; and the qubit bit is written to the address i of simulation vector M—the last one can be defined by coordinates with the symbol X, which allows us to perform multivalued simulation, test synthesis, and verification of digital systems. (4) Go back to step 2 if there are not processed ($i < n$) primitives. (5) Return to step 1 if there are not processed input test patterns ($t < m$). (6) The qubit simulation ends.

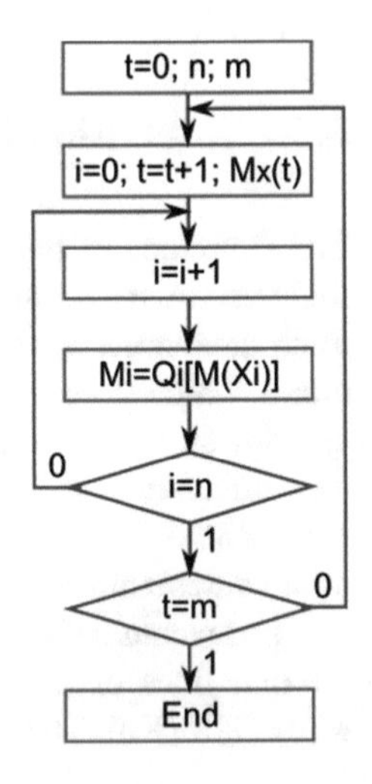

Fig. 6.2 Qubit simulation method for digital systems

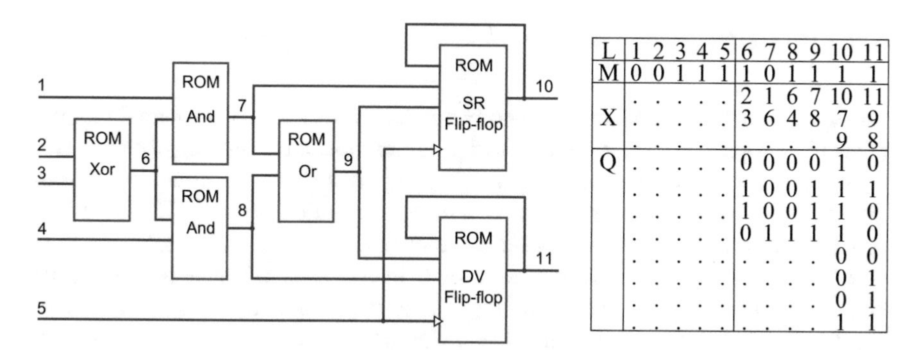

L	1	2	3	4	5	6	7	8	9	10	11
M	0	0	1	1	1	1	0	1	1	1	1
X	.	.	.	.	.	2	1	6	7	10	11
	.	.	.	.	.	3	6	4	8	7	9
	.	.	.	.	.	.	.	.	.	9	8
Q	.	.	.	.	.	0	0	0	0	1	0
	.	.	.	.	.	1	0	0	1	1	1
	.	.	.	.	.	1	0	0	1	1	0
	.	.	.	.	.	0	1	1	1	1	0
	.	.	.	.	.	.	.	.	.	0	0
	.	.	.	.	.	.	.	.	.	0	1
	.	.	.	.	.	.	.	.	.	0	1
	.	.	.	.	.	.	.	.	.	1	1

Fig. 6.3 Memory-based digital circuit

The characteristic universal equation of the digital system quantum simulation allows us to draw the conclusion that the proposed <MAT> (memory–address–transaction) processor should be defined as an interpretative structure of functional memory-driven primitives, which are free of wired connections. The goal of addressable memory data transactions is innovative computing in time and cyber physical space.

The digital circuit with flip-flops and logical elements is shown in Fig. 6.3, where all components are implemented in the memory, which stores the Q coverages of each logical primitive. The qubit data structures required for the digital circuit simulation are represented in the table, where the basic components are the following: M is the simulation vector, which in this unit has five input and six no-input lines, the states of which have to be determined; and X is a vector of primitive input numbers. The last ones are required to generate an address for reading the state of the primitive output, the logic function of which is defined by the Q_i vector.

Before the beginning of a circuit simulation, all primitives should be ordered on the rule that the next logical element is analyzed if all its predecessors have been processed. In the simulation process the addressed bit of the current Q coverage is read and entered in the corresponding bit M_i of the M vector. Sequential processing of all Q vectors of the circuit structure creates the bits of the M vector. For the mentioned circuit the talk is about 6–11 bits of the M vector. The initial states of the flip-flop pseudoinputs are defined by unknown signals, zero or one, depending on the internal technological culture of the company producing industrial tools for design, testing, simulation, and verification. The input variable number q defines the Q vector length of the primitive by the equation $\text{card}(Q) = 2^q$. The correctness of the qubit simulation method has been verified by using the Active-HDL 9.1 simulator (Aldec Inc., Henderson, NV, USA) and opened benchmarks. The innovative feature of the digital system's data structures is description of all primitives by leveraging memory elements where Q vectors of output states are written.

6.4 Fault-Free Simulation of Synchronous Circuits

The sync signals are the cause of inconvenience for describing sequential components (flip-flops, registers, counters) and programming simulation algorithms. This is connected with the existence of rising–falling edge control that requires execution of transactions between two neighboring simulation vectors (M^{t-1},M^{t}) The synchronous primitive has two master-slave-connected elements controlled by low- and high-level signals for recording 1 or 0 in the first and then in the second element, respectively. However, the logic simulation method, taking into account the details of circuit synchro-signals, can significantly reduce the analysis time of digital systems. Therefore, it is necessary to create logically adequate models of real components, focused on high-performance processing in terms of the simulation method. At that, the imposed constraint has to be associated with address analysis of the circuit components. The proposed simulation method considers the sync input as the logical variable forming a bit address of the Q vector, where the sequential primitive is divided into two functions: (1) a logic qubit, generating an enable 1 signal when the rising edge in two time cycles is formed; and (2) a qubit of the sequential component such as flip-flops, registers, and counters. Taking into account the abovementioned rules, a synchronous D flip-flop can be described by two Q coverages to address-based simulation of the output states: <CLK(t – 1) CLK(t) (0100) C(t)>, <Q(t – 1) D(t)C(t) (00010011) Q(t)>. The digital circuit model with two sync signals represented in Fig. 6.4 has a data structure consisting of a Q coverage set, which forms the current simulation vector M depending on the M(t – 1) defined in the previous timing cycle. Increasing the variable number by introducing two sync logic functions allows a reduction in the total dimension of the qubit vector tables, which, for the abovementioned circuit, have 56 coordinates at 7 variables.

If we do not introduce two additional variables (two synchronization elements), the memory volume of Q coverage will be increased up to 80 cells. Such a circuit implementation is most focused on data structures of industrial tools for modeling and verification.

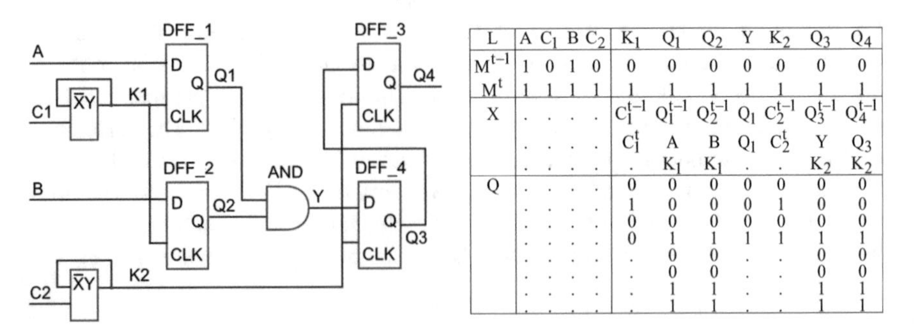

L	A	C_1	B	C_2	K_1	Q_1	Q_2	Y	K_2	Q_3	Q_4
M^{t-1}	1	0	1	0	0	0	0	0	0	0	0
M^{t}	1	1	1	1	1	1	1	1	1	1	1
X	.	.	.	.	C_1^{t-1}	Q_1^{t-1}	Q_2^{t-1}	Q_1	C_2^{t-1}	Q_3^{t-1}	Q_4^{t-1}
	.	.	.	.	C_1^{t}	A	B	Q_1	C_2^{t}	Y	Q_3
	.	.	.	.	.	K_1	K_1	.	.	K_2	K_2
Q	.	.	.	.	0	0	0	0	0	0	0
	.	.	.	.	1	0	0	0	1	0	0
	.	.	.	.	0	0	0	0	0	0	0
	.	.	.	.	0	1	1	1	1	1	1
	.	.	.	.	.	0	0	.	.	0	0
	.	.	.	.	.	0	0	.	.	0	0
	.	.	.	.	.	1	1	.	.	1	1
	.	.	.	.	.	1	1	.	.	1	1

Fig. 6.4 Circuit with sync D flip-flops

The reduction in the total circuit Q coverages is connected to the number of variables creating the addresses of Q vector bits. Of course, partitioning of the variables into two equal subsets usually allows a reduction in the memory volume for storing already two Q vectors. The functional dependence of decreasing the dimension of the initial Q vector, defined on n variables, when dividing by two subcircuits with an equal number of outcome variables ($n/2$), is followed:

For instance, partitioning the vector of eight variables into two Q coverages of four variables allows us to decrease the memory volume by eight times. But each partition of a functional module on k subcircuits creates k internal lines and requires d additional cycles to calculate the states of the primitive outputs of the circuit, where the simulation time is $T = d(k + 1)$. Decreasing the performance of the divided functionality is considered as the cost of the memory reduction of the digital system data structure. Partitioning the functionality into k equal parts allows a memory volume saving depending on the number of partitions of the Q vector defined on the input variables:

$$Q = \frac{2^n}{k \times 2^{n/k}} = \frac{1}{k} \times 2^{n-n/k}.$$

The formula consists of $k = 2, 4, 8, 16$ as the partitioning parameter. The value k should be no more than $n/2$. In general, m partitions may exist for the vector of input variables; each of them has more than one line. The partitioning condition is true if the sum of all the variables involved in the partitions is not more than n):

$$Q = \frac{2^n}{2^{k_1} + 2^{k_2} + \ldots + 2^{k_i} + \ldots + 2^{k_m}}, k_1 + k_2 + \ldots + k_i + \ldots + k_m = n.$$

The formula shows the benefit from functionality partitioning as the relationship between the dimensions of the original Q vector and the Q vectors obtained after the split. The effectiveness of functionality splitting on the circuit fragments depends not only on the memory volume decreasing. It is necessary to take into account the negative effects connected to increasing the simulation time, which depends on the number of obtained circuit components d:

$$Q = \frac{2^n}{d \times m \times 2^{k_1} + 2^{k_2} + \ldots + 2^{k_i} + \ldots + 2^{k_m}}.$$

The qubit-driven modification of the fault-free interpretative simulation method of digital systems allows transformation of the Moore automaton for the description of synchronous digital devices to functional relations by leveraging addressable (A) operations:

$$S(t) = A[X(t), S(t-1)];$$
$$Y(t) = A[X(t), S(t)].$$

There are the input (X); internal automaton states in two time frames S(t), S (t −1); and the rules for address-based calculation of the states S(t) and the output values Y(t).

The addressable qubit automaton allows the following: (1) avoidance of wired interconnections between the logic and sequential elements in their hardware/software implementation by using the memory; (2) achievement of the property of online replacement of digital system elements due to their addressability; (3) simplification of all processes of simulation, verification, and testing by leveraging of the addressable transaction procedures for circuit components, represented by qubit coverages; (4) unification of SoC design processes through use of transactions on addressable memory components; (5) increased efficiency of the simulation procedures of digital circuits by introducing interpretative addressable qubit models of primitives; (6) execution of all simulation procedures based on the leverage of qubit coverages and simulation vector M, given in two adjacent automaton clock cycles; and (7) leverage of the standard BIST infrastructure [6, 7] of testable design (IEEE 1687, 1500, 1149) for online testing, diagnosis, and repair of addressable functional blocks.

The computational complexity C of the qubit fault-free simulation method is defined by linear dependency of the primitive or line number n:

$$C = (R + W) \times n + \sum_{i=1}^{n} m_i,$$

The formula consists of parameter R (W) as the time of reading–writing the memory cell; m_i is the number of i element inputs to be concatenated for an address definition. The linear dependency of computational complexity from the primitive amount creates the market's attractor for this technology in the practice of SoC design and testing.

6.5 QuaSim: A Cloud Service Simulator of Digital Devices

QuaSim is designed for analysis, testing, and verification of small-dimension digital projects; it is intended for leverage in the educational process as a cloud service, available to students via any mobile device or computer. To create a commercial cloud service based on the proposed technology, it is necessary to gather a team of five programmers and funding of $800,000 for 2 years.

The goal is a significant improvement in the quality of the educational process through provision of technologically advanced microservices for digital device analysis with simultaneous visualization of circuits, testing, simulation results, and qubit coverages of functional elements.

The objectives are the following: (1) creation of a structure for cloud simulation of digital devices based on the Google platform of computing services; (2) development of the module (microservice) Q element, implementing circuit creation and visualization of logical primitives; (3) design of a software module for qubit primitive generation and more complex digital devices, and also tools for their visualization; (4) development of a module or control panel integrating a simulator, generator of logic elements, circuit designs, and input–output ports; (5) development of a module for analysis of digital circuits based on recursive processing of the logic elements; (6) creation of a library for storage of the descriptions of functional elements, complex digital circuits, test patterns, the results of their analysis, and service information; and (7) testing and verification of the QuaSim cloud service, designed for interpretive simulation of digital devices.

The substance of the quantum method for digital circuit fault-free analysis is the address implementation of digital system components, which allows us to significantly improve the performance of the interpretive simulation method and the quality of SoC handling through rapid replacement of faulty logic elements by their readdressing.

The structure of the cloud service includes the following. (1) A Q element module generates quantum descriptions of logic elements of the digital functionality structure. (2) A View module visualizes the circuit elements and input and output ports on the monitor. (3) A Collapse module controls the display windows and their sizes by using the appropriate icons. (4) A Split module images the operation of all controllers in the assembly of the circuit on the screen, and also scales the project components. (5) An Evaluate module generates the output state of the current element by reading the contents of the qubit cell by using the corresponding address. (6) A QuaSim module implements the simulation algorithm of all circuit lines by way of developing a recursive model on the first step for serial-to-parallel processing of the elements. In the second step, the states of all outputs of logic elements are calculated. Modeling ends after processing of the entire test input patterns. If a circuit has a global or local feedback, the simulation is carried out up to fixing the same signal values on all lines of the scheme. If the circuit is not set in a stable state at the input patterns, the generator mode is fixed after performing n (=20) iterations. In this case, all the changing lines are set to binary uncertainty $X = \{0,1\}$. (7) A Library Control module for elements, circuits, and projects realizes reading, writing, and connecting the fragments. (8) A module of the waveform visualizes tests and output signals on a monitor in the form of continuous signals divided into clock cycles in absolute or modeling time. (9) All the modules of the cloud service are programmed in Swift language, the operating system OSX 10.9, and the compiler XCode 7. The number of source files is 36, and the total number of code lines is 1450.

A visualization of the graphical design results of a circuit with flip-flops is shown in Fig. 6.5. This circuit is fully consistent with the functionality shown in Fig. 6.3. It includes four input ports for input operating or test inputs and two output ports. The

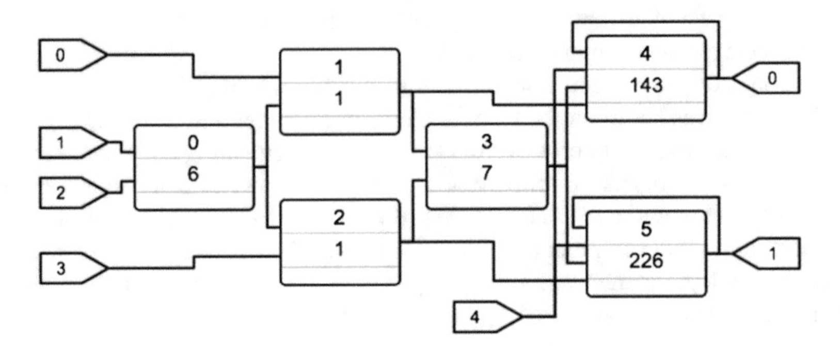

Fig. 6.5 Screenshot of flip-flop structure visualization

structure contains four logical elements and two flip-flops. Mnemonic description of the circuit components is reduced to the universal form of a rectangle and differs only in the number of device primitives, as well as in the type of functionality that is set by a qubit vector represented by a decimal number. In the circuit shown in Fig. 6.5, the elements have ordinal numbers (at the top) and integers to identify functionalities: 0/6—0110, 1/1—0001, 2/1—0001, 3/7—0111, 4/143—11110001, 5/226—01000111. Here the binary vector corresponds to the decimal equivalent of the number to set the functionality. Because a qubit vector does not have an explicit specification of input patterns, it can be seen as an implicit or compact form of a set theoretic truth table. It is not necessary to explicitly specify the input values if they are strictly sequential addressing of the output values. Thus, the truth table, as a collection of inputs and corresponding output values, always concedes to the qubit vector form of functionality representation in the volume and speed of data analysis. There are no fundamental differences between the descriptions of the combinational element, circuit, or sequential primitive, since qubit vectors, which are written in the addressable memory, formally present them. Moreover, all the primitives are also addressable ones, and the circuit structure can also be described as a qubit vector.

Thus, it is possible to come to a realization of a computing device where there is nothing but an address memory or qubit vectors of different lengths and the functionality is determined by an ordered set of zero and unit signals. The advantages of the QuaSim service for qubit description and modeling of digital devices are as follows. (1) All functional components and circuits are defined by Q vectors unifying the procedures for synthesis and analysis of digital devices. (2) It is technologically simple to change or modify the functionality of the circuit or any primitive by replacing the individual bits of the Q vector. (3) Unification of the qubit form for describing circuit primitives allows us to apply a single procedure for analyzing functionality, which is reduced to calculating an address $M_i = Q_i[M(X_i)]$, making the process of programming the QuaSim cloud service technologically simple in terms of implementation and not dependent on the functional and structural complexity of the digital structures. (4) The simple and intuitive user graphical interface makes the cloud service competitive in the market of

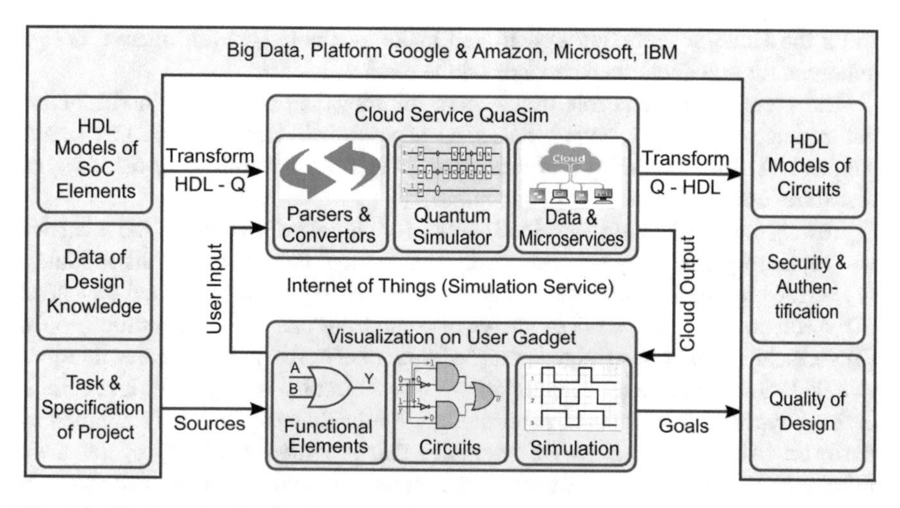

Fig. 6.6 Cloud service for simulating digital devices

educational services, where complex and heavy simulation tools from leading world companies are not available in universities because of high cost and, for students, because of time-consuming and complex preparation of HDL specifications for small educational projects. (5) Unification of the form for describing primitives creates conditions for technologically solving the problems of synthesis, fault simulation, testing, verification, and diagnostics based on operations with qubit vectors. (6) The disadvantage of qubit or quantum technology for the description and analysis of digital structures is some reduction in simulation speed compared with existing commercial compilers for ASIC and VLSI projects, where the volume of reusable logic is dominant in order to achieve high performance.

The structure of the interacting components of the QuaSim cloud service is shown in Fig. 6.6. Quantum or qubit representation of the digital device model with an interpretive simulator constitutes the core of the system, which is integrated into the big data of cyberspace or the Internet. This makes it possible to use as input data open specifications and test benches described in the VHDL language Verilog. Such data and/or test cases are available in almost all leading companies, universities, and subject conferences of the IEEE, TTTC, and ISCAS. In addition, loading the QuaSim service in the Internet space will provide unloading of its operating results related to the analysis and synthesis of educational or market-focused projects in data storage services based on the Google, Amazon, Microsoft, IBM, and Facebook platforms. Naturally, the integration of the cloud service with cyberspace requires parser microservices for conversion specifications from the hardware description languages to the QuaSim internal language, and also there must be reverse data conversion of a qubit representation in HDL languages. Parsers provide the opportunity to use open Internet projects for study and comparison by using the QuaSim modeling system and make available QuaSim internal design solutions for all educational services on the market.

The security unit controls user access for statistical accounting and requires authentication of every person based on a password, surname, and first name, supplemented by any valid (corporate) attribute from the following list: {digital signature, e-mail, digital key, phone number}.

Testing and verification of a cloud service for digital system simulation is carried out separately for each microservice, and for the interaction of all modules: (1) verification of the generation of more complex logic and functional elements; (2) verification of structural synthesis of digital circuits and visualization tools; (3) verification and testing algorithms for binary and ternary synchronous interpretive fault-free simulation of input patterns on 40 circuits of sequential and combinational types; at that, we used patterns algorithmically generated and combined by the user; (4) checking of service modules that provide operability of the main microservices: libraries of elements and circuits, user authentication, and modules for generating statistical data for projects and users; and (5) verification of interface microservices providing interaction between the cloud back-end and custom front-end modules.

6.6 Conclusion

A qubit method for digital device simulation is proposed; it is characterized by the following features: (1) implementation of address transactions between all memory-only components included in the operation and control mechanisms of computing; (2) the quantum analysis method of digital circuits, based on superposition and parallelism of primitive qubit vector execution for significantly increasing the performance of modeling, simulation, testing, and diagnosis; (3) compact data structures for digital system simulation, which simplify design and testing algorithms and improve performance due to address-parallel transactions between memory components; and (4) software implementation of computing structures and processes based on leverage of the qubit address automaton, with testable design standards to improve yield due to online repair of SoC primitives.

The practical aim of the quantum-driven investigation lies in the design of a cloud service for interpretative simulation, based on the leverage of memory-only primitives that make a homogeneous computing structure. The addressability of qubit vectors delivers technological convenience to design, manufacturing, and exploitation, including simulation, online testing, diagnosis, and repair through leverage of the addressable spare memory field.

Quantum simulation of digital systems using the address qubits of components creates high performance of design computing due to regular data structures and simple transaction on the memory. The use of qubit memory-based models for describing digital elements and systems is connected with increasing the SoC yield, improving the quality and reliability of computing products, and reducing the costs of design, manufacturing, and online repair.

References

1. Metodi, T., & Chong, F. (2006). *Quantum computing for computer architects*. Morgan & Claypool: Synthesis lectures on computer architecture. University of Wisconsin: Madison.
2. Stig S., & Suominen K.-A. (2005). Quantum approach to informatics. John Wiley & Sons, Inc., Hoboken, New Jersey.
3. Hahanov, V. I., Gharibi, W., Litvinova, E. I., & Shkil, A. S. (2015). Qubit data structure of computing devices. *Electronic Modeling Journal, 1*, 76–99.
4. Hahanov, V., Bani Amer, T., Hahanov, I. (2015). *MQT-model for virtual computer design*. In: Proc. of Microtechnology and Thermal Problems in Electronics (Microtherm), 2015.
5. Hahanov, V.I., Litvinova, E.I., Chumachenko, S.V., et al. (2012). *Qubit model for solving the coverage problem*. In: Proc. of IEEE East–West Design and Test Symposium, 2012.
6. Zorian, Y., & Shoukourian, S. (2013). *Test solutions for nanoscale systems-on-chip: Algorithms, methods and test infrastructure*. In: Computer Science and Information Technologies (CSIT), 2013.
7. Zorian, Y., & Shoukourian, S. (2003). Embedded-memory test and repair: Infrastructure IP for SoC yield. *IEEE Design & Test of Computers Journal, 20*(3), 58–66.
8. Dugganapally, I.P., Watkins, S.E., Cooper, B. (2014). *Multi-level, memory-based logic using CMOS technology*. In: 2014 I.E. Computer Society Annual Symposium on VLSI (ISVLSI), 2014.
9. Yueh, W., Chatterjee, S., Zia, M., Bhunia, S., & Mukhopadhyay, S. (2015). A memory-based logic block with optimized-for-read SRAM for energy-efficient reconfigurable computing fabric. *IEEE Transactions on Circuits and Systems II: Express Briefs Journal, 62*(6), 593–597.
10. Matsunaga, S., Hayakawa, J., Ikeda, S., Miura, K., Endoh, T., Ohno, H., Hanyu, T. (2009). *MTJ-based nonvolatile logic-in-memory circuit, future prospects and issues*. In: Design, Automation & Test in Europe Conference & Exhibition, 2009.
11. Harada, S., Bai, X., Kameyama, M., Fujioka, Y. (2014). *Design of a logic-in-memory multiple-valued reconfigurable VLSI based on a bit-serial packet data transfer scheme*. In: IEEE 44th International Symposium on Multiple-Valued Logic (ISMVL), 2014.
12. Hahanov, V. I., Amer, T. B., Chumachenko, S. V., & Litvinova, E. I. (2015). Qubit technology analysis and diagnosis of digital devices. *Electronic Modeling Journal, 37*(3), 17–40.
13. Melikyan, V. S. (2009). A method of eliminating false paths during statistical static analysis of timing delays of digital circuits. *Elektronica i svyaz Journal, 2–3*(1), 93–96.
14. Melikyan, V. S., & Vatyan, A. O. (1997). Interconnections model delays for the logic analysis of ECL circuits. *SUAB, Computer Engineering, Moscow Journal, 2*, 187–194.

Chapter 7
Computing for Diagnosis of HDL Code

Vladimir Hahanov, Eugenia Litvinova, and Svetlana Chumachenko

7.1 State of Assertion-Driven Verification

The technology for testing and verification of Hardware Description Language (HDL) code in a digital system on a chip (SoC) is mutually enriched to create (1) a scalable verification environment [1]; and (2) an embedded SoC IP infrastructure [2, 3]. The ultimate goal of both design redundancies is fast and accurate diagnosis of faulty HDL blocks and correction of bugs. To achieve this goal, there are practical solutions that are aimed at partial automation of testing and verification of HDL code blocks without specifying types of functional errors (faults) [4, 8, 14]. This makes it possible to significantly reduce the amount of diagnostic information. Numerous other published decisions are related to avoiding time-consuming debugging tools, and a transition to assertion-based verification has been shown [5–13]. The basic idea of the abovementioned publications is that assertion-based verification successfully solves many practical problems of design testing, namely: (1) observation of essential variables in time, when HDL code is tested [1–3]; (2) automatic control of the verification process to reduce the time needed for diagnosis of functional errors [1–3]; (3) minimization of the test bench by optimal placement of assertion statements in the design HDL code body [7, 12]; (4) increasing the diagnosis depth of faulty HDL blocks by inserting additional assertions [4, 14–16, 18]; (5) leveraging the assertion mechanism at different (system, transaction, RT) levels of digital design descriptions [5, 11–13]; (6) automatic generation of smart assertions in time and space of the design verification process to increase the diagnosis quality [7, 8]; (7) creation of a smart infrastructure

V. Hahanov (✉) • E. Litvinova • S. Chumachenko
Design Automation Department, Kharkov National University of Radio Electronics, Ukraine, Nauka Avenue 14, Kharkov 61166, Ukraine
e-mail: hahanov@icloud.com; litvinova_eugenia@icloud.com; svetachumachenko@icloud.com

V. Hahanov, *Cyber Physical Computing for IoT-driven Services*,
DOI 10.1007/978-3-319-54825-8_7

of assertion-based verification, diagnosis, and automatic error correction of SoC HDL code [7, 9, 12]; (8) language services—because of the large design appearance, verification is becoming more and more complex, where the specification is written in English, the design is written in HDL (typically Verilog/VHDL), and the verification model is written in HDL or some proprietary verification language (System Verilog, SystemC, Superlog, PSL-Sugar, OVA and OVL) [6, 8]; (9) transaction-level assertions [5, 12, 13] as an attractive way to create a transaction-driven verification (TDV) environment and design under verification (DUV) using SVA [in a conventional class-based transaction-driven verification environment (e.g., OVM, UVM), System Verilog temporal assertions are possible only in design elements like a module; for modeling system-level assertions, transactions are needed to pass from the class environment to the module/program block where the assertions are implemented; one of the proposed ideas is doing transaction-level assertions by exploiting the concept of method ports and System Verilog scoping rules]; and (10) considering the state of current trends in the field of testing and verification of SoC HDL models, formulation of further market-attractive goals and problems of the proposed research oriented for assertion-based diagnosis of the faulty HDL code blocks.

The main goals are a significant reduction in the designing time to market and increased yield by creating a verification infrastructure for SoC HDL code based on a transaction-level assertion graph that can improve the diagnosis depth of faulty HDL code blocks.

The objectives are (1) development of a transaction-level assertion graph of SoC HDL code based on leveraging of simulation tools; (2) creation of a matrix model of a Transaction-Level Test Assertion Blocks Activated Graph (TABA graph) for diagnosing faulty blocks of the SoC HDL model; (3) development of metric units of diagnosability assessment for a TABA graph or Test Assertion Blocks Activated Matrix (TABA matrix) [18]; (4) creation of an assertion-based diagnosis algorithm for faulty HDL code blocks by using an SoC HDL hierarchical model; and (5) implementation of the developed models and algorithm in the testing and verification infrastructure of Riviera (Aldec, Inc., Henderson, NV, USA) [17], which enables ultimate test bench productivity, reusability, and automation by combining a high-performance simulation engine, advanced diagnosis at different levels of abstraction, and support for the latest Language and Verification Library Standards.

7.2 TABA Model for Diagnosis of Faulty HDL Blocks

The first purpose is to develop a transaction-level TABA matrix structure and diagnosis method to decrease the time needed for HDL code testing and the memory needed for storage of the ternary relationship represented by <test pattern–assertion engine–functional blocks> in a single table. The investigation subjects are (1) development of a digital system HDL model in the form of a transaction-level graph for faulty block diagnosis of HDL code by leveraging an assertion engine [1, 4, 5, 15, 16]; (2) a development method for TABA matrix analysis to detect faulty blocks [15, 16]; (3) synthesis of the logic structure for embedded faulty block diagnosis that can be integrated with the IEEE standards 1500, 11.49, 1687 [15].

The structure for testing HDL code is represented by an *xor* operation between the components <test pattern–gold block model–faulty block model B^*>:

$$T \oplus B \oplus B^* = 0; B^* = T \oplus B = \{T \times A\} \oplus B,$$

which have to be transformed to the relationship of the HDL code diagnosis components in the TABA matrix:

$$M = \{\{T \times A\} \times \{B\}\}, M_{ij} = (T \times A)_i \oplus B_j.$$

A coordinate of the TABA matrix is equal to 1 if the test–monitor pair $(T \times A)_i$ detects any fault in the functional block $B_j \in B$.

An analytical model for verification by leveraging a temporal assertion (additional observation statements or code lines) allows achievement of an ordered diagnosis depth in the set of faulty blocks:

$$\begin{aligned} &\Omega = f(G, A, B, S, T), G = (A^*B) \times S; S = f(T, B); \\ &A = \{A_1, A_2, \ldots, A_i, \ldots, A_h\}; B = \{B_1, B_2, \ldots, B_i, \ldots, B_n\}; \\ &S = \{S_1, S_2, \ldots, S_i, \ldots, S_m\}; T = \{T_1, T_2, \ldots, T_i, \ldots, T_k\}. \end{aligned}$$

Here, $G = (A^*B) \times S$ is the structure of the functionality, in the form of a Code-Flow Transaction (CFT) graph (Fig. 7.1); $S = \{S_1, S_2, \ldots, S_i, \ldots, S_m\}$ are nodes of software variable states during the test segment simulation. Otherwise the structure is considered as a TABA graph. The states $S_i = \{S_{i1}, S_{i2}, \ldots, S_{ij}, \ldots, S_{ip}\}$ correspond to the values of design variables (Boolean, register, memory). The graph arcs correspond to the software–hardware blocks:

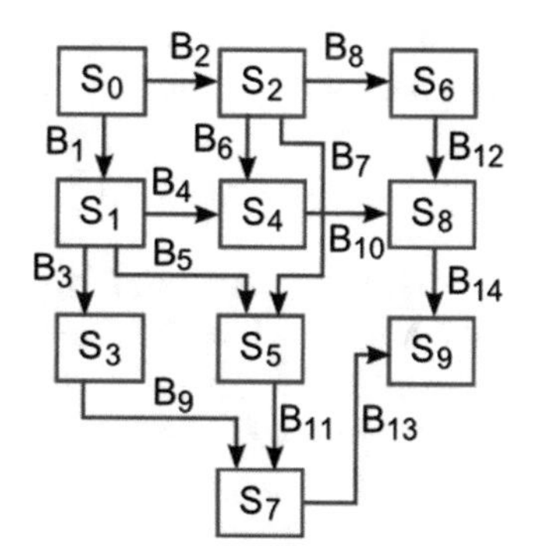

Fig. 7.1 Example of a TABA graph for HDL code

$$B = \{B_1, B_2, \ldots, B_i, \ldots, B_n\}; \bigcup_{i=1}^{n} B_i = B; B_i \bigcap_{i \neq j} B_j = \varnothing.$$

The assertion $A_i \in A = \{A_1, A_2, \ldots, A_i, \ldots, A_n\}$ can be inserted into the end of each block B_i—the code statement sequence that creates the state of the graph node $S_i = f(T, B_i)$ during simulation of the test pattern $T = \{T_1, T_2, \ldots, T_i, \ldots, T_k\}$. The monitor gathering an assertion of incoming arcs $A(S_i) = A_{i1} \vee A_{i2} \vee \ldots \vee A_{ij} \vee \ldots \vee A_{iq}$ can be put in every graph node.

The structure of the HDL program, represented as a TABA graph, describes the software and test segments of the functional coverage, generated by leveraging software blocks incoming to the considered node. The last one defines the relationship between achievement on the test variable space and potential (maximal) space, which forms the functional coverage of the graph node state $Q = \text{card}S_i^r / \text{card}S_i^p$. In the aggregate, all graph nodes have to get a complete state coverage space of the software variables, which defines the test quality, equal to 100%:

$$Q = \text{card} \cup_{i=1}^{m} S_i^r / \text{card} \cup_{i=1}^{m} S_i^p = 1.$$

The assertion engine <A,S>, existing in the graph, monitors arcs (HDL code coverage) $B = \{B_1, B_2, \ldots, B_i, \ldots, B_n\}$ and also graph nodes (functional coverage) $S = \{S_1, S_2, \ldots, S_i, \ldots, S_m\}$. The assertions on arcs $B_i \in B$ are designed for diagnosis of the functional failures in software blocks. The assertions on graph nodes $S_i \in S$ carry information about the quality of the test and assertion set for improvement or complement. The TABA graph makes possible the following operations: (1) an increase in software quality through design diagnosability; (2) minimization of the time needed for test generation, faulty block diagnosis, and correction by leveraging the assertion engine; and (3) optimization of synthesis of the test pattern, covering all arcs and nodes by a minimum set of activated logical paths.

For diagnosis, test segments $T = \{T_1, T_2, \ldots, T_r, \ldots, T_k\}$ activate transaction (logical) paths in the graph model covering all nodes and arcs. The testing model can be represented by the Cartesian product $M = T \times A \times B$ with the dimension $Q = k \times h \times n$ accordingly. To reduce the amount of test diagnosis data, a separate assertion point is assigned to each test segment. This action makes visible the functional block activation and decreases the matrix dimension to $Q = n \times k$ retaining all features of the relationship $M = T \times A \times B$. The pair <test–monitor> generates three forms:

$$< T_i \rightarrow A_j >, < \{T_i, T_r\} \rightarrow A_j >, < \{T_i\} \rightarrow \{A_j, A_s\} >.$$

The diagnosis method for the faulty functional blocks leverages the prebuilt TABA matrix $M = [M_{ij}]$, where the row defines the relationship between the test segment and a subset of the activated blocks $T_i \rightarrow A_j \approx (M_{i1}, M_{i2}, \ldots, M_{ij}, \ldots, M_{in}), M_{ij} = \{0, 1\}$, monitored by A_j. The column of the matrix defines the relationship between the functional blocks, detected on test segments, by using monitors $M_j = B_j\,(T_{j,}\, A_j)$.

For faulty block diagnosis in the testing procedure, the real assertion response (vector) $A^* = \{A_1^*, A_2^*, \ldots, A_i^*, \ldots, A_n^*\}$ on the test pattern T is determined by calculating the function $A_i^* = f(T_i, B_i)$. The faulty block detection is based on an *xor* operation between the real assertion response vector and TABA matrix columns:

$$A^* \oplus [M_1(B_1) \vee M_2(B_2) \vee \ldots \vee M_j(B_j) \vee \ldots \vee M_n(B_n)].$$

The faulty functional block is defined by a vector B_j, which delivers results with a minimal number of 1-unit coordinates in the column:

$$B = \min_{j=\overline{1,n}} \left[B_j = \sum_{i=1}^{h} (B_{ij} \oplus A_i^*) \right].$$

With the redundancy in the diagnosis model of HDL code, it is necessary to describe the next useful features of the TABA matrix:

$$\begin{aligned}
&(1)\ M_i = (T_i \times A_j);\ \ (2) \bigvee_{i=1}^{m} M_{ij} \to \bigvee_{j=1}^{n} M_j = 1;\\
&(3)\ M_{ij} \bigoplus_{j=1}^{n} M_{rj} \neq M_{ij};\ (4) M_{ij} \bigoplus_{i=1}^{k} M_{ir} \neq M_{ij};\\
&(5)\ \log_2 n \leq k \leftrightarrow \log_2 |B| \leq |T|\\
&(6)\ B_j = f(T, A) \to B \oplus T \oplus A = 0.
\end{aligned}$$

The math expressions mean that (1) the matrix row appears as a subset of the Cartesian product between the test and monitor; (2) the logical union of all matrix rows results in a 1-unit vector over all coordinates; (3) the matrix rows are pairwise distinct, which eliminates the test redundancy; (4) the matrix columns are pairwise distinct, which excludes the existence of equivalent faulty blocks; (5) the number of matrix rows has to be greater than the binary logarithm of the number of columns that provides the maximal HDL code diagnosability up to every block; and (6) the function of diagnosis of every block depends on the number of test segments and monitors, which have to be minimized without diagnosability reduction. In accordance with the six test segments, there are six activated graph node paths loaded to the assertion point S_9:

$$\begin{aligned}
T = {} & S_0S_1S_3S_7S_9 \vee S_0S_1S_4S_8S_9 \vee S_0S_1S_5S_7S_9 \vee \\
& \vee S_0S_2S_4S_8S_9 \vee S_0S_2S_5S_7S_9 \vee S_0S_2S_6S_8S_9.
\end{aligned}$$

The leveraging HDL code graph creates the opportunity to define the minimal number of test-activated paths completely covering the functional blocks (oriented arcs):

$$B = B_1B_3B_9B_{13} \vee B_2B_7B_{11}B_{13} \vee B_1B_5B_{11}B_{13} \vee$$
$$\vee B_1B_4B_{10}B_{14} \vee B_2B_6B_{10}B_{14} \vee B_2B_8B_{12}B_{14}.$$

The assertions $\{A_9 \subseteq S_9, A_3 \subseteq S_3, A_6 \subseteq S_6\}$ can divide all blocks into three groups of components, which create logical equations for monitoring HDL code blocks:

$$A_9 = T_1(B_1B_3B_9B_{13}) \vee T_2(B_2B_7B_{11}B_{13}) \vee T_3(B_1B_5B_{11}B_{13}) \vee$$
$$\vee T_4(B_1B_4B_{10}B_{14}) \vee T_5(B_2B_6B_{10}B_{14}) \vee T_6(B_2B_8B_{12}B_{14});$$
$$A_3 = T_1(B_1B_3); A_6 = T_6(B_2B_8).$$

The next action creates six rows of TABA matrix $M_{ij}(G_1)$, defining the relationship between the test segments and activated HDL blocks:

$M_{ij}(G_1)$	B_1	B_2	B_3	B_4	B_5	B_6	B_7	B_8	B_9	B_{10}	B_{11}	B_{12}	B_{13}	B_{14}
$T_1 \to S_9$	1	·	1	·	·	·	·	·	1	·	·	·	1	·
$T_2 \to S_9$	1	·	·	1	·	·	·	·	·	1	·	·	·	1
$T_3 \to S_9$	1	·	·	·	1	·	·	·	·		1	·	1	·
$T_4 \to S_9$	·	1	·	·	·	1	·	·	·	1	·	·	·	1
$T_5 \to S_9$	·	1	·	·	·	·	1	·	·	·	1	·	1	·
$T_6 \to S_9$	·	1	·	·	·	·	·	1	·	·	·	1	·	1
$T_1 \to S_3$	1	·	1	·	·	·	·	·	·	·	·	·	·	·
$T_6 \to S_6$	·	1	·	·	·	·	·	1	·	·	·	·	·	·

The TABA matrix shows the existence of equivalent faulty blocks: {3,9}, {8,12} defined with six test segments and one assertion point as node 9 in the HDL graph. Really, the columns {3,9}, {8,12} are equivalent. To resolve indistinguishability of two pairs of equivalent faulty blocks, it is necessary to add two monitors in graph nodes S_3 and S_6 for test segments T_1 and T_6, respectively. As a result, three additional assertions in the graph nodes $A = (S_9, S_3, S_6)$ provide complete distinction of the faulty blocks of software HDL code. So, the HDL code graph enables us to synthesize the optimal test to define the minimal number of assertion monitors in the graph nodes, to detect faulty blocks with a given or maximal diagnosis depth.

Leveraging the created TABA matrix, the diagnosis procedure can be defined by the following equation of a vector *xor* operation between the real eight assertion values and the B columns:

$$\left\{[A_9(T_1, T_2, T_3, T_4, T_5, T_6), A_3(T_1), A_6(T_6)] \oplus B_j = 0\right\} \to$$
$$\to \left(B_j - \text{failed}\right).$$

7.3 Design for Diagnosability

The diagnosability of the HDL program (hardware/software functionality) is the relationship $D = N_d/N$ between the recognized number of faulty blocks N_d (when there are not equivalent components, or the diagnosis depth is equal to 1) and the total number N of HDL blocks.

The expense criterion E evaluates the TABA matrix model for faulty block detection via the test assertion efficiency for a given diagnosis depth. Criterion E functionally depends on the relationship between the "gold"$]\log_2 N[\times N$ and real $|T| \times |A| \times N$ dimensions of the memory sizes, where $|T|$ – test length, $|A|$ – number of assertions, for the corresponding TABA matrices, which compose the relative expenses reduced to a 0–1 interval:

$$E = \frac{]\log_2 N[\times N}{|T| \times |A| \times N} = \frac{]\log_2 N[}{|T| \times |A|}.$$

The general diagnosis quality criterion depends on the time–money expense E and diagnosability D:

$$Q = E \times D = \frac{]\log_2 N[}{|T| \times |A|} \times \frac{N_d}{N}.$$

For instance, the diagnosis quality of the TABA matrix $M_{ij}(G_1)$ before and after adding two rows is equal to:

$$Q_1[M(6 \times 1 \times 14)] = \frac{]\log_2 14[}{|6| \times |1|} \times \frac{10}{14} = 0.47;$$

$$Q_2[M(8 \times 1 \times 14)] = \frac{]\log_2 14[}{|8| \times |1|} \times \frac{14}{14} = 0.5.$$

It means the first matrix dimension is a little bit less than the second, but diagnosability is better in the second variant of the matrix, which becomes the winner of the whole. Comparing this to the well-known solutions [12], when every cell of a matrix contains all existing assertions $|M_{ij}| = |A|$ the second version evaluates the following low value:

$$Q_2[M(6 \times 3 \times 14)] = \frac{]\log_2 14[}{|6| \times |3|} \times \frac{14}{14} = 0.2.$$

So, the TABA matrix operating by the selected pair test assertion concurrently allows having essential advantages in memory size reducing in $|A| - 1$ times with the same diagnosability value. The TABA matrix diagnosis quality is the ratio of the bit number needed for identification (recognition) of all blocks $]\log_2 N[$ related to the real number of code bits, presented by the product of the test length and number of assertions $|T| \times |A|$. If the first part E of quality criterion Q is equal to 1 and every block with functional failures is recognized in the field of the rest components $N_d = N$, it means the test and assertions are optimal, giving the best quality criterion of diagnosis model $Q = 1$. The goal of the TABA graph analysis is evaluation of assertion engine structure, which can deliver the maximal diagnosis depth of faulty blocks. The diagnosability of the TABA graph is a function depending on the number N_d of transit internal nodes where there exist only two adjacent arcs; one of them is incoming and the other arc is outgoing. Such graph arcs create paths though the node without fan-in and fan-out branches (N is the total arc number in the TABA graph): $D = (N - N_n)/N$. The estimation N_n is the number of unrecognizable or equivalent functional blocks, which can be faulty. The additional assertion point for improving the diagnosability of the faulty blocks is transit nodes of the TABA graph. The diagnosis quality criterion of the TABA graph is represented below:

$$Q = E \times D = \frac{]\log_2 N[}{|T| \times |A|} \times \frac{N - N_n}{N}.$$

This equation creates some practical advice for the synthesis of a diagnosable HDL code structure. (1) The test pattern has to create a minimal number of single activation paths, which cover all the nodes and arcs in the TABA graph. (2) The initial number of monitors has to be equal to the end node number of the TABA graph. (3) An additional assertion point can be placed on each internal graph node, which has one incoming and one outgoing arc. (4) The (n) parallel independent HDL code blocks have to have (n) monitors and a single concurrent test, or one integrated monitor and (n) serial tests. (5) Serially connected HDL code blocks have one activation test for one path and ($n - 1$) assertion points, or (n) tests and (n) monitors. (6) The graph nodes, which have more than 1 number of input and output arcs, create quasioptimal conditions for the diagnosability of the considered graph fragment by single path activation tests without additional assertion points. (7) The generated test pattern has to have 100% functional coverage for the TABA graph nodes. (8) The diagnosis quality criterion as a function depending on the TABA graph structure, test, and assertion points can always be increased up to the 1 value. For this goal, there are two alternative ways. The first one proposes increasing the test segment amount by activating new paths for recognition of equivalent faulty blocks without increasing the number of monitors, if the software graph structure allows creation of the new single activation path. The second way suggests adding assertion points on transit nodes of the TABA graph. A third hybrid variant is possible, when the superposition of the two abovementioned techniques creates the design for diagnosability.

7.4 Diagnosis Method for a Hardware–Software System

A hierarchical model in the form of multitree B is focused on the software or hardware system components, with redundancy in the form of a three-dimensional activation TABA matrix of functional subcomponents. The outcoming graph arcs from the node are connected to the lower and more detailed level of diagnosis when replacing the faulty block is too expensive:

$$B = \left[B_{ij}^{rs}\right], \mathrm{card}B = \sum_{r=1}^{n}\sum_{s=1}^{m_r}\sum_{i=1}^{p_{rs}}\sum_{j=1}^{k_{rs}} B_{ij}^{rs}.$$

The formula description contains the following: n is the level amount of the diagnosis multitree of the digital system; m_r is the number of functional components at the level r; k_{rs} (p_{rs}) is the number of components (test length) in the table B^{rs}; and $B_{ij}^{rs} = \{0, 1\}$ is a component of the activation table, which is defined by 1-unit, identifiying the detected faulty block under the test segment T_{i-A_i} relative to the observed assertion point A_i. Each table as a node has a number of outcoming down arcs equal to the number of functional components, which are represented by activation TABA matrices.

The TABA method for faulty block diagnosis of the HS system, based on the multitree model, creates the universal engine algorithm (Fig. 7.2) for traversal of tree branches on the prior specified depth:

$$B_j^{rs} \oplus A^{rs} = \begin{cases} 0 \rightarrow \left\{B_j^{r+1,s}, R\right\}; \\ \left\{1 \rightarrow B_{j+1,}^{rs}, T\right\}. \end{cases}$$

Here the *xor* vector operation is executing the matrix columns with the assertion (output) response vector A^{rs}, which is determined by the real (m) and gold (g) functionality responses under test patterns based on the *xor* operation: $A_i^{rs} = m_i^{rs} \oplus g_i^{rs}, i = \overline{1, k_{rs}}$. If all coordinates of the vector *xor* sum $B_j^{rs} \oplus A^{rs} = 0$ are equal to zero, then one of the following operations is performed: transition to the activation matrix of the next lower level $B_j^{r+1,s}$ or repair of the functional block $B = B_j^{rs}$. One of two operations is executed, which is the most optimal: (1) the time ($t > m$, block 10)—then repairing the faulty block is performed; or (2) the money ($t < m$)—then a transition down specifies the fault location more exactly, because replacement of the smaller block decreases the cost of repair. If one coordinate of the *xor* sum vector is equal to one, $B_j^{rs} \oplus A^{rs} = 1$, then transition to the next matrix column is performed. When all coordinates of the assertion vector are equal to zero $A^{1s} = 0$, then a fault-free state of the HS system is declared. If all resulting vectors by *xor* executing the TABA matrix column are not equal to zero, $B_j^{rs} \oplus A^{rs} \neq 0$, it means a test pattern, generated for faulty block detection, has to be corrected. If more than

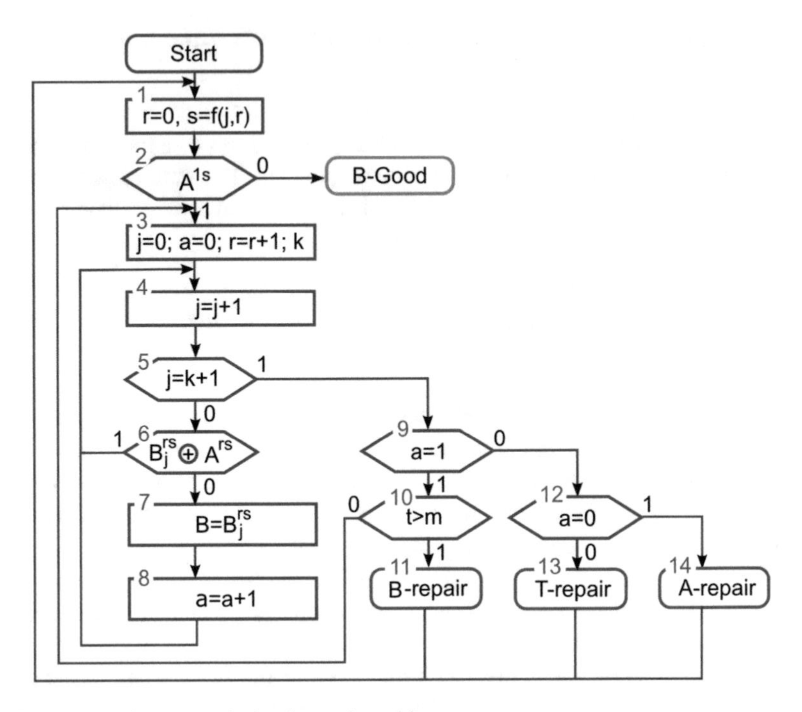

Fig. 7.2 Engine for traversal of a diagnosis multitree

one vector sum obtained by *xor* executing the TABA matrix column is equal to zero, $B_j^{rs} \oplus A^{rs} = 0$, it means an assertion engine created for faulty block detection on the represented test has to be supplemented with extra assertion points. So, the TABA engine algorithm has four end nodes, where one of them is B good, which indicates successful finishing of the testing. The other three end nodes mean intermediate results in the testing, which needs to take into account the increasing test quality and diagnosis depth by using extra assertions points and/or additional test pattern generation. Thus, the multitree B appears to be an efficient infrastructure IP for complicated hardware–software digital systems. The advantage of the TABA engine, which is a scalable hierarchical diagnosis model, is the simplicity of preparation and presentation of diagnostic information as an activation table of functional blocks on the test segments with the assertion monitors.

Practical implementation of the models and verification methods is integrated into the Riviera simulation environment (Fig. 7.3). New assertion and diagnosis modules, added into the system, have improved the existing verification process, allowing a 15% reduction in the design time of the digital device. Actually, application of assertions makes it possible to decrease the length of the test bench code and considerably reduce in three times the verification period (Fig. 7.4), which is the most expensive aspect. The assertion engine allows an increase in the diagnosis depth of functional failures in software blocks up to the level of 10–20

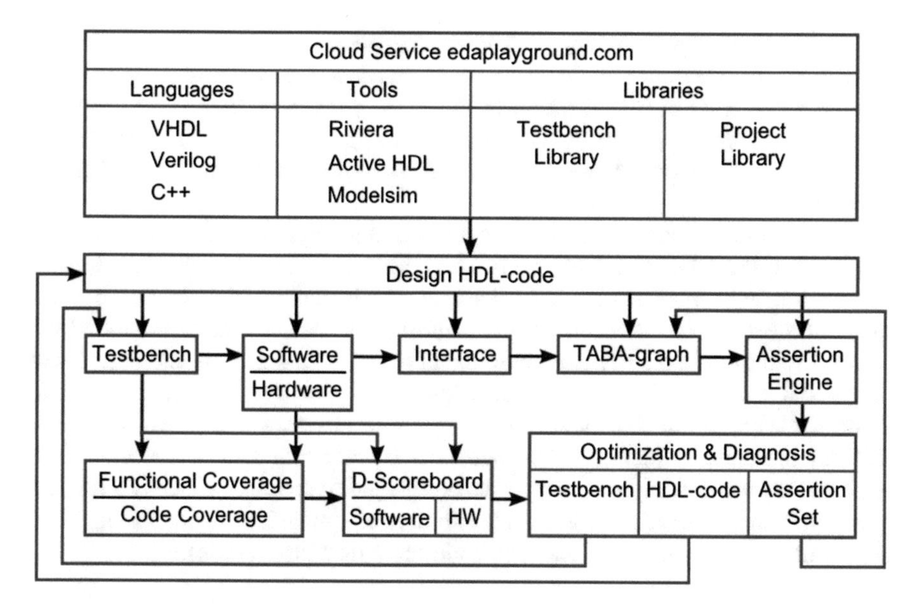

Fig. 7.3 Implementation of results in the cloud EDA system

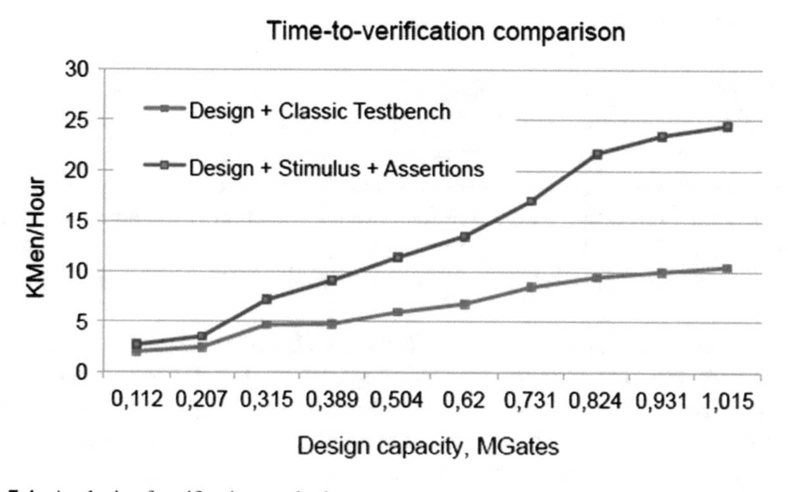

Fig. 7.4 Analysis of verification methods

HDL code statements. The interaction of the simulation tools and assertion engine creates access of diagnosis tools to all internal signals of HDL code. This allows quick identification of the location and type of the functional failure, as well as a reduction in the time of error detection in the evolution of a product with top-down design. Application of assertion for 50 real-life designs (from 5000 up to 5 million gates) allows hundreds of dedicated solutions to be obtained, which are included in the verification template library (VTL), which generalizes the temporal verification limitations that are most popular in the electronic design automation (EDA) market for the broad class of digital products. Software implementation of the proposed system for analyzing assertions and diagnosing HDL code is part of the multifunctional integrated environment of Riviera for simulation and verification of HDL models.

The high performance and technological combination of the assertion analysis system and the HDL simulator from Aldec have been largely achieved through integration with internal simulator components, including HDL language compilers. Processing of the results of the assertion analysis system is provided by a set of visual tools in the Riviera environment to facilitate diagnosis and removal of functional failures. The assertion analysis model can also be implemented in hardware with certain constraints on a subset of the supported language structures. Riviera products including the components of assertion temporal verification, which allow a 3–5% improvement in the design quality, currently occupy a leading position in the world IT market, with the number of system installations being 5000 a year in 200 companies and universities in more than 20 countries.

7.5 Conclusion

1. The infrastructure and technology for HDL code diagnosis are presented. The proposed transaction-level TABA graph is considered as the redundancy to the HS functionality, which creates an opportunity for assertion-based verification for diagnosis of digital systems on chips. The proposed method is focused on considerably reducing the time of faulty block detection and memory for storing the diagnosis TABA matrix describing the ternary relationship in the format of monitor-focused test segments that detect faulty HDL blocks of SoC under design.
2. The innovative diagnosis quality criterion as a function depending on the graph structure, test, and assertion engine is represented. For this goal, there are two alternative ways to make a good choice to improve TABA graph diagnosability: by increasing the test segment set for recognition of equivalent faulty blocks or by adding assertion points on transit nodes of the activated HDL blocks.
3. An improved TABA engine or algorithm for faulty block detection in HDL code is proposed. Leveraging the *xor* operation makes it possible to improve the diagnosis performance for single and multiple faulty blocks on the basis of parallel analysis of the TABA matrix and leveraging of vector (*and*, *or*, *xor*) operations.
4. A model for diagnosing the functionality of SoC HDL code in the form of a multitree and a method for tree traversal, implemented in the engine for detecting faulty blocks with a given depth, are developed. This considerably increases the performance of the testing and diagnosis infrastructure.

5. An assertion-based verification and diagnosis method is performed with three real case studies, presented by SoC HDL blocks of a cosine transform filter, which show the consistency of the results in order to minimize the timing period of faulty block detection and memory for storing diagnostic information, as well as increasing the diagnosis depth of SoC HDL blocks.
6. Practical implementation of the assertion-based verification and diagnosis method is able to improve the quality of HDL code design by 3%, reduce the time of verification by 15%, and enlarge the diagnosis depth by 30% without increasing the test set.
7. Further investigations are connected with automatic generation: (1) TABA graph and TABA matrix; (2) placement of assertion monitors into minimal nodes of a transaction-level graph; and (3) diagnosis and correction of faults in HDL blocks automatically, forming an assertion-based verification environment as a cyber physical system in the cloud.

References

1. Bergeron, J. (2006). *Writing Testbenches Using SystemVerilog*. Springer. New York.
2. Benso, A., Di Carlo, S., Prinetto, P., & Zorian, Y. (2008). IEEE standard 1500 compliance verification for embedded cores. *IEEE Transactions VLSI Systems, 16*(4), 397–407.
3. Benabboud, Y., Bosio, A., Girard, P., Pravossoudovitch, S., Virazel, A., Bouzaida, L., & Izaute, I. (2009). A case study on logic diagnosis for system-on-chip. In *International symposium on quality electronic design*, 2009.
4. Ubar, R., Kostin, S., & Raik, J. (2009). Block-level fault model-free debug and diagnosis in digital systems. In *DSD '09, 12th Euromicro conference*, 2009.
5. Sohofi, H., & Navabi, Z. (2014). Assertion-based verification for system-level designs. In *15th international symposium on quality electronic design (ISQED)*, 2014.
6. Datta, K., & Das, P. P. (2004). Assertion based verification using HDVL. In *Proceedings of the 17th international conference VLSI design*, 2004.
7. Lingyi, L., Sheridan, D., Athavale, V., & Vasudevan, S. (2011). Automatic generation of assertions from system level design using data mining. In: *9th IEEE/ACM international conference on formal methods and models for codesign (MEMOCODE)*, 2011.
8. Piccolboni, L., & Pravadelli, G. (2014). Simplified stimuli generation for scenario and assertion based verification. In *15th Latin American test workshop, LATW*, 2014.
9. Di Guglielmo, G., Di Guglielmo, L., Fummi, F., & Pravadelli, G. (2012). Enabling dynamic assertion-based verification of embedded software through model-driven design. In *Design, automation & test in Europe conference & exhibition (DATE)*, 2012.
10. Ruan, A. W., Wang, Y., Shi, K., Zhu, Z. J., Wu, Q., Han, X., & Liao, Y. B. (2011). SOC HW/SW co-verification technology for application of FPGA test and diagnosis. In *IEEE international conference on computational problem-solving (ICCP)*, 2011.
11. Bombieri, N., Fummi, F., Guarnieri, V., Pravadelli, G., Stefanni, F., Ghasempouri, T., Lora, M., Auditore, G., & Marcigaglia, M. N. (2014). On the reuse of RTL assertions in SystemC TLM verification. In *15th Latin American test workshop, LATW*, 2014.
12. Sudhish, N., Raghavendra, B. R., & Yagain, H. (2011). An efficient method for using transaction level assertions in a class based verification environment. In: *International symposium on electronic system design*, 2011.
13. Niemann, B., & Haubelt, C. (2006). Assertion based verification of transaction level models. In *ITG/GI/GMM workshop*, Vol 9, Dresden, 2006.

14. Cook, A., Hellebrand, S., Imhof, M. E., Mumtaz, A., & Wunderlich, H. (2012). Built-in self-diagnosis targeting arbitrary defects with partial pseudo-exhaustive test. In *13th Latin American test workshop*, 2012.
15. Hahanov, V. I., Hahanova, I. V., Litvinova, E. I., & Guz, O. A. (2010). Digital system-on-chip design and verification. Novoye Slovo, Kharkov.
16. Ngene, C. U., & Hahanov, V. (2011). A diagnostic model for detecting functional violation in HDL-code of SoC. In *Proceedings of IEEE east-west design and test symposium*, 2011.
17. https://www.aldec.com/en/products/functional_verification/riviera-pro
18. Bondarenko, M. F., Hahanov, V. I., & Litvinova, E. I. (2012). Logical associative multiprocessor structure. *Automation and Remote Control, 73*, 1648–1666.

Chapter 8
Qubit Computing for Digital System Diagnosis

Vladimir Hahanov, Svetlana Chumachenko, and Eugenia Litvinova

8.1 Introduction

Evolution of world cyberspace is divided into the following periods [1–3]: (1) the 1980s—creation of personal computers; (2) the 1990s—introduction of Internet technologies into production processes and people's lives; (3) the 2000s—improvement of the quality of life through introduction of mobile devices and cloud services; (4) the 2010s—creation of a digital infrastructure for monitoring, control, and interaction of moving objects (air, sea, ground transportation, and robots); (5) 2015–2020s—creation of a global digital infrastructure of cyberspace, where all processes and phenomena are identified in time and in three-dimensional space, and become smart.

In recent years, qubit structures for creating cloud Internet services have become interesting for parallel computing of unordered data, which is explained by their alternativeness to the existing time-consuming models of computing processes focused on sequential processing of set theoretic structures and considerable increases in memory [1]. But now they are acceptable because of nanoelectronic technologies proposing up to one billion gates, located on a chip with the dimensions of 2 × 2 cm with a substrate thickness of 5 μ. That modern technology allows creation of a package (sandwich) containing up to seven dies. Practically wireless connection of such chips is based on through-silicon vias (TSVs)—the technological capability to drill about 10,000 through vias in 1 cm^2 of wafer or die. In addition, the emergence of FinFET transistors and 3D technology based on them for implementation of digital systems provides new opportunities for creating high-

V. Hahanov (✉) • S. Chumachenko • E. Litvinova
Design Automation Department, Kharkov National University of Radio Electronics, Ukraine, Nauka Avenue 14, Kharkov 61166, Ukraine
e-mail: hahanov@icloud.com; svetachumachenko@icloud.com; litvinova_eugenia@icloud.com

V. Hahanov, *Cyber Physical Computing for IoT-driven Services*,
DOI 10.1007/978-3-319-54825-8_8

speed devices by reducing delays in parallel computing elements [2–7]. So, it is necessary to use hardware-focused models and methods for creating high-speed tools for parallel solving of real world problems. The discreteness and multivaluedness of the alphabets for describing information processes, with parallelism inherent in quantum computing, are particularly important when developing effective and intelligent engines for cyberspace, cloud structures, and Internet services, improving the reliability of digital devices, and testing and simulation of digital systems on chips. We do not consider the physical basis of quantum computing, originally described in the works of scientists, focused on the use of nondeterministic quantum interactions within the atom. We do not address the physical foundations of quantum mechanics, concerning nondeterministic interactions of atomic particles [1], but we use the concept of the qubit structure as a vector form for joint definition of the power set (the set of all subsets) of the states in the discrete cyberspace area that provides high parallelism and superposition for processing the proposed qubit models and methods.

Quantum simulators on classical computers are effectively used for solving optimization problems by way of a brute force method through the use of set theory [1, 8]. A set of elements in a traditional computer is orderly, because each bit, byte, and other component has its own address. Therefore, set theoretical operations are reduced to an exhaustive search of addresses of primitive elements. The address order of data structures is useful for applications where model components can be strictly ranked, which makes it possible to carry out their analysis in a single pass (a single iteration). If there is no order in the structure, for instance, the set of all subsets, the classical model of memory and computational processes, does not improve the analysis time of primitive elements of the circuit, or processing of associative groups is ineffective.

What can be offered for unordered data instead of a strict order? A processor where the unit cell is the image or pattern of the universal set of n primitives, which generates $Q = 2^n$ possible states of a cell as a power set or the set of all subsets. A direct solution for creating such a cell is based on unitary positional coding states of primitives that form the set of all subsets and the limit of the universal set of primitives by superposition of the last ones [8, 9].

The n qubit is a vector form of unitary encoding of the universal set of n primitives to specify the power set of states 2^{2^n} by using 2^n binary variables.

For example, if $n = 2$, then the 2 qubit sets 16 states by using four variables. If $n = 1$, the qubit sets four states in the universal set of two primitives by using two binary variables (00,01,10,11) [1]. Herewith, the superposition (simultaneous existence) of 2^n states in a vector is supposed. The qubit (n qubit) allows use of logical operations instead of set theoretic ones to significantly speed up the analysis of discrete systems. Further, the qubit is identified with the n qubit or vector if this does not prevent understanding of the presented material. As quantum computing is related to the analysis of qubit data structures, further, we use the definition “quantum” for identifying technologies based on two properties of quantum mechanics: concurrency of processing and superposition of states.

8.2 Qubit Analysis for Digital System Diagnosis

A matrix method for diagnosing functional failures and stuck-at faults in software or hardware units is proposed. It is based on qubit or multivalued data structures for defining diagnostic information, which allows a significant reduction in the computational complexity of simulation and diagnosis due to the introduction of parallel logical operations on the matrix data. A qubit method for fault-free simulation of digital units with the possibility of online repair of digital system components is presented. It is characterized by significantly improved performance due to the addressable processing procedure of the functional primitives, defined by qubit vectors of output states.

A model for diagnosing an object is represented as a graph of a digital system, which has functional elements connected by communication lines. Among them, there are assertion points, needed for verification, testing, and diagnosis of faults [2]. Diagnostic information is provided by the following components: (1) a verification test for diagnosing faults of a given class—in this case they are considered single stuck-at faults of circuit lines {≡0,≡1}; (2) a fault detection table [6], the rows of which define vectors of faults, checked by each test pattern and associated with circuit lines; (3) a reachability matrix, which determines the reachability of each assertion point by using the set of input (previous) lines [8]; and (4) a matrix of assertion engine states or an output response matrix, which determines the status of each assertion on test patterns by comparing the reference response at a given point to the real signal during the execution of the diagnostic experiment [2, 7].

The base model for diagnosing the digital unit, discrete process, or phenomenon is represented by the components, which create four dimensions in the feature space:

$$
\begin{cases} D_b =< S, A, F, T > \\ D = \{< S, A >, < F, T >\}; \end{cases}
\begin{cases} V_b = (|S| \times |A| \times |F| \times |T|); \\ V = (|S| \times |A|) + (|F| \times |T|); \\ V_b >> V; \end{cases}
\begin{cases} S^* = f(S, A, T); \\ A^* = g(T, A); \\ F^* = h(S, A, F, T); \end{cases}
$$

At that, the amount of diagnostic information V is formed by the Cartesian product of four components (powers) in the order specified above: (1) the object structure; (2) an assertion or monitoring engine; (3) a set of faults or modules, which can fail; and (4) test patterns or segments to diagnose faults or a set of these modules. A significantly reduced amount of diagnostic information can be achieved by reducing the dimension of the feature space by partitioning the base model into two disjoint subsets $<S,A>$,$<F,T>$. In this case, the estimate of the diagnostic information volume is not multiplicative but additive with respect to the power of subsets, derived from partitioning, without reducing the diagnosis depth. Here, the first component of the diagnosis model is presented by a reachability matrix, which allows minimizing the mask of possible faults by analyzing the structure of the circuit by comparing the true and real results of simulation of output signals for each test pattern or segment. The number of rows of this matrix is the number of observed outputs or assertions.

In realizing the diagnosis method, a binary matrix of structural fault activation is created, which is a mask for a substantial reduction in the set of suspected faults, when concurrently analyzing the fault detection table. In this case, the symbols of single stuck-at faults $\{0,1,X,\varnothing\}, X = \{0,1\}$ in the cells of the fault detection table [6] are encoded by the corresponding qubits (10,01,11,00) of the multivalued Cantor alphabet $A^k = \{0,1,X,\varnothing\}$, which makes it possible to exclude the set theoretic procedures from the calculation processes and replace them with vector logical operations.

A fragment of a digital circuit shown in Fig. 8.1 is used for consideration of the method. There are three assertion points *A*, *B*, *C* to monitor the status of all circuit lines under test (during the diagnostic experiment) by input of five test patterns specified by a fault detection table *F*(*T*). The coordinates of the table define faults 0 and 1 detected by the test vectors, as well as the states of coordinates ∅(.), which means lack of detectable faults, and *X*—detecting constants 0 and 1 on the line simultaneously. The right side of the table is a matrix of assertion engine states as a comparison of the results of the reference and actual responses of the digital device to the test patterns. A value of 1 means a negative result (incomparable), and 0 means a match of the responses.

The circuit structure is not taken into account in a fault detection table to enhance the diagnosis depth, based on calculating the actual state matrix of the assertion engine, which together with the reachability matrix creates a structure mask, minimizing the set of suspected faults. For the fragment of a digital circuit shown in Fig. 8.1, the reachability matrix is as follows:

$S = \lvert S_{ij} \rvert$	1	2	3	4	5	6	7	8	9	*A*	*B*	*C*
1	1	1	·	·	·	·	·	·	·	1	·	·
2	1	1	1	1	·	1	1	1	·	·	1	·
3	·	·	1	1	1	1	1		1	·	·	1

Here, assertion outputs *A*, *B*, *C* are monitors of the technical condition of the diagnosis object. Each of them can have two states: $A_{ij} = \{0,1\}$, defining the output

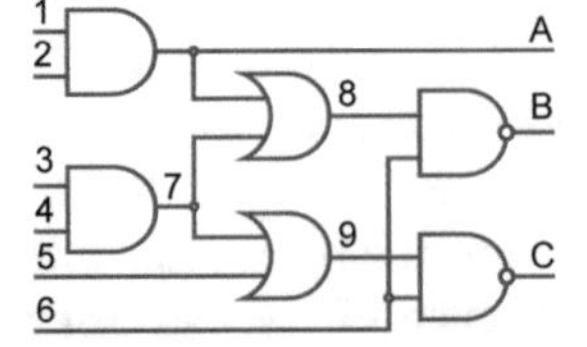

T\F	1	2	3	4	5	6	7	8	9	A	B	C	A_a	A_b	A_c
111101	0	0	0	0	.	0	0	0	0	0	1	1	1	0	0
010101	1	.	1	.	1	.	1	1	1	1	0	0	0	1	1
101001	.	1	.	1	1	.	1	1	1	1	0	0	0	0	0
000011	.	.	.	.	0	0	1	1	0	1	0	1	0	0	0
111110	0	0	.	.	.	1	.	.	.	0	0	0	1	1	1

Fig. 8.1 Fragment of a digital circuit and fault detection table

response matrix $A = |A_{ij}|$ by comparing reference states $T = |T_{ij}|$ and real states $U = |U_{ij}|$ of lines under test or output lines: $A_{ij} = T_{ij} \oplus U_{ij}$, which form a mask of possible faults by using the following expression: $S_i = S(T_i) = \left(V_{A_{ij}=1} S_{ij}\right) \wedge \left(\overline{V_{A_{ij}=0} S_{ij}}\right)$. Each test vector (segment) activates its own structure of possible faults that is functionally dependent on the mask, assertions (states of observed outputs), and test patterns: $S = f(S, A, T_i)$. Assuming that in the matrix $S = |S_{ij}|$ on the first test vector, the states of the assertion outputs are equal to $A_{1A} = 0; A_{1B} = 1; A_{1C} = 1$, where the value 1 identifies the appearance of a fault in the unit; the mask of possible faults corresponds to the following functional $S_1 = S(T_1) = \left(V_{A_{1j}=1} S_{1j}\right) \wedge \left(\overline{V_{A_{1j}=0} S_{1j}}\right)$ and has the view:

$$
\begin{aligned}
S_1 \; &= S(T_1) = (S_2 \vee S_3) \wedge \left(\bar{S}_1\right) = (111101110010 \vee 001111101001) \\
&\times \wedge \left(\overline{110000000100}\right) = (111111111011) \wedge (001111111011) \\
&= (001111111011).
\end{aligned}
$$

The resulting mask is applied to the first row of the fault detection table that defines a set of suspected faults $F_i = T_i \wedge S|_{i=1} \rightarrow F_1 = T_1 \wedge S_1$, which form assertion output response $A_{1(A,B,C)} = (011)$ of the device on the first test vector:

Faults	1	2	3	4	5	6	7	8	9	*A*	*B*	*C*
T_1	0	0	0	0	.	0	0	0	0	0	1	1
S_1	0	0	1	1	1	1	1	1	1	0	1	1
$F_1 = T_1 \wedge S_1$	.	.	0	0	.	.	0	0	0	.	1	1

Under the proposed procedure for obtaining a mask of one line building a matrix of structural fault activation $S(T)$ is performed based on the use of output response table $A = |A_{ij}|$ defining the states of the assertion engine during test execution $S(T) = S \bigotimes A$:

$S=\|S_{ij}\|$	1	2	3	4	5	6	7	8	9	*A*	*B*	*C*
1	1	1	.	.	.	.	.	.	.	1	.	.
2	1	1	1	1	.	1	1	1	.	.	1	.
3	.	.	1	1	1	1	1	.	1	.	.	1

$\otimes \rightarrow$

$A=\|A_{ij}\|$	*A*	*B*	*C*
T_1	1	0	0
T_2	0	1	1
T_3	0	0	0
T_4	0	0	0
T_5	1	1	1

=

S(T)	1	2	3	4	5	6	7	8	9	*A*	*B*	*C*
T_1	1	1	0	0	0	0	0	0	0	1	0	0
T_2	0	0	1	1	1	1	1	1	1	0	1	1
T_3	0	0	0	0	0	0	0	0	0	0	0	0
T_4	0	0	0	0	0	0	0	0	0	0	0	0
T_5	1	1	1	1	1	1	1	1	1	1	1	1

To form data structures suitable for computer processing, it is necessary to translate the symbols of the faults of the fault detection table into the two-digit code in accordance with the rules of $\triangleright$ coding: $\triangleright = (0 = 10; 1 = 01; X = 11;$ $\varnothing = 00\}$, which applied to the fault detection table $F(T)$ gives the following result:

F(T)	1	2	3	4	5	6	7	8	9	A	B	C
T_1	0	0	0	0	.	0	0	0	0	0	1	1
T_2	1	.	1	.	1	.	1	1	1	1	0	0
T_3	.	1	.	1	1	.	1	1	1	1	0	0
T_4	.	.	.	.	0	0	1	1	0	1	0	1
T_5	0	0	.	.	.	1	.	.	.	0	0	0

$\triangleright$ $\longrightarrow$

F(T)	1	2	3	4	5	6	7	8	9	A	B	C
T_1	10	10	10	10	00	10	10	10	10	10	01	01
T_2	01	00	01	00	01	00	01	01	01	01	10	10
T_3	00	01	00	01	01	00	01	01	01	01	10	10
T_4	00	00	00	00	10	10	01	01	10	01	10	01
T_5	10	10	00	00	00	01	00	00	00	10	10	10

After obtaining the structural matrix $S(T)$, intended to mask the real faults in the fault detection table, it is necessary to perform # superposition of two matrices: $F\ (T)\ F = S(T) \# F(T)$, which is reduced to performing an # operation with the coordinates of the same name $F_{ij} = \bar{F}_j \leftarrow (F_j = 00) \vee (S_{ij} = 0)$ meaning modification of coordinate codes in the table $F(T)$ when the predetermined conditions are fulfilled. Otherwise, the operation is reduced to the negation of the matrix cells of the fault codes, masked by zero signals of the structural activation matrix, and also all zero codes in the fault detection table. The truth table of # operation in the symbol and coded form is shown below:

$\# = S_{ij} \setminus F_{ij}$	∅	1	0	X
0	X	0	1	∅
1	X	1	0	X

$\# = S_{ij} \setminus F_{ij}$	00	01	10	11
0	11	10	01	00
1	11	01	10	11

The truth table is changed with respect to a negation of the state 00 in 11 in the presence of a 1 value of the fault activation signal because such a code (00) indicates the presence of an empty set of detectable faults in a circuit line, which is impossible. But the code 00 blocks all the calculations of the conjunction on the column, making the result 00. The negation of the code makes it possible to not mask faults of any signs when logical multiplying. It is assumed that checking circuit faults of different signs on the same line by a test vector is impossible.

Performing the superposition procedure of the structural matrix with the coded fault detection table $F(T) = S(T) \# F(T)$ gives the following result:

S (T)	1	2	3	4	5	6	7	8	9	A	B	C
T_1	1	1	0	0	0	0	0	0	0	1	0	0
T_2	0	0	1	1	1	1	1	1	1	0	1	1
T_3	0	0	0	0	0	0	0	0	0	0	0	0
T_4	0	0	0	0	0	0	0	0	0	0	0	0
T_5	1	1	1	1	1	1	1	1	1	1	1	1

$\overset{\#}{\rightarrow}$

F (T)	1	2	3	4	5	6	7	8	9	A	B	C
T_1	10	10	10	10	00	10	10	10	10	10	01	01
T_2	01	00	01	00	01	00	01	01	01	01	10	10
T_3	00	01	00	01	01	00	01	01	01	01	10	10
T_4	00	00	00	00	10	10	01	01	10	01	10	01
T_5	10	10	00	00	00	01	00	00	00	10	10	10

=

F (T)	1	2	3	4	5	6	7	8	9	A	B	C
	10	10	01	01	11	01	01	01	01	10	10	10
	10	11	01	11	01	11	01	01	01	10	10	10
	11	10	11	10	10	11	10	10	10	10	01	01
= $\wedge$	11	11	11	11	01	01	10	10	01	10	01	10
	10	10	11	11	11	01	11	11	11	10	10	10
$F(T) = \bigwedge_{i=1}^{n} F_i$	10	10	01	00	00	01	00	00	00	10	00	00
$F =$	0	0	1	.	.	1	.	.	.	0	.	.

In the final stage of diagnosis a single vector operation of a logical *and* operation for all the rows of the modified encoded truth table $F(T)$ is performed:

$$F(T) = \left(\bigvee_{A_i=1} F_i\right) \wedge \left(\overline{\bigvee_{A_i=0} F_i}\right) = \left(\bigwedge_{A_i=1} F_i\right) \wedge \left(\overline{\bigvee_{A_i=0} F_i}\right) = \left(\bigwedge_{A_i=1} F_i\right) \wedge \left(\bigwedge_{A_i=1} \bar{F}_i\right)$$
$$= \left(\bigwedge_{i=1}^{n} F_i\right).$$

This makes it possible to exactly detect all the faults presented in the diagnosis object, which are shown in the two lower rows of the above encoded fault detection table $F(T)$: $F = \{1^0, 2^0, 3^1, 6^1, A^0\}$.

The theoretical proof of the matrix diagnosis of single and multiple faults is presented below in the form of two theorems.

Theorem 1 Single stuck-at faults of a digital circuit, defined by qubits on the test patterns of a multivalued fault detection table, are determined by using a vector *and* operation, masked in the rows by output response vector $A = |A_{ij}|$ for all assertion points:

$$F(T) = \left(\bigvee_{A_i=1} F_i\right) \wedge \left(\overline{\bigvee_{A_i=0} F_i}\right) = \left(\bigwedge_{A_i=1} F_i\right) \wedge \left(\overline{\bigvee_{A_i=0} F_i}\right) = \left(\bigwedge_{A_i=1} F_i\right) \wedge \left(\bigwedge_{A_i=1} \bar{F}_i\right)$$
$$= \left(\bigwedge_{i=1}^{n} F_i\right).$$

The expression is true because (1) the second multiplier is pure mathematics—the negation of the disjunction is the conjunction of negations—which means the *and* operation of table codes by their preliminary negation; and (2) the first multiplier is focused on detection of consistent faults, so it is replaced by $(\wedge_{A_i=1} F_i)$. Indeed, in one line or a variable, two detected faults of opposite sign cannot be present simultaneously. Therefore, in the basic formula the fault disjunction $(\vee_{A_i=1} F_i)$ is largely focused on detecting multiple faults, but not associated with a single line. The faults $X = \{0,1\}$ in the line, as well as the negation of the empty fault set, theoretically create conditions for *and* operation with other coordinates of a column in order to obtain in each line a detected fault or an empty set of faults.

Theorem 2 Multiple stuck-at faults of a digital circuit, defined by qubits on test patterns of a multivalued fault detection table, are defined by using vector *or* operations and masked in rows by an output response vector $A(T)$ for all assertion points:

$$F(T) = \left(\underset{A_i=1}{\vee} F_i \right) \wedge \left(\overline{\underset{A_i=0}{\vee} F_i} \right) = \left(\underset{A_i=1}{\vee} F_i \right) \wedge \left(\underset{A_i=0}{\wedge} \bar{F}_i \right).$$

The expression is true because (1) the second multiplier is the negation of the disjunction or the conjunction of negations, which means the logical *and* operation of table codes with their preliminary negation; and (2) the first operand is focused on detection of the multiple faults on the assumption that two detectable faults of opposite sign may be present simultaneously in a line or variable. This formula is more focused on detection of multiple faults in blocks of digital systems not connected with a single line. Multiple faults of a digital system theoretically create conditions for *or* operation with other column cells to form a fault set, defining an output response vector; detectable faults on a test that do not affect the appearance of incorrect responses on outputs must be subtracted from them.

Detecting multiple faults based on a Hasse multiprocessor [4, 5], which is focused on solving the coverage problem, by leveraging the output response vector and the columns of the fault detection table, is based on the formula:

$$F(T) = \left(\underset{i}{\vee} F_i \right) \oplus A = 0$$

This solution is a combination of columns involved in vector operations of logic addition, which provides the equality to the output response vector. Because the operation is time consuming, it should use a Hasse multiprocessor, focused on calculating the power set in the almost parallel mode.

To sum up, it should be noted that a model for diagnosing digital devices contains transducers, focused on the implementation of the following steps (Fig. 8.2):

1. Preprocessing: generating the initial diagnostic information in the form of a diagnostic test, fault detection table, and reachability matrix of a digital system.
2. Testing the real unit, based on the use of industrial simulators to compare the actual responses and reference values on the observed assertion lines, which enables forming a matrix of output responses or output response vectors in a binary alphabet.

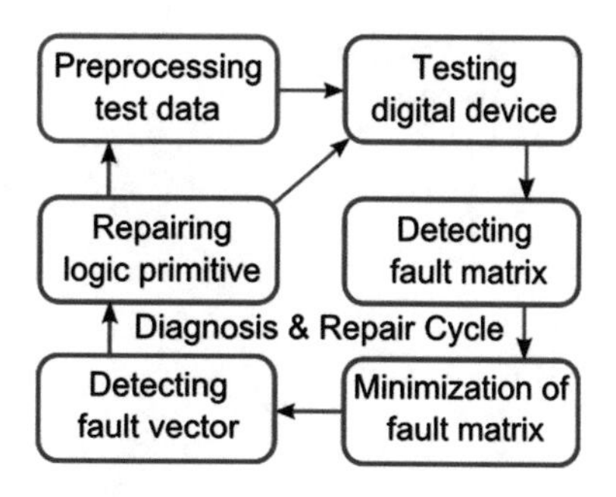

Fig. 8.2 Diagnosis and repair cycle for logic blocks

3. Calculating the activity matrix of the graph structure for each input test pattern, which is equal to the dimension of the fault detection table. by using the output response matrix and reachability matrix, which allows a significant reduction in the area of suspected faults.
4. Modification of the content of the fault detection table by means of their masking by using an activity matrix of the graph structure in order to detect only those faults, which really form an output response matrix in the diagnostic process.
5. Executing the procedure of a logical *and* operation of the fault detection table rows to obtain the vector of suspected faults.
6. Repairing the digital unit by means of readdressing the faulty logic components to their analogs from spares and repeating the diagnostic process.

Thus, the novelty of the proposed method for fault diagnosis lies in usage of a single parallel logical *and* operation, which in combination with structural fault masking offers advantages over analogs in terms of increasing the performance and depth of diagnosis.

8.3 Qubit Modeling of Digital Systems

The data structures, effectively from the viewpoint of software or hardware implementation of fault-free interpretative modeling of discrete systems, described in the form of qubit vectors of primitive output states, are considered. To describe the digital circuit shown in Fig. 8.3, the structure of interrelated elements and cubic coverage (truth tables) of logic elements are used.

The aim of the proposed method for qubit simulation is to replace the truth tables of digital device components with vectors of output states. Let a functional primitive with the number P_6 have the following truth table:

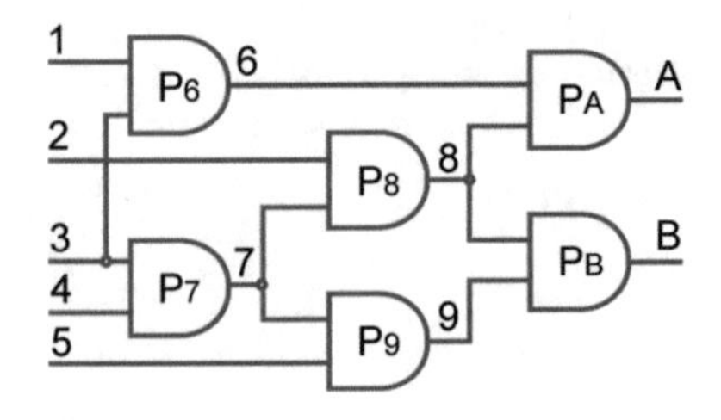

Fig. 8.3 Fragment of a digital circuit

$$P_6 = \begin{array}{|cc|c|} \hline X_1 & X_2 & Y \\ \hline 0 & 0 & 1 \\ 0 & 1 & 1 \\ 1 & 0 & 1 \\ 1 & 1 & 0 \\ \hline \end{array}$$

This coverage of a logic element can be transformed by unitary encoding of input vectors based on the use of a two-frame alphabet [4–7]. Symbols and their codes for describing automaton variables are the power set (the set of all subsets) on the universal set of four primitives that corresponds to the vector format containing two qubits:

$$\begin{aligned} B^*(Y) &= \{Q=(1000), E=(0100), J=(0001), O=\{Q,H\} \\ &= (1010), I=\{E,J\}=(0101), A=\{Q,E\}=(1100), B=\{H,J\} \\ &= (0011), S=\{Q,J\}=(1001), P=\{E,H\}=(0110), C=\{E,H,J\}=(1110), F \\ &= \{Q,H,J\}=(1011), L=\{Q,E,J\}=(1101), V=\{Q,E,H\} \\ &= (1110), Y=\{Q,E,H,J\}=(1111), U=(0000)\}. \end{aligned}$$

By using a two-frame alphabet, any coverage of the functional primitive can be represented by one or two cubes through encoding of input sets and subsequent combining of symbols, given that the cubes are mutually negated:

$$P_6 = \begin{array}{|c|c|} \hline 00 & 1 \\ 01 & 1 \\ 10 & 1 \\ 11 & 0 \\ \hline \end{array} = \begin{array}{|c|c|} \hline Q & 1 \\ E & 1 \\ H & 1 \\ J & 0 \\ \hline \end{array} = \begin{array}{|c|c|} \hline V & 1 \\ J & 0 \\ \hline \end{array} = \begin{array}{|c|c|} \hline 1110 & 1 \\ 0001 & 0 \\ \hline \end{array} \rightarrow \begin{array}{|cccc|} \hline 1 & 1 & 1 & 0 \\ \hline \end{array}$$

Two cubes show not only all the solutions but also the negated output signal that is interesting from the point of activation of all logical paths in the circuit structure, when synthesizing tests. For example, to change the output state it is necessary to create a pair of consecutive terms of the inputs, where the first three vectors (addresses) should be in the first cycle, and the fourth vector formed by two input variables in the second one.

To simulate fault-free behavior, it is enough to have a single cube (zero or unit), since the second one is always a complement to the first cube. Consequently, focusing, for example, on the unit cube, forming 1 at the output, we can remove a bit of the primitive output state, reducing the dimension of the cube or primitive model up to the number of addressable primitive states, where the address is a vector composed of the binary values of the input variables, which identifies the primitive state of the output.

Qubit Q coverage is the vector form of interpretative description of the functionality, where the coordinate value determines the state of the function output corresponding to the binary input word, forming a cell address. Q coverage of a 1-output primitive is always represented by two mutually negated cubes (vectors) whose dimension is equal to the power of two of the number of input variables, where a single coordinate value defines the usage of the address of the considered bit in the formation of the corresponding (0,1) state of the primitive output. Qubit models of primitives require the creation of a novel theory for modeling, forward propagation and backward implication, test synthesis, fault simulation, and fault detection. Here and below, we present the main procedures for fault-free simulation based on manipulating addresses implicitly represented in cube coordinates of Q coverage.

The model for analyzing a digital system based on the use of qubit data structures can be described by four components:

$$\begin{aligned}
&F = < L, M, X, Q > ,\\
&L = (L_1, L_2, \ldots, L_j, \ldots, L_n);\\
&M = (M_1, M_2, \ldots, M_j, \ldots, M_n);\\
&X = (X_{nx+1}, X_{nx+2}, \ldots, X_{nx+i}, \ldots, X_n);\\
&Q = (Q_{nx+1}, Q_{nx+2}, \ldots, Q_{nx+i}, \ldots, Q_n).
\end{aligned}$$

The following designations are used here: L—a vector of identifiers for equipotential lines of a digital circuit, which, because of its triviality, can be excluded from the model, but it is necessary to know a number of input variables of a device and the total number of lines; M—a state modeling vector for all circuit lines; X—the ordered set of input variable vectors for each circuit primitive associated with the output numbers; Q—a set of Q coverage for the primitives, strictly associated with the output numbers and the input variables of the primitives; n—the number of lines in the circuit; and n_x—the number of input variables.

As an example of a qubit model of the digital device $F = <L,M,X,Q>$ represented in Fig. 8.3, a variant of the circuit description table for analyzing fault-free behavior (fault-free simulation) is given below:

L	1	2	3	4	5	6	7	8	9	A	B
M	1	1	1	1	1	0	1	0	1	1	0
X	.	.	.	.	.	13	34	27	75	68	89
Q	.	.	.	.	.	1	0	1	1	1	1
	.	.	.	.	.	1	1	0	0	0	0
	.	.	.	.	.	1	1	0	0	1	1
	.	.	.	.	.	0	1	0	1	0	1

The method of qubit fault-free simulation is reduced to the definition of the output value for the element at the address generated by concatenation of binary states of input variables for each primitive of the digital circuit:

$$M(Y_i) = Q_i\left[M\left(X_{i1}{}^*X_{i2}\ldots{}^*X_{ij}\ldots{}^*X_{ik_i}\right)\right].$$

Here, k_i is a number of input lines in the primitive i.

If the variables create a nonbinary address, in this case, there is the possibility of forming a nonbinary output state of the primitive, which is defined in the ternary alphabet by the symbol X. The output states are formed by consistent modeling, based on simple iterations or Seidel iterations [6, 8]. In the second case, a preprocessor procedure for ranking lines and circuit primitives is necessary, which considerably reduces the number of passes on the circuit primitives to achieve convergence when the equality of the states of all circuit lines in two adjacent iterations is fixed. In addition, ranking of primitives at the levels of forming outputs allows significant improvement of the performance of simulation due to parallel processing of the functional elements of one level. For example, for the circuit shown in Fig. 8.3, we can handle concurrently the elements with the numbers 6, 7, and then 8, 9, and beyond—A, B. In the first case, when simple iterations are used, ranking is not required, but the cost for the simplicity of the simulation algorithm is a significantly greater number of iterative passes through the circuit primitives to achieve the convergence criterion.

Because the outputs of the processed primitives uniquely identify the numbers of noninput lines for the vector L, the formula for modeling can be reduced to the loop determining the status of all noninput variables:

$$M_i = Q_i\left[M\left(X_{i1}{}^*X_{i2}\ldots{}^*X_{ij}\ldots{}^*X_{ik_i}\right)\right] = Q_i[M(A_i)],\ \ i = \overline{n_X + 1, n}$$

Here, the modeling process is associated with obtaining the bit address in a functionality qubit by concatenation and determining the status of the primitive or noninput line of a digital structure, starting with the number $i = n_x + 1$. If the variables create a nonbinary address, in this case, there is the possibility of forming the output state of the logic element in the ternary alphabet by the symbol X. The output states are formed by the primitive procedure for processing a primitive qubit $M_i = Q_i[M(X_i)]$ based on simple iterations or Seidel iterations [6, 8]. In the second case, a preprocessor procedure for ranking lines and circuit primitives is necessary, which considerably reduces the number of passes on the circuit primitives to achieve convergence when the equality of the states of all circuit lines in two adjacent iterations is fixed. The computational complexity of the proposed Q method for modeling based on qubit functionalities is determined by the procedures for generating an address (input vector), containing k_i variables for each i-th primitive $[(r + w) \times k_i]$, reading a bit from the qubit vector at the concatenated address and writing $(r + w)$ for a given bit to the modeling vector:

$$\eta = \sum_{i=n_X+1}^{n} \{[(r+w) \times k_i] + (r+w)\} = \sum_{i=n_X+1}^{n} [(r+w) \times (k_i+1)]$$
$$= (r+w) \times \sum_{i=n_X+1}^{n} (k_i+1).$$

The modeling time for a test vector by the Q method, on the condition that a digital circuit composed of 900 four-input primitives is characterized by the parameters $r = w = 5$ ns, $k_i = 4$, $n_x = 100$, $n = 1000$, is equal to 45 μs:

$$\eta = (r+w) \times \sum_{i=n_X+1}^{n} (k_i+1) = (5+5) \times 900 \times (4+1) = 10 \times 900 \times 5$$
$$= 45{,}000 \ \text{ns} = 45 \ \mu s.$$

This means that the performance of an interpretative Q method for modeling allows processing of 22,222 input vectors per second for a given circuit. At that, a digital device has a significant advantage—a service function for repairing failure of primitives online by readdressing on the spare element.

For the synthesis of quasioptimal data structures of a combinational device, it is necessary to use the following rules:

1. To simulate by using Seidel's method, the ranked circuit of a digital device on the structural depth must have the same type of primitives as possible at each level (layer) of operation.
2. It is desirable to have the same number of primitives at each level, meaning that synthesis of the digital device has to be focused on creating a rectangular or matrix-like structure of similar logic elements.
3. Implementation of the combinational primitives provides for using addressable memory elements existing in programmable logic devices (FPGA, CPLD), widely used for prototyping.
4. Spare primitives are provided for each level of the combinational device for online repair—one spare element for each type of component that is used at that level.
5. The hardware cost for implementing a high-performance combinational device should be determined by the sum of all the primitives associated with the levels of the combinational device, extended by set of spares—one for each layer (assuming the existence of the same primitives in each layer):

$$Q = \sum_{i=\overline{1,n}}^{j=\overline{1,m}} P_{ij} + n.$$

6. Implementation of the combinational device based on minimizing the hardware cost is determined by the sum of all types of primitives, which are invariant to the levels of the combinational device and extended by a set of spares—one for each type:

$$Q = \sum_{i=1}^{m} P_i + m$$

7. The matrix of the combinational elements is processed by using the processor line of primitives, the number of which is equal to the power of the maximum level or layer in the rectangular structure, which provides the possibility for parallel processing of all primitives at each element level in order to improve the performance of the combinational prototype, implemented in the PLD.

Thus, the novelty of the proposed Q method for interpretative fault-free simulation of digital circuits lies in significantly increasing the performance and reducing the volume of data structures through replacing the truth tables by Q coverage, providing the competitiveness of the proposed development in comparison with compilation simulation.

8.4 Repair of Logic

The few works devoted to repairing logic circuits [9–13] describe two ideas. The first one provides for reconfiguring the structure of logical components offline, which provides the possibility of replacing each of the faulty primitives. The second one creates the conditions for replacement of faulty components through the use of spare components and extension of multiplexers for readdressing failed primitives.

Qubit data structures are modified by way of adding a row of primitive types $F = \langle L, M, X, P, Q \rangle, P = (P_1, P_2, \ldots, P_i, \ldots, P_m)$, used when synthesizing the digital system if necessary to repair the system during operation by using spare primitives, which like basic elements are implemented on the basis of memory.

Figure 8.4 illustrates an example of a circuit structure, composed from addressable elements and three spare components. The data structures corresponding to the circuit with three additional elements are presented below:

The table (see Fig. 8.4) uses the numbers of structural primitives, which gives the opportunity to replace any failed element with a spare by changing the address number in the line of primitives P. Repair elements in this table begin with the number 7.

The following table shows the line of logic elements, and also the addresses of the types of these primitives, marked with numbers:

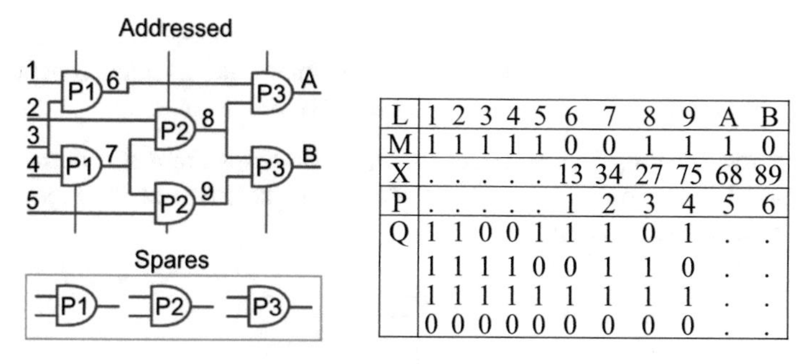

L	1	2	3	4	5	6	7	8	9	A	B
M	1	1	1	1	1	0	0	1	1	1	0
X	.	.	.	.	.	13	34	27	75	68	89
P	.	.	.	.	.	1	2	3	4	5	6
Q	1	1	0	0	1	1	1	0	1	.	.
	1	1	1	1	0	0	1	1	0	.	.
	1	1	1	1	1	1	1	1	1	.	.
	0	0	0	0	0	0	0	0	0	.	.

Fig. 8.4 Circuit structure of addressable and spare elements

L	1	2	3	4	5	6	7	8	9	*A*	*B*
M	1	1	1	1	1	0	0	1	1	1	0
X	.	.	.	.	.	13	34	27	75	68	89
P	.	.	.	.	.	1	2	3	4	5	6
Q	1	1	0	0	1	1	1	0	1	.	.
	1	1	1	1	0	0	1	1	0	.	.
	1	1	1	1	1	1	1	1	1	.	.
	0	0	0	0	0	0	0	0	0	.	.

This data structure is focused on software simulation, and spare primitives begin with the number 2. If it is possible to reprogram the logic in the memory element with the same number of input variables, then the procedure should be performed after fixing the faulty element, where it is known—which element in the structure refused and the type of it. This repair procedure is focused on PLD implementation of digital systems. If the qubit models of circuits have no spare primitives, the corresponding tables will be as follows:

L	1	2	3	4	5	6	7	8	9	*A*	*B*
M	1	1	1	1	1	0	0	1	1	1	0
X	.	.	.	.	.	13	34	27	75	68	89
P	.	.	.	.	.	1	1	2	2	3	3
Q	1	0	1	1	0	1	.	.	.	.	.
	1	1	0	1	1	0	.	.	.	.	.
	1	1	1	1	1	1	.	.	.	.	.
	0	0	0	0	0	0	.	.	.	.	.

Thus, quantum data structures are focused on compact description of digital product functionality by means of qubit vectors, speeding-up of simulation procedures by using addresses of output states of primitives, and repair of logic gates due to their implementation in PLD memory elements or software modules. This is very important—in the last case the storage of spare primitives is not necessary, because the interpretative structures of the tabular data provided herein are focused on fault correction during the operation of a digital device prototype.

The processing circuit on a chip is reduced to determining the address, compiled from binary bits of the simulation vector, which determines a logical function. Each primitive has a processing loop that contains three procedures:

1. An address reading the numbers of input variables from the corresponding column of the matrix X to form the address of the states of the input variable for the simulation vector:

$$A = X_{ij}, i = \overline{1, n}; j = \overline{1, s_p - 1};$$

2. Generation of the address (binary code) to compute the logic function by concatenating the corresponding states of the input variables in the simulation vector:

$$A = M(X_{ij})^* M(X_{ir});$$

3. Saving of the result of executing the logical function (the status of the output) to the corresponding bit of the simulation vector:

$$M(X_{is_p}) = P\left[M(X_{ij})^* M(X_{ir})\right].$$

The processing of all circuit primitives in this case is strongly consistent, resulting in a considerable slowdown of the procedure for forming the states of output variables. However, the decrease in performance can be considered as the cost for embedded and autonomous repair of the functionality of the digital structure, which is one of the stages of the functioning system-on-a-chip (SoC) infrastructure IP, shown in Fig. 8.5.

The combinational circuit becomes an operating unit, where there are operational and control automata. Replaceable components in the operating automaton are primitive types—functional elements or structural primitives.

The operational unit for implementation of the element-addressable combinational circuits contains the following components: a counter for processing the current primitive C_1; a memory for storing the types of primitives, corresponding to the structural elements P; a counter for reading the numbers of input and output variables of the current primitive C_2; a decoder of primitive types DC; a memory for storing the simulation vector M; a matrix memory for storing the numbers of inputs–outputs of the structural primitives X; a memory line, realizing the functional primitives $P(Q)$; a register for generating the input address word for the processed primitive RG; and a logic element *or* for switching the results of processing functional primitives.

A flowchart of the control algorithm for simulating the structure of the combinational circuit is shown in Fig. 8.6 and includes the following items:

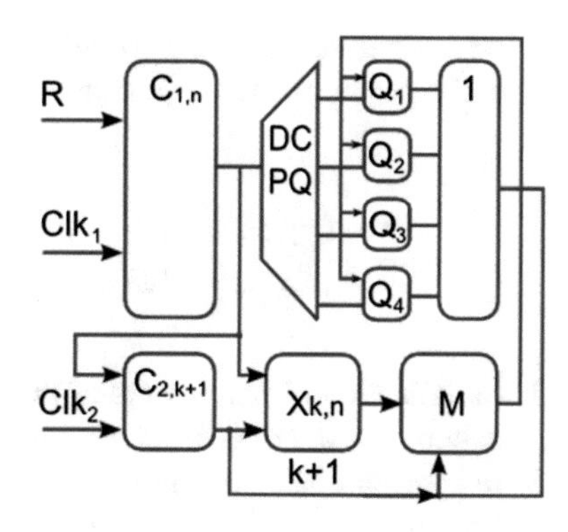

Fig. 8.5 Operational structure of a logic circuit

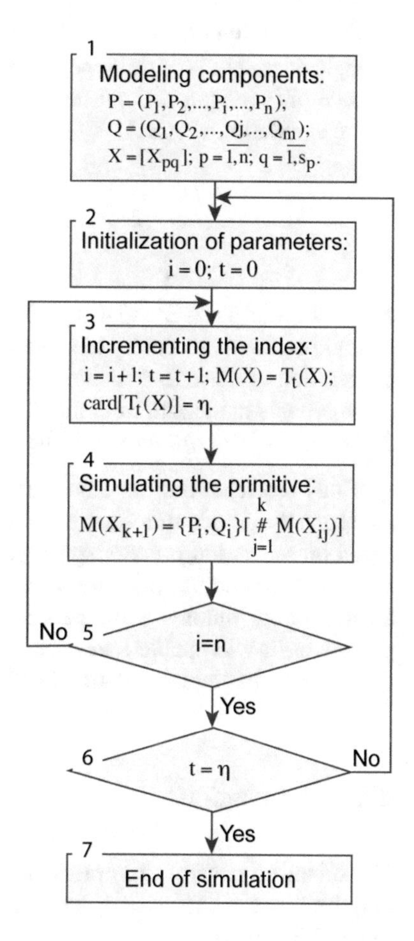

Fig. 8.6 Flowchart of a control algorithm for simulation

1. Initialization (modeling) of all components of the circuit structure (elements P and types of elements Q, concatenating lines X for input vector forming):

$$P = (P_1, P_2, \ldots, P_i, \ldots, P_n); \quad Q = (Q_1, Q_2, \ldots, Q_j, \ldots, Q_m);$$
$$X = [X_{pq}]; p = \overline{1, n}; \quad q = \overline{1, s_p}.$$

2. Initialization of parameters of the processed primitive and the number of input pattern $i = 0$, $t = 0$ for its simulation in the binary alphabet $M_r = \{0,1\}$
3. Incrementation of the index of the primitive, the numbers of the test, and initialization of the input test pattern: $i = i + 1, t = t + 1, M(X) = T_t(X), |T_t(X)| = \eta$
4. Concatenation (#) of word bits for generating the input stimulus $\#_{j=1}^{k} M(X_{ij})$ for logic element P_i (of the type Q_i), performing the procedure for determining the state of its output and subsequent writing into the corresponding coordinate of the simulation vector $M(X_{k+1})$:

$$M(X_{k+1}) = \{P_i, Q_i\} \left[\mathop{\#}_{j=1}^{k} M(X_{ij}) \right].$$

5. Iteration of items 3 and 4 in order to obtain the states of the outputs for all the logic elements to satisfy the condition $i = n$
6. Iteration of items 2–4 for simulating all input test patterns to the equality $t = \eta$, where η is the length of the test
7. The end of the simulation of the digital device

Thus, the scientific novelty of the proposed model of a digital system lies in adding to the device structure redundant spare components and a control automaton based on sequential processing of logic primitives that allows realizing the procedure for readdressing primitives in a case of failure of one of them. It is easy to create similar automata for parallel processing of the layers of ranged circuit primitives; it will enable a maximal increase in the simulation performance of the device up to its implementation in PLD chips.

8.5 Conclusion

The scientific novelty and practical significance of this chapter is based on the use of qubit structures focused on parallel computation of set theoretic data, represented in the following items:

1. The method for fault diagnosis of digital systems is improved through the use of a single parallel operation of a logical *and* operation, which in combination with structural fault masking offers advantages over analogs in terms of data

compactness, improvement of performance, and enhancement of the diagnosis depth of functional failures and stuck-at faults in software or hardware units; it is focused on the qubit structures of diagnostic information allowing a significant reduction in the computational complexity of the simulation and diagnosis through the introduction of parallel logic operations on matrix data.
2. A new Q method for interpretative fault-free simulation of digital circuits is proposed, which is characterized by using compact Q coverage instead of truth tables; it allows a significant increase in performance through forming addressable outputs of functional primitives and reducing the volume of data structures, providing competitiveness of the proposed method in comparison with compilative simulation.
3. The model of digital systems is improved by way of adding to the device structure redundant spare components and a control automaton based on sequential processing of logic primitives that allow realizing the procedure for readdressing faulty primitives in online operational mode.
4. Examples of using qubit data structures and forming addresses for simulation of digital circuits and solving diagnosis problems through the use of vector parallel logic operations and repair of faulty modules based on addressable logic primitives are considered.
5. The main innovative idea of quantum (qubit) computing is to move from computational procedures of the byte operand, defining a single solution (point) in the discrete space to quantum parallel processes of the qubit operand, concurrently forming a power set of solutions.

References

1. Nielsen, M. A., & Chuang, I. L. (2010). *Quantum computation and quantum information*. Cambridge University Press: Cambridge.
2. Hahanov, V. I., Litvinova, E. I., & Guz, O. A. (2009). *Digital system-on-chip design and test*. Kharkov: Novoye Slovo.
3. Hahanov, V., Gharibi, W., Litvinova, E., & Chumachenko, S. (2011). Information analysis infrastructure for diagnosis. *Information an International Interdisciplinary Journal Japan, 14* (7), 2419–2433.
4. Hahanov, V. I., Murad, A. A., Litvinova, E. I., Guz, O. A., & Hahanova, I. V. (2011). Quantum model of computing processes. *Radioelectronics and Informatics, 3*, 35–40.
5. Bondarenko, M. F., Hahanov, V. I., & Litvinova, E. I. (2012). Logical associative multiprocessor structure. *Automation and Remote Control, 73*, 1648–1666.
6. Hahanov, V. I. (1995). *Technical diagnosis of digital and microprocessor structures*. Kiev: ISIO.
7. Hahanov, V. (2011). Infrastructure intellectual property for SoC simulation and diagnosis service. In M. Adamski, A. Barkalov, & M. Węgrzyn (Eds.), *Design of digital systems and devices* (pp. 289–330). Berlin: Springer.
8. Gorbatov, V. A. (1986). *Basics of discrete mathematics*. Moscow: Higher School.
9. Hahanov, V. I., Litvinova, E. I., Hahanova, I. V., & Murad, A. A. (2012). Infrastructure of built-in logic PLD-circuits repair. *Radioelectronics and Informatics, 2*, 54–57.
10. Hahanov, V., Litvinova, E., Gharibi, W., & Murad, A. A. (2012). Qubit models for SoC synthesis. Parallel and cloud computing. *USA, 1*(1), 16–20.

11. Hahanov, V. I., Litvinova, E. I., Chumachenko, S. V., Baghdadi, A. A. A., & Mandefro, E. A. (2012). Qubit model for solving the coverage problem. In *Proceedings of IEEE east-west design and test symposium*, 2012.
12. Zheng, G., Manning, E., & Metz, G. (1972). *Diagnosis of digital computers system failures*. Moscow: Mir.
13. Koal, T., Scheit, D., & Vierhaus H. T. (2009). A comprehensive scheme for logic self repair. In *Conference of the proceedings on signal processing algorithms, architectures, arrangements, and applications*, 2009.

Chapter 9
Cloud Service Computing: The "Smart Cyber University"

Vladimir Hahanov, Oleksandr Mishchenko, and Eugenia Litvinova

9.1 The Smart Cyber University in the Big Data of Cyberspace

A crisis in the higher education system in recent decades has encouraged scientists to look for disruptive decisions related to structural reforms of universities through use of computing services and Internet of Things (IoT) technology when creating cyber social relationships, which eliminate corruption by developing a smart cyber physical system for digital monitoring of scientific and educational processes, and also for cloud management of resources and personnel, free of officials.

Big data is a technological culture of cyberspace, aimed at creating the infrastructure of the cyber physical ecosystem of the planet in order to carry out online management of social, physical, and virtual processes, based on precise digital monitoring of large data flows and subsequent metric analysis of them by using intelligent cloud filter services to enhance the quality of life and preserve the planet's ecosystem. The market feasibility of big data, as a new technological culture, is defined as follows: (1) instead of retrieving data—exhaustive information; (2) instead of storing data—reuse of data for prediction and management of processes and phenomena; (3) instead of data structures with hard connections—addressable organization of physical and virtual objects and processes; (4) instead of manual data input and output—use of the Internet as an input and output for the cyber system; (5) instead of passive representation of the real and virtual worlds—cyber physical systems for monitoring and analysis of data for management of physical and virtual processes; (6) instead of the chaos of static data and knowledge in the cyberspace of the Internet—gradual semantic structuring of dynamic big data

V. Hahanov (✉) • O. Mishchenko • E. Litvinova
Design Automation Department, Kharkov National University of Radio Electronics, Ukraine, Nauka Avenue 14, Kharkov 61166, Ukraine
e-mail: hahanov@icloud.com; santific@gmail.com; litvinova_eugenia@icloud.com

V. Hahanov, *Cyber Physical Computing for IoT-driven Services*,
DOI 10.1007/978-3-319-54825-8_9

flows of cyber physical processes and phenomena for their effective monitoring, analysis, and management; (7) instead of unsorted data that are difficult to understand and use by human or cyber systems—smart, metrically ranked information structures focused on creating optimal solutions; and (8) instead of separate development of the real and virtual spaces—creation of a closed cyber physical ecosystem of the planet for harmonious development of the real and virtual worlds.

Cyberculture is the level of development of social and technological relationships between society, the physical world, and cyberspace determined by implementation of online cyber services for precise digital monitoring and reliable metric cloud management in all processes and areas of human activity, including education, science, production, and transport, in order to improve the quality of life of the people and preserve the planet's ecosystem.

Cyberspace is a set of addressable and metrically interacting digitized processes and phenomena in the global telecommunication infrastructure of computer networks with distinct functions for monitoring, computing, storage, transaction, and management to achieve goals.

Cybersecurity is a branch of knowledge aimed at infrastructural support of normal functioning of the object in cyberspace, which includes access authentication, vulnerability management, crypto security of transactions, testing, diagnosis, and protection from destructive penetrations. Cybersecurity (in the narrow sense) is a metric property of a digitized process or phenomenon in cyberspace that lies in the ability to withstand destructive penetrations, retaining all operating parameters in accordance with the specification.

Science is an area of human activity aimed at collection and analysis of facts to obtain objective knowledge about reality in order to predict natural phenomena, with management of social and cyber physical processes to ensure the quality of life of people and preserve the planet's ecology.

Education is an area of human activity aimed at the continuous process of formation of spiritual, physical, emotional, intellectual, and professional human culture through meaningful accumulation of conventional values, knowledge, and skills by using a multilevel education and training system that exists in time and space, allowing each individual to acquire social importance in the process of human development, as well as improving the quality of life and preserving the ecosystem of the planet.

Competence is a metric evaluation of the spiritual, physical, emotional, intellectual, and professional culture of an individual that defines his integral importance for possible application of knowledge and skills in performance of his social role, aimed at improving the quality of life and preserving the ecosystem of the planet.

A *metric* is a way of measuring distance in the space–time continuum between processes or phenomena by comparing their parameters.

A *quality* is a set of properties of a process or phenomenon that determine its suitability to meet certain requirements in accordance with the purpose.

A *university* is a community of teaching and support staff, united by an infrastructure and managed by legislation, statutes, regulations, orders, and moral and ethical relations, aimed at implementation of actual research and training of professionals with academic degrees demanded by the market, and also providing a

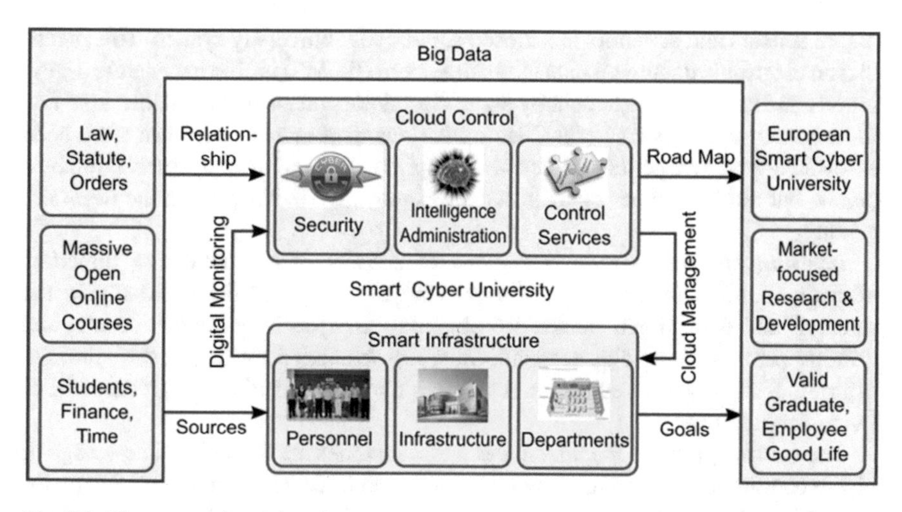

Fig. 9.1 The smart cyber university

high quality of life for staff by attracting foreign investment through the sale of educational services and scientific and technical products.

A *system* is a set of relationships between components and the external environment, with monitoring and control functions to achieve goals.

A *cyber physical system* (CPS) is a set of communicatively connected addressable virtual and real components in a digitized metric space with the functions of adequate physical monitoring and optimal cloud management in real time to achieve goals (Fig. 9.1).

Smart is a characteristic of a process or phenomenon, associated with network interaction of addressable system components in time and space between themselves and the environment, based on self-learning technologies to achieve goals.

A *smart cyber physical system* is a set of communicatively network–connected addressable virtual and real components in a digitized metric space with the functions of adequate physical monitoring and optimal cloud management and self-learning in real time to achieve goals.

A *smart cyber university* (SCU) is a metric culture of social and technological relationships, integrating personnel and a smart infrastructure in a network to perform actual research and train professionals with academic degrees demanded by the market through adequate monitoring and cloud management of digitized scientific and educational processes and phenomena in order to attract investment and achieve a high quality of life for employees (Fig. 9.1).

Electronic document management involves intelligent and legitimate transactions of digitized document streams (sensory signals and regulatory impacts) in a smart logically distributed data network, designed to implement paperless relationships with the outside world and direct monitoring and management of scientific and educational processes and departments of the university. Digitized documents (accessible for understanding by computers and humans) can be considered as digital sensors and actuators in a closed smart cyber university system. This means that an electronic document management system (EDMS) is able to generate digital reports and realize management by using digital documents that are understandable for a human-free cyber system. Electronic document management has often been associated with transactions of electronic copies of papers for visual perception by a person but not by a cyber system; that was innovative technology in the twentieth century.

International cooperation is an area of activity of the collective, aimed at attracting foreign investment and improving the image of the university in the research and educational market by obtaining competitive scientific results and training professionals with academic degrees demanded by the market, through offering of high-quality educational services aimed at ensuring a high standard of living for employees.

The *metric of a valid scientist* in the world includes the following: knowledge of high technologies and foreign languages, and scientific and technical achievements, proven by publications in scientometric index databases. The metric of the university image is defined as follows: (1) absence of corruption and high salaries for employees; (2) market feasibility of scientific achievements, innovations, and educational services, and quality of graduates; (3) the presence of a critical mass (20%+) of the world's authoritative scientists; (4) a high rate (20%+) of scientometric publications (books, articles, and reports) by scientists at leading international conferences (e.g., the IEEE); (5) receipt of international awards and prizes by scientists and students; and (6) a high level of foreign investment (50%+) in science and education.

Public relations is cyber informational management of public opinion, based on digital monitoring of market preferences in order to create and maintain the image and positive interaction of the subject in economic activity with the outside world to promote their own interests, processes, or phenomena in a particular segment of the market.

Legitimate relationships are a set of social and technological relations between management, staff, and offices, forming a system structure of the university for monitoring and management of scientific and educational processes, human resources, infrastructure, and financial and time resources, based on existing legislation, statutes, regulations, orders, and traditions in order to do the following: (1) implement actual research and create scientific and technical products that are competitive in the market; (2) offer high-quality educational services and training of professionals with academic degrees demanded by the market; (3) improve the image of the university and attract foreign investment; and (4) ensure a high quality of life for employees and create a positive moral and ethical climate.

A *metric of relationships* is a digitized set of informing documents and regulatory actions (orders, regulations, statutes, customs, and laws) forming the basis of cooperation, operational and strategic digital monitoring, and cyber management of research and educational processes, human resources, infrastructure, and financial and time resources, aimed at creating an external image and an internal moral and ethical climate in order to achieve a high level of research, trained professionals, and quality of life for employees.

Metric ability of informative documents, laws, regulations, and orders becomes the main condition for the appearance and formation of the digital space of social–metric relations, specifying cyber services for monitoring and management of government agencies—in particular, universities. In the higher education system, relationships constitute the main system-forming component that defines the success of a university in the education market. Everything else (goals, personnel, infrastructure, management, science, and education) is directly dependent on relationships.

A popular action of all managers is staff rotation that does not create a qualitatively new system (replacement of a faulty component does not lead to a new service but repairs the old functionality). To create a top European university it is necessary de facto and de jure to implement three axioms of legitimate (cyber) relationships: (1) the head (officer), and then the cyber system, estimates an employee through the metric of constructive, productive activity of a scientist, and measures him adequately; (2) a researcher receives from the (cyber) manager adequate service (according to his metric) that provides career development, with moral and material compensation; (3) the success of a scientist in a moral society is a matter of admiration and imitation, but not of envy and meanness.

The relationship is a fundamental principle of all phenomena and processes. Georg Cantor and many scholars of mathematics and practice have put forward this thesis. Formally, relationships are a defined set of connections within a finite set of components forming the structure. Actually, the structure of carbon can make diamond or graphite. Genome relationships can create humans or monkeys in four dimensions. The constitution–relationships determine the success or collapse of the state. Statutes, orders, and regulations form the genome of successful development or decay of the university.

9.2 Computing as a New Trend in the Cyber Service Market

The most significant market-oriented innovations are created through the use of a cyber physical system as an automaton model of computing for monitoring and management of processes and phenomena in all areas of human activity and nature. Implementation of structural and understandable cyber physical models of computing in nature, manufacturing, biology, sociology, ecology, and technology almost always gives a positive result. Simple and proven models can be applied to complex natural and biosocial objects and processes, which gives a more useful perspective than an attempt to replicate the complex entity artificially in computing. The following global technologies are proof of this, scaling the computing model: (1) cyber physical systems; (2) the Internet of Things and Everything; (3) web, cloud, mobile, service, network, automotive, big data, and quantum computing; and (4) smart objects and infrastructure: enterprises, universities, cities, and government.

The Internet of Things is a structure of cyber physical systems, combining large data centers, knowledge, services, and applications, focused on monitoring and management of smart processes and phenomena in a digitized physical space through sensors and actuators in order to ensure a high quality of human life and the preservation of the planet's ecology.

Computing is a branch of knowledge focused on research, design, and implementation of computer systems, networks and cloud–mobile services for monitoring and managing cyber physical processes and phenomena. Computing includes the following: computer engineering and management, software engineering and artificial intelligence, computer science, and information systems and technologies. The concept of computing is associated with the classical automaton model of a calculator, with control and execution mechanisms, monitoring and actuation signals, inputs for instructions and data, and outputs of the system status and results. Computing is a closed scalable system for monitoring and management of processes and phenomena in order to achieve goals.

Computing differs from information technology by active management of processes and phenomena in the real and virtual worlds. We can put an equals sign between the concepts of form and content: Internet = information technology; Internet of Things = computing. The development of computing, the main function of which is cyber management, should only be considered in conjunction with the real or physical world, a part of which is humanity. There exists an interaction between the two worlds, the real and the virtual ones: (1) mankind always poorly manages the real world and creates computing for assistance; (2) computing is a more advanced mechanism and takes control of technological processes—this is taking place now; (3) in order to save mankind from self-destruction, computing should pick up the rest of the operational management of social processes within the next 10 years; and (4) humanity and computing are united in the desire to change the cyber physical continuum for peaceful coexistence.

9.3 Motivation and Technological Solutions of a Smart Cyber University

Instead of information technology, the Internet of Things (IoT) has appeared. Instead of passively monitoring information, this means human-free active cloud management in a digitized cyber physical space, based on sensor–actuator fog-network monitoring of physical processes and phenomena [1–5]. It is difficult to propose new knowledge to a human and even more difficult to load in his brain a new metric of a vision of reality. We are talking about digitization of scientific and educational processes [6, 7] to create cyber physical systems not only for monitoring but also for management. The future of humanity is connected with the idea of creating human-free cloud cyber management of social institutions [8, 9], aimed at realization of open and objective regulation of digitized processes, where, instead of a subjective head (officer), an impartial cyber system works [10, 11]. The format of a cyber system cycle of management—"fact, measurement, evaluation, action"—related to monitoring, measurement, and management, is based on the postulate: "No measurement—no management" [2]. The use of digital competence metrics for transparent cyber distribution of moral and material incentives among participants in metric evaluation is realized in accordance with the ratings of processes or phenomena.

Management of a smart university is characterized by changing the disruptive paradigm of passive information technology monitoring into active IoT management of physical processes through the use of big data analytics [2]. Creation of the cyber physical system of a smart cyber university for monitoring and management is based on an automaton computing model, the feature of which is the use of cloud services as a control mechanism and smart fog networks as monitoring and execution mechanisms [12, 13]. The decision-making methods of the cyber system are focused on big data analysis [14–16] by using human-free filters of metric relations, where a head (officer) performs a decorative representative function.

Computational methods use virtual cloud processors, which are also based on nonarithmetic metrics of object measurement in cyberspace [19]. The digital cyberspace of science and education is a platform to create scalable human-free cloud cyber services [20]. Digitization of the physical and virtual components of scientific and educational processes is a necessary condition of cyber physical monitoring and management of the university. To do this, it is necessary to generate competence metrics in order to measure the quality of structural university components: (1) relationships; (2) road maps; (3) management; (4) infrastructure; (5) personnel; (6) resources; (7) production: educational services, graduates, and scientific achievements; (8) science; and (9) education.

The purpose of the SCU system is to improve the quality of educational services and scientific achievements of the higher education system through creation of a metric system of relations defining the rules of the digital monitoring and active cloud cyber management of scientific and educational processes, which makes it possible to eliminate corruption, attract foreign investment, and increase productivity and the quality of life of scientists and professors, creating market-demanded products in the form of graduates and scientific achievements. Creation of a smart cyber university is associated with integration of the following technologies: big data, cloud computing, mobile services, and cyber physical systems within the IoT culture through the use of service-oriented software platforms offered by leading organizations such as IBM, Google (www.google.com), Microsoft, NASA, Amazon, and Facebook.

The effectiveness of the social system (university) is defined by three purely economic assessments at the following levels: consumption, exports, and investment (Fig. 9.2). Leaders in the scientific and educational markets are formed secondarily by personnel and the economy, and primarily by the relationships defined by the leaders, laws, history, culture, and traditions. In the cause–effect cycle of the system economy, investment is a consequence and not the cause of efficient and stable work of the social system—money likes silence and

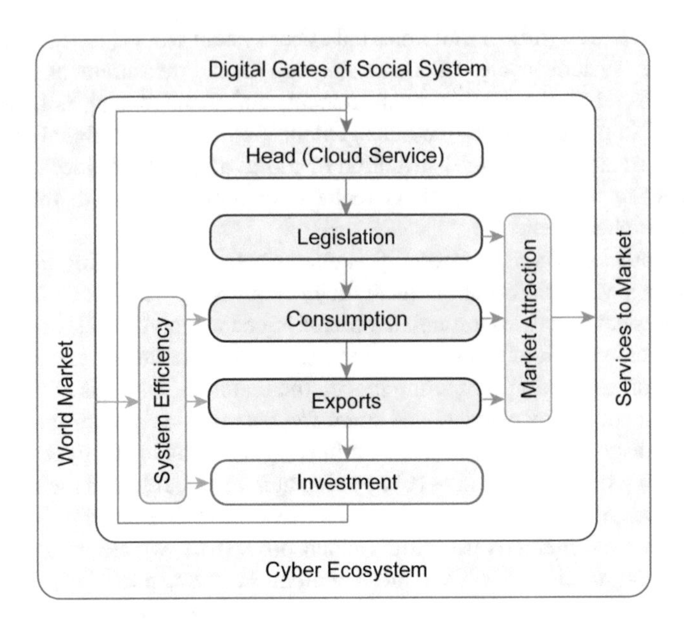

Fig. 9.2 Effectiveness of the social system

sustainability of creative relationships. The roots of corruption lie not in personnel, creating a tree of the social hierarchy of officials and executors, but in the relations between them, defined by legislation that knowingly permits subjective allocation of public funds and positions. The solution to the problem is cyber cloud management of financial and human resources based on digitized metric relationships defined by laws.

Analysis of research related to the creation of a smart cyber university [5–10, 14–18] shows the absence of system understanding of cloud–mobile cyber services, based on the use of metric relations, defining rules not only for digital monitoring but also for active human-free cyber management of scientific and educational processes to eliminate corruption, attract foreign investment, improve productivity, and improve the living standards of scientists and professors. This white spot in research exists in spite of the education market appeal of the SCU, which is about US$1 billion in the world. This can be explained by the difficulties of power transfer from traditionally corrupt officials to a human-free cyber management service. However, there are some publications [6, 7, 9–11, 13, 16, 18] considering human-free management of physical and virtual processes within a cyber system (cloud services + fog networks) emerging on the planet.

The work published by the authors on the subject [12, 19, 20] is devoted to solving the following problems: (1) creation of nonarithmetic metrics and multiprocessors for parallel analysis of big data in cyberspace using only logical operations [19]; (2) adaptation of quantum data structures and methods to improve the performance of virtual processors when performing information retrieval and decision-making processes; and (3) implementation of cyber systems for precise digital monitoring and human-free active management of physical and social objects and processes [12–20].

9.4 Innovative Services in the Smart Cyber University

Innovative cloud services, which constitute a smart cyber university as a structural prototype of a global scientific and educational virtual cyberspace—a global smart cyber university—are presented below:

1. Electronic document management for digital monitoring and intelligent cyber management of scientific and educational processes in the closed-cycle format of "fact–measurement–evaluation–action" to eliminate paper flow through the use of cloud–mobile service computing, databases, digital signatures, ID cards, e-mail, and mobile phones
2. E-voting for monitoring of public opinion, implementation of student surveys, and decision making at operational and academic council meetings, and employee conferences; election of experts, the student senate, and the governing scientific and teaching staff for the filling of vacant positions
3. Human resource management based on online monitoring, measurement, rating, and accumulation of digital competence metrics for evaluating the activity of students and all categories of staff in order to develop transparent regulatory moral and material incentives
4. Management of the structural department based on online monitoring, measurement, and accumulation of digital competence metrics of a department associated with scientific and educational processes for the generation of regulatory management actions and documents necessary for life
5. Evaluation of the quality of educational processes and components, and online testing of knowledge and skills, to exclude illegitimate relationships between a teacher and student during exams and tests
6. Management of scientific processes based on digital assessment of activities of scientists and departments, research results, projects, and proposals by using metrics developed by experts in order to ensure legitimate and transparent distribution of financial, human, and time resources between departments and employees
7. Provision of educational services in the form of massive open online courses (MOOCs) and onsite courses, and also management of educational processes based on clear distribution of financial and time (credit) resources between departments and employees in strict accordance with the metric estimation of the contribution of each subject to the asset and the university's image
8. Monitoring and management of scientific and educational processes of the student in real time, with generation and storage of electronic documents for his support in time and space through creation of a personal virtual cabinet associated with the mobile devices and e-mail

9. Measurement and support of undergraduate and postgraduate works and dissertations, and also competitive projects based on the integration of international metrics for evaluating the scientific novelty and practical significance of the results of the research with an internal quality system developed by experts
10. Licensing and accreditation of education programs based on measurement of the scientific and educational activities of departments and subsequent generation of packages of documents required for external evaluation of the quality of educational processes
11. Electronic 24–7 access and monitoring of staff and students present in the classrooms of the university, based on use of mobile devices and ID cards, and also electronic banking for paying for educational services and use of corporate credit cards to purchase goods and services within the earned funds of the department
12. Protection of the information and physical space of the university and electronic access authorization for all cyber physical components and processes associated with the life of the university

These services use the following main principles to build cyber physical systems: (1) mobile monitoring and human-free cloud management of addressable components of scientific and educational processes; (2) digitization of the system of metric relationships in the university (statutes, regulations, and orders) for measurement of scientific and educational processes in order to rate departments, students, and staff; (3) creation of digital competence metrics for evaluating job positions throughout the range of employees, as well as digital passports for all departments of the university; (4) electronic document management to eliminate paper through use of electronic digital signatures, keys, ID cards, e-mail and mobile phones; (5) an electronic voting system that operates regardless of the location of the expert, freeing him from the need to attend numerous meetings for decision making, academic council meetings, and employee conferences; (6) an electronic system of elections for vacant positions of rectors, other leaders, and scientific and educational positions, using the digital competence metrics of the applicants; (7) a cyber physical expert system to minimize the risk of making wrong decisions due to introduction of cloud services for metric evaluation of projects and proposals for management of scientific and educational processes; (8) digital measurement of all processes and phenomena in the university, as this is the only way to transform the chaos of errors of authoritarian rule into exact regulatory actions of cloud cyber management; (9) modeling of suggested actuator cyber solutions by the cloud service in order to predict their effectiveness and/or waste for the university; (10) distribution of financial resources between departments and employees in strict accordance with the metric estimation of the contribution of each subject to the final budget and the university's image; and (11) blended (online–onsite) scientific and educational processes, synchronized with the speed of understanding of each student individually, giving individual access to knowledge (via lectures and MOOCs) 24–7.

9.5 Metrics of Employees and University Departments

The university rating is determined by metric measurement of the quality of the following entities: relationships, human resources, management, infrastructure, resources, goals, graduates, research results, educational services, and the lives of the staff. Each process or phenomenon has its own metric. The basic metric primitive for measuring the activity of the university (for a scientist) is one A4-sized page (single spaced and in 11-point font) extracted from the original text, an article in a journal, or a 3000-character (without spaces) piece of text, which is estimated as 1 point. All other estimates are given to this primitive by scaling the complexity of tasks.

The symbolic designations of the process parameters are as follows: k is the number of pages; $f = 2$ (1) is the foreign factor; F is the impact factor of the journal; $i = 2$ (1) is the scientometric index of the publication (Scopus, Web of Science); n is the total number of attributes; $\$$ is the money equivalent in the local currency; and L is the number of lectures in the course.

The metric of the employee is represented by numerical values of parameters (the values or formulae for their calculation are represented in parentheses) for honors/awards/prizes and for teaching, science/research, volunteering, and education activities within the university [21]:

1. Honors, awards, and prizes ($300f$); membership in the National Academy of Sciences ($300f$) and in public academies of sciences ($20f$)
   (a) Teaching and prepared courses ($10nLf$)
   (b) Technological culture: programming languages and cloud services ($n10$)
   (c) Knowledge of foreign languages and certification ($n100$)
   (d) Number of publications (kfi); preparation of bachelor's ($5f$), master's ($10f$), PhD ($100f$), and DSc theses/dissertations ($300f$)
2. Science/research activities
   (a) Market products: software applications, devices, and models ($100f$)
   (b) Patents, inventors' certificates, and diplomas ($10f$); exhibitions ($10f$)
   (c) Monographs: foreign and national (kf)
   (d) Journal articles with HM indices: foreign, national, and other (ikf)
   (e) Reports at conferences with online–onsite representation and HM indices: foreign and national (ikf)
   (f) International grants, national projects, and research contracts ($f\$/1000$)
   (g) Training of scientists: DSc ($300ft$) and PhD students ($100ft$)
   (h) Defense of theses: DSc ($300t$) and PhD ($100t$) students
3. Teaching activities
   (a) Textbooks ($2kf$); manuals and training materials (kf)
   (b) Numbers of PhD students ($5f$), master's degree students ($2f$), and bachelor's degree students (f) in the current year

(c) Foreign internships and invited lectures (20*f*)
(d) Agreements with companies and universities: foreign and national (20*f*)

4. Volunteering activities

(a) Organization of conferences (200*f*), seminars (5*f*), and olympiads (20*f*)
(b) Membership of specialized scientific boards (10*f*)
(c) Critique of theses: foreign and national (10*f*)
(d) Review of articles: foreign and national (2*f*)
(e) Participation in Program Committee (PC) of conferences: foreign and national (5*f*)
(f) Individual grants: foreign and national (20*f*)
(g) Lectures at universities, and television and media appearances (10*f*)
(h) Sport and cultural activities (5*f*)
(i) Publication of websites, magazines, newspapers, brochures, videos, and conference proceedings (20*f*)

5. Educational activities

(a) Research and educational workshops for pupils, students, and PhD students (3)
(b) Organization of student attendances at conferences, exhibitions, and olympiads (10*f*)
(c) Excursions for students (physical and aesthetic culture) (3)
(d) Lectures and clubs focused on professional guidance for pupils and students (3)

These parameters form the positive numerator of the integral criterion for employee efficiency of the department for n parameters for a certain period.

In addition, the numerator is formed by the following integral parameters of the department:

1. Quality of the students, as the sum of scores of all tests divided by the number of them (10n)
2. Quality of scientific and teaching staff (D – number of DSc, K – number of PhD), normalized for the number of them 100 $(3D + K)/n$
3. Cooperation with companies and universities (agreements) (20*f*)
4. Research work by students (Q/n)
5. Profits of undergraduate and graduate students (N – number of students) ($/(1000$N$))
6. Graduation of bachelor's and master's degree students ($B/b + M/m$) (1000/2)
7. Employment of bachelor's and master's degree students ($B/b + M/m$) (1000/2)

where Q is the total number of points earned by the students of the department, and B/b, M/m is the ratio between the number of employed/graduated bachelor's/master's degree students.

There is also the denominator, consisting of parameters that determine the financial and time costs necessary for implementation of the scientific and educational processes of the department: integral financing of all employees of the department; expenditure on infrastructure maintenance and production areas of the department; equipment costs of the equipment in departmental laboratories, purchased by the university; and other expenses for preparation (education and diploma defence) of bachelor’s, master’s, PhD, and DSc.

Implementation of a rating system for each employee provides an opportunity for employees to protect themselves from authoritarian rule by first-level managers who can subjectively distribute time, money, and positions. It is proposed that this be weighted and normalized in the range (0–1) for a quality criterion Q of the integral activity of the department for the current year (month) Y, taking into account the average collective activity within the previous m years (months), having staff S at n parameters P_i. Each of them is given the maximum or reference value P_i (max) in the structure of the university [22]. The essentiality of the parameter is determined by its multiplication by a coefficient of scientific–educational and socioeconomic importance, approved by a council of scientists–experts: $k_i = \{[0,1], [1,q]\}$:

$$Q_Y = \frac{1}{m+1}\left[\frac{1}{S \times n} \times \sum_{i=1}^{n} \frac{k_i \times P_i}{P_{i(\max)}} + \sum_{j=1}^{m} Q_{Y-j}\right].$$

The quality criterion for each department and employee must be represented in the language of moral and material incentives for scientific and educational achievements. With regard to the integrated metrics for estimating the effectiveness of scientific and educational activities of the scientist–professor, it is difficult to offer more understandable and simple quality criteria, taking into account the story of current results and achievements:

$$Q_Y = \frac{1}{m+1}\left[\frac{1}{n} \times \sum_{i=1}^{n} \frac{k_i \times P_i}{P_{i(\max)}} + \sum_{j=1}^{m} Q_{Y-j}\right].$$

In fact, the metric defines the average over n parameters of the performance of a scientist in the department, faculty, or university. The numerator represents personal achievements, and the denominator represents the best numerical values of the achievements of scientists in each of n categories in the department or university. Zero values do not have a fatal impact on the evaluation of a scientist or department, because they are compensated for by high values of parameters in other areas of scientific and educational activities. The criterion also takes into account the total cumulative activity of the scientist over the previous m years, which is especially important for elderly employees who should receive a decent financial reward for their achievements in previous years. Incidental peaks and troughs in activity should not have a significant impact on the value of stimulation. If scientists and departments neglect significant parameters for the university, the cyber physical system should attract the attention of employees by raising the profile of the respective expert factors. In accordance with the quality criteria for the activity of each scientist, the cyber system, not the head (officer), allocates bonuses and allowances within the university or faculty. It is important that this information is available to all employees in order to avoid the spreading of rumors about inequitable distribution of benefits.

It is advisable to perform extra special and separate ratings for the following positions: (1) vice rector; (2) deans; (3) heads of department; and (4) heads of infrastructure and secondary departments, whose ratings should evaluate their performance as managers. In addition, all employees of infrastructure units should be evaluated based on the results of their activities, in accordance with the competence metric developed for them. The quality of the employee during the hiring process and in the workplace should be checked for compliance with the official competencies.

9.6 Leveraging of Cyber Democracy in the University

A cyber democracy is a metric culture of social–technological relationships, formed by experts, combining social groups and intelligent infrastructure in cyber physical space for comprehensive digital monitoring of public opinion and cloud management of digitized social processes and phenomena in order to save the planet and achieve a high quality of life. More popularly, cyber democracy is a combination of a comprehensive democratic discussion and digital monitoring of social problems with exact cyber management of society by cloud regulations. A process model of the cyber democratic interaction of university staff and cloud management is shown in Fig. 9.3.

The structure of the cyber democracy are as follows: (1) the social group of the university elects experts democratically by using their transparent metrics and delegates them powers to prepare competency metrics for digital measurement of a process or phenomenon; (2) the experts create digital metrics for assessment of a process or phenomenon in the form of a specification that is approved democratically by the academic council or administration and, after that, it is inserted into the cloud service as the rules for decision making; (3) the decision with the highest rating among all proposals is generated by the cloud cyber service of management based on the digital metric, approved by the social group, for measurement of the input data of the process or phenomenon. The additional components F and S mean the filtering of input request data and signature of the social group head approving the decision.

It is important to adopt new rules of the game: the metric for decision making is subject to voting—not the decision itself, which is generated by the cloud service. The winner is defined by the cloud service as the manager, scientist, or professor with the best metric value. Each member of the team has equal rights to moral and material rewards according to his activity, independent of subjective leader opinion. Within a short time, the new relationship between the cloud service and the team

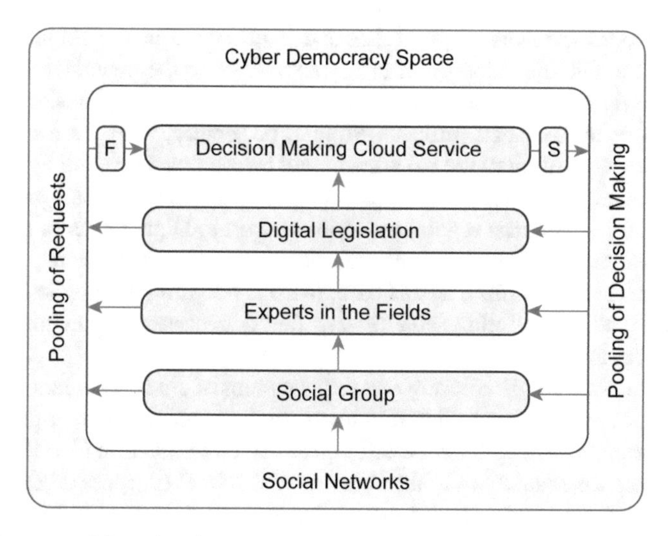

Fig. 9.3 Structure of the cyber democracy

leads to a change in the structure of the university from the ratio of 20%/80% (active/passive employees) to the ratio of 80%/20% that characterizes private companies. The new relationship makes the university attractive to the outside world, attracting a flow of investment and enrollees from highly developed countries. The image of the university is gradually transformed with a significant ranking in the world market of scientific and educational services, which increases the level of the employees' salaries at least two times.

The principles of a cyber democracy consistent with the structure of the proposed system are as follows: (1) the initiation of the cyber democracy is fulfilled by a leader or social group; (2) the choice of experts is made through metric evaluation of the competence matrices of candidates, and it is possible to withdraw experts, laws, regulations, orders, and decisions in the case of their negative impact on the functioning of the social system; (3) digitization of all physical and virtual social processes and phenomena—and the relationships between them—in the cyber physical (cyber social) space is a necessary condition, and creation of competence metrics in the cyber physical space for measurement of all social processes, events, and the relationships between them is necessary; (4) a cloud cyber service based on a digitized cyber physical space of social–technological relationships provides comprehensive digital monitoring of public opinion and also metric evaluation of processes and phenomena in order to generate management regulations; (5) data storage (concerning processes, phenomena, and relationships) and regulatory actions (concerning decisions, orders, regulations, statutes, and laws) take place in the intellectual containers of big data in cyberspace; (6) transparency of monitoring and management of all processes and phenomena for all members of the social group is required; and (7) digital modeling, simulation, and forecasting [11] of possible consequences of processes and phenomena in the social system are necessary to define the response of the cyber service to the generated decisions.

The innovations of cyber democracy are the following: decision making is performed not by the head (officer), experts, or society, but by a corruption-free cloud service, according to the rules generated by experts delegated by the society. Cyber democracy operates in real time, allowing monitoring of all processes and phenomena for generating regulatory actions on time. Digitization of all members in social groups is possible if each individual has a mobile phone, e-mail, and digital signature (for adults). In this case, there is a legitimate interactive relationship between state institutions and citizens that is necessary for monitoring, measurement, and management.

A cloud service for monitoring and management of student careers is described below. The innovation is creation of a mathematical model of career growth in the form of an equation describing the difference-driven interaction [16, 18, 21, 22] of three competence matrices: (1) the future social role P (purpose); (2) the current achievements C (current); (3) and the current activity A (activity). The equation—PCA—is universal and determines the distance between the three components/matrices: goal–competence–activity by using an *xor* operation. The PCA equation allows three problems to be specified for student activity: (1) the P problem: who do you want to be—the potential reachability of the desired social role; (2) the C problem: how are you clever today—an assessment of the current level of competence; and (3) the A problem: what are you doing to achieve the desired future—a road map in the educational space during the time interval. Thus, the proposed mathematical model describes and assesses all the processes of formation of the individual as a socially significant personality during the time interval. In accordance with the PCA equation, it is simple to create a cloud service for monitoring and management of the process of student education in the form of a computing model that answers the most difficult question—how to achieve the desired future (universities, courses, professors, and companies). The model includes the cloud service, which gives the student recommendations on a gadget: the road map—what, where, and when to study in response to the conditions (syllabus) in the form of matrices for the future social role and current competence. The competence PCA model is scaled to all social, cyber, and technological processes and objects, wherever there exists a need to address any of three abovementioned problems, if two of the three components (goal, road map, status) are known.

9.7 Conclusion

1. A cyber physical system—a smart cyber university—is proposed; it is characterized by use of a digitized space of regulatory rules, exact monitoring and active cyber management of addressable components of scientific and educational processes, automatic generation of regulatory signals, cyber decision

making (independent of head officers) to manage financial and human resources, and elimination of paper documents from scientific and educational processes.
2. The idea of the smart cyber university is formulated to improve the quality of educational services and scientific achievements of the higher education system through creation of a metric system of relations defining the rules for digital monitoring and active cloud cyber management of scientific and educational processes, which makes it possible to eliminate corruption, attract foreign investment, and increase the productivity and living standards of researchers and professors, creating market-demanded products.
3. The essential components of a smart cyber university are defined as the infrastructure, personnel, relationships, management, road map, and resources, which have digital representations in cyberspace to perform scientific and educational processes based on exact monitoring and cloud–mobile management.
4. Innovative cloud–mobile services, implementing the smart cyber university, are offered as the prototype of a global virtual research and education cyberspace and the future of higher education.
5. The structure of cloud monitoring and active cyber management of digitized scientific and educational processes within the IoT culture is represented; it includes the hierarchy of the cloud–fog networks–mobile gadgets, scalable to universities in the higher education system, and eliminates paper flows and dependence on subjective officials.
6. The market feasibility of the SCU service, with a market capitalization on the order of US$1 billion, is determined by the trend toward global penetration of cyber services into state scientific and educational structures, and is aimed at initiating constructive activity of scientists and professors, with the ability to improve the performance of scientific work by at least two times.

References

1. Barnaghi, P., Sheth, A., Singh, V., & Hauswirth, M. (2015). Physical–cyber–social computing: Looking back, looking forward. *IEEE Internet Computing, 19*(3), 7–11.
2. *Proceedings of IEEE SERVICES. BigData Congress CLOUD/ICWS/SCC/MS*, New York City, 2015.
3. Dameri, R. P., & Rosenthal-Sabroux, C. (Eds.). (2014). *Smart city: How to create public and economic value with high technology in urban space*. Cham: Springer.
4. Clohessy, T., Acton, T., & Morgan, L. (2014). Smart city as a service (ScaaS): A future roadmap for E-government smart city cloud computing initiatives. In *IEEE/ACM 7th international conference on utility and cloud computing (UCC)*.
5. Bueno-Delgado, M. V., Pavon-Marino, P., De-Gea-Garcia, A., & Dolon-Garcia, A. (2012). The smart university experience: An NFC-based ubiquitous environment. In *Sixth international conference on innovative mobile and internet services in ubiquitous computing (IMIS)*.
6. Owoc, M., & Marciniak, K. (2013). Knowledge management as foundation of smart university. In *Federated conference on computer science and information systems (FedCSIS)*.
7. Moon, J., Kim, C., & Cho, K. W. (2008). CFD cyber education service using cyber infrastructure for e-science. In *Fourth international conference on networked computing and advanced information management*.

8. Gilani, S. M. M., Ahmed, J., & Abbas, M. A. (2009). Electronic document management: A paperless university model. In *2nd IEEE international conference on computer science and information technology*.
9. Forell T., Milojicic D., Talwar V. (2011) Cloud management: Challenges and opportunities. In *IEEE international symposium on parallel and distributed processing workshops and Phd forum (IPDPSW)*, 2011.
10. Bellini, P., Cenni, D., & Nesi, P. (2015). A knowledge base driven solution for smart cloud management. In *IEEE 8th international conference on CLOUD computing (CLOUD)*.
11. Da Fonseca, N., & Boutaba, R. (2015). *Cloud services, networking, and management*. Hoboken: Wiley-IEEE Press.
12. Hahanov, V., Gharibi, W., Zhalilo, A., & Litvinova, E. (2015). Cloud-driven traffic control: Formal modeling and technical realization. In *4th Mediterranean conference on embedded computing (MECO)*.
13. Kalvet, T. (2009). Management of technology: The case of e-voting in Estonia. In *International conference on computer technology and development*, vol 2.
14. Priyogi, B., Nan Cenka, B. A., Paramartha, A. A. G. Y., Rubhasy, A. (2014). Work in progress—Open education metric (OEM) developing metric to measure open education service quality. In *1st international conference on information technology, computer and electrical engineering (ICITACEE)*.
15. Tsay, D., & Matthews, E. P. (2005). Metrics for comparative analysis of operations competency. *Bell Labs Technical Journal, 10*(1), 175–179.
16. Jani, H. M. (2011). Intellectual capacity building in higher education: Quality assurance and management. In *5th international conference on new trends in information science and service science (NISS)*, vol 2.
17. Zhang, L., Wu, C., Li, Z., Guo, C., Chen, M., & Lau, F. C. M. (2013). Moving big data to the cloud: An online cost-minimizing approach. *IEEE Journal on Selected Areas in Communications, 31*(12), 2710–2721.
18. Tantatsanawong, P., Kawtrakul, A., & Lertwipatrakul, W. (2011). Enabling future education with smart services. In *Annual SRII global conference (SRII)*.
19. Bondarenko, M. F., Hahanov, V. I., & Litvinova, E. I. (2012). Logical associative multiprocessor structure. *Automation and Remote Control, 73*, 1648–1666.
20. Hahanov, V., Litvinova, E., Gharibi, W., & Chumachenko, S. (2015). Big data driven cyber analytic system. In *IEEE international congress on big data*, New York City.
21. Hahanov, V., Mischenko, A., Chumachenko, S., & Zaichenko, S. (2014). Cyber services of active management of an university. *Radioelectronics and Informatics, 4*, 56–61.
22. Hahanov, V., Chumachenko, S., Hahanova, A., Mishchenko, A., Hussein, M. A. A., & Filippenko, I. (2015). CyUni service—Smart cyber university. In *2015 I.E. east–west design & test symposium (EWDTS)*, Batumi.

Chapter 10
Transportation Computing: "Cloud Traffic Control"

Vladimir Hahanov, Artur Ziarmand, and Svetlana Chumachenko

10.1 Introduction

Integration of interdisciplinary research yields the most significant market-focused discoveries and innovations, and also implementation of the achievements of a technological culture in related disciplines. Computing models are increasingly being used for monitoring and management of processes and phenomena in all areas of human activity and nature [1–4]. The integration of technical, natural, biosocial cultures and cyber technological solutions for monitoring and control often leads to an original system (cyber physical, bioinformational) of scientific results and innovations in traditionally conservative areas of knowledge, such as nature, biology, sociology, ecology, transport, technology, and industry. Proof of this can serve the global fashion in technology, using a scalable computing model [2, 3]: (1) cyber physical systems; (2) the Internet of Things and Everything; (3) web, cloud, mobile, service, network, automotive, big data, and quantum computing; (4) Internet-driven smart infrastructures: enterprises, universities, cities, and government; and (5) cybersecurity computing.

The concept of computing has evolved from a classical automaton computing model that combines control and execution mechanisms, monitoring and actuation signals, inputs of instructions and data, and outputs of the system status and results. Computing is the process of achieving a goal through use of control and execution mechanisms in a closed-cycle system with given inputs and outputs, monitoring, and actuation signals. In a narrow sense, computing is an activity focused on research, design, and application of intelligent software and hardware systems and networks for monitoring and control of cyber physical processes and

---

V. Hahanov (✉) • A. Ziarmand • S. Chumachenko
Design Automation Department, Kharkov National University of Radio Electronics, Ukraine, Nauka Avenue 14, Kharkov 61166, Ukraine
e-mail: hahanov@icloud.com; artziarmand@gmail.com; svetachumachenko@icloud.com

V. Hahanov, *Cyber Physical Computing for IoT-driven Services*,
DOI 10.1007/978-3-319-54825-8_10

phenomena. The area of computing covers computer engineering and management, software engineering and artificial intelligence, computer science, information systems, and technologies.

As a result of human interaction with the cyber physical world, a new concept of cyberculture is being formed [5]. It reflects the level of social and technological relations between society, the physical world, and cyberspace, defined by implementation of Internet services for precise digital monitoring and reliable metric control in all processes and areas of human activity, including education, science, production, and transport, in order to improve the quality of life of people and preserve the planet's ecosystems. Today, Google cars drive about 10,000 miles a week on an innovative urban road infrastructure. More than 10.2 million smart clothing items will be produced by 2020. By that year, the market for RFID tags for digital identification of objects and processes will have significantly increased from US$11.1 billion to US$21.9 billion. As a result, the financial impact of IoT in 2025 on the world market will amount to US$11 trillion, and the levels of IoT capitalization will be US$4.6 trillion and US$14.4 trillion in the public and private sectors, respectively. Susan Galer of *Forbes* believes that the cloud service will be the most interesting model of IT business in the next 10 years [6]. By the end of 2017, two thirds of the Forbes Global 2000 companies will have transformed their activities with digital processes for monitoring and control.

Computing is a technology of interactive monitoring and control of processes and phenomena in order to achieve a goal using a given program (free of direct human actions). The basis of computing is a transactional interaction of addressable data (memory components) to achieve the goal. The computing model is the following: <memory, address, transaction>. Memory is a substance that can store data (information). The address is the substance that determines the structure by the component's coordinates in virtual or real space. The transaction is a purposeful process of data reception and transmission between the addressable components of the structured memory to implement the functionality or service.

Virtual cyber computing manipulates data instead of memory; it is distributed in cyberspace and presented by fashionable structural directions: (1) cloud computing; (2) fog networking; (3) mobile computing; (4) service computing (web services); (5) social computing; (6) automotive computing; (7) Internet computing—smart everything; (8) cyber physical systems or the Internet of Things (embedded microsystems); (9) big data computing; and (10) quantum computing (security). Computing is described by an automaton structure of controlled physical and virtual processes, where the main component is memory as any substance that can store information, on which the structural components of computing (control and execution) are formed. Operation and development of the system are carried out through monitoring (M signals) and control/actuation (A signals) by the components of the system. The inputs and outputs of the system connect with a single external virtual–physical (virical) space (a virtual–physical continuum), as Josh Linkner of *Forbes* has described.

10.2 Goal and Objectives of This Investigation

The status of the transport system of the planet depends on the following components: (1) the human and his skills as the main control unit; (2) the level of technological implementation of the road infrastructure [7]; and (3) the production quality of vehicles. In this triad of components, the driver plays a major role. Safety depends on road users, and the driver is also a major culprit in virtually all road incidents on land, in the water, and in the air. So it is proposed to successively reduce—and eventually eliminate—the role of the vehicle driver as the most unreliable link in the transport system.

The goal of this investigation is to improve the quality of people's life in the process of their movement and preserve the ecology of the planet through creation of an innovative human-free transport system for monitoring and control of moving objects due to integration of the following components into a unified transport system of the planet: (1) a digitized road geoinfrastructure; (2) smart and intelligent vehicles with precise digital global positioning and authentication in space; and (3) cloud services for digital monitoring and precise human-free cyber control of each vehicle.

The basic principles of such system are (1) interactive communication of cloud services and the car (vehicle) to monitor and control it on the route; (2) gradual exclusion of the driver from the driving cycle, to be turned into a passenger, defining the route for the movement; (3) removal of vehicle license plates to save the considerable material resources involved in their production and use, through global digital identification of the vehicle; (4) gradual removal of all traffic lights in the road infrastructure through creation of an accurate cloud infrastructure, which is able to adequately reflect the real transport infrastructure of the planet; and (5) gradual removal of road signs, to be transferred into the cloud infrastructure to make the planet greener and allow considerable reductions in the costs of maintaining traffic lights and road signs.

The objectives are to create a cyber physical system in the form of an intelligent cloud for traffic management in real time (a cyber physical system–smart cloud traffic control), based on development of a cloud traffic infrastructure, integrated with virtual traffic lights and road signs, with mobile tools for vehicle identification in order to improve the quality and safety of vehicle movement, minimizing the time and costs involved in the realization of the specified routes.

This innovative proposal is the following: smart cloud traffic control is designed to include the time parameter in a digital map of the world, and also gradual transfer of road signs and traffic lights to the cloud, allowing the traffic infrastructure on the ground to be made more green and creating the potential for savings in the thousands of tons of metal used for manufacture of traffic lights and unnecessary vehicle license plates, the millions of kilowatt hours of electricity used to maintain the operation of the system, and the millions of dollars spent on installation and operation of traffic lights and road signs, as well as reducing the time needed for installation and updating of traffic lights in the city's virtual infrastructure from a few days to several minutes.

10.3 Global Cyber Physical System for Digital Monitoring and Cloud Traffic Control

Figure 10.1 shows the cyber physical system for digital monitoring and cloud traffic control, which is implemented in the Internet of Things technology. The innovative feature is determined by the absence of inputs and outputs, which are identified with big data. This means that the input data are associated with the information arrays of cyber physical space. The data analysis result provided by the CPS cloud services is obligatorily written into the specialized arrays of big data. This makes it possible to use specialized big data and services for each user, located inside the vehicle. Naturally, the cyber physical system has finite monitors and smart sensors located in the physical space of the planet. These monitors are modern vehicle computers and drivers' or travelers' own gadgets. At the same time, the above tools are actuators too, which control the routing of vehicles; in the future, driving will be human free, whereas today the human performs the role of the driver. The compulsory condition of human-free movement is global positioning of the vehicle with a digital ID or any other gadget in the digitized physical space—represented, for example, by a digital Google map with precision of up to 10 cm.

The advantage of this system is minimization of traffic police across the planet, who currently perform the roles of registration and analysis of traffic accidents. Global digital cyber physical monitoring of vehicles in time and space allows us to

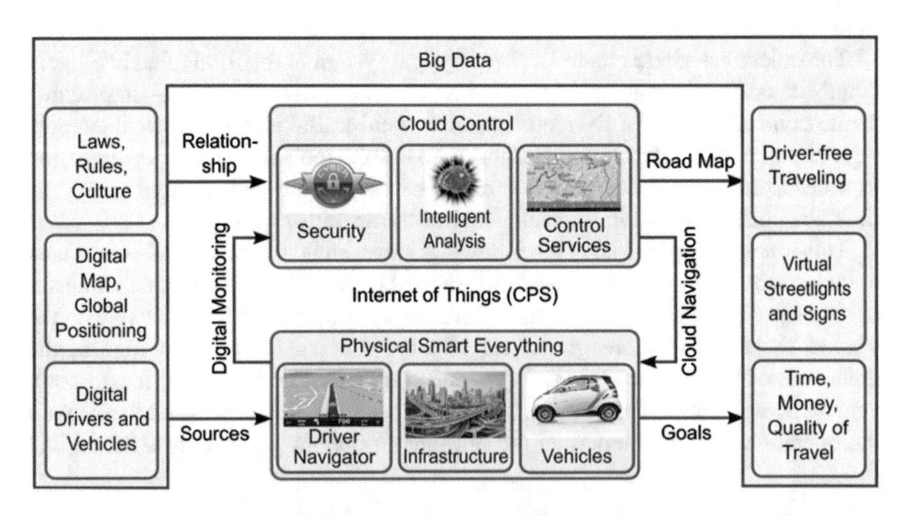

Fig. 10.1 Global cyber physical system for digital monitoring and cloud control of traffic

register all movement of the vehicle and fix all disorders associated with excessive speed, movement through red traffic lights, intersection of lanes, and also all emergency situations, analyzed online by the corresponding cloud services with generation of regulations for insurance companies, banks, and other organizations related to participation in the investigation.

10.4 Register-Based Structure of Smart Car–Driven Traffic Lights

The register structure of smart traffic lights is characterized by optimal control of signals, which depends on the number of vehicles in the movement lanes at the intersection. Each movement lane consists of parallel nonintersecting flows of vehicles directed toward each other. These flows form the states of the register, where each bit is associated with a car on an intersection lane; when forming the traffic light signal the preference is referred to the register, which has the maximum number of unit values. Taking into account that the register may have zero and unit digits in each control cycle, a left shift of register bits is performed. This operation eliminates the arithmetic instructions of multiplication, addition, division, and subtraction from the control process of the smart traffic light, which considerably simplifies virtual (physical) implementation of the traffic light system and increases the speed of its work.

To shift data in one clock cycle, we use an original patented solution [8]. A circuit for computing the control signal as a game of two transport flows is presented in Fig. 10.2. There are two flows, *A* and *B*, which inherently conflict (contradict each other). Input m defines artificially created preferences that should take place in the urban infrastructure. By default, these preferences are determined by zero signals of the register m. The procedure of traffic light control contains three steps: (1) shift left crowding $A = \text{SLC}(A)$, $B = \text{SLC}(B)$; (2) winner lane definition $a = (A \,\&\, B) \oplus A$, $b = (A \,\&\, B) \oplus B$; and (3) forming of streetlight signals $Q^a = V_n^{i=1} a_i$, $Q^b = V_n^{i=1} b_i$, $i = \overline{1, n}$, where *n* is the number of register bits (the number of cars in all nonintersected lanes).

Register variables (*a b*), designated as vectors of compressed-to-the-left unit values, are united and inverted for simultaneous implementation of the *xor* operation. The results in the form of states of the registers are inputs of two *or* gates,

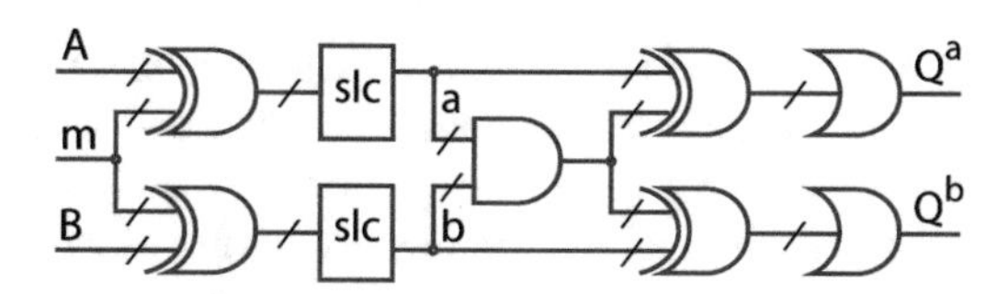

Fig. 10.2 Circuit control of a crossroad

which form the states of two Boolean variables to create three combinations: 00, 01, and 10. The unit value (1) of one of the two variables is the best solution we have to select. Two zero states mean that both solutions are equivalent in the level of preference, the choice in this case being given to flow A. A unit combination of Boolean variables is impossible.

Thus, precise control of smart traffic lights according to traffic flows at the intersection can and should be carried out on the basis of only the logical operations *and*, *or*, *not*, *xor*, and *slc* without use of arithmetic functions to allow designing of high-speed vector logical physical and/or virtual multiprocessors in order to significantly reduce the execution time required for cloud-based crossroad control.

In the case of control of a complicated intersection, where n conflicting traffic flows exist, a more complex control circuit is used, which is realized in the form of a vector logic processor, where the input register m forms a preference function, which has a unit default value for all digits in the register. At the outputs of the control system of traffic light signals there can be a single unit value, permitting traffic flow with the maximum number of cars. Setting a traffic signal in the unit value is a compromise solution that depends not only on the number of vehicles in the movement lane but also on the number of vehicles in the other lanes crossing the intersection.

10.5 Cyber Physical Monitoring of Transport—Technical Conditions

The quality of transport services depends on the reliability and satisfactory operation of all its components. Therefore, a remote online mode to obtain information about the condition and positioning of each car is one of the essential services for trouble-free operation of a cyber physical transport system constructed with Internet of Things technologies.

The innovative idea of an intelligent cyber physical transport system (ICTS) is not informed after the accident fact by using contents of black boxes; it needs to prevent collisions on the basis of continuous monitoring. In order to ensure the timeliness and accuracy of parametric information delivery, it is necessary to diversify telecommunication channels of data from each vehicle to the control cloud.

It is necessary to use the channels of satellite and differential positioning systems (i.e., GPS or GLONASS), mobile channels, and telematics, including online Wi-Fi communication of cars with each other and with the road infrastructure transponders (wireless protocol Wi-Fi, 802.15.4, BT, Lora, cellular).

Thus, ICTS allows remote monitoring of the car's parameters and the transport route, performing diagnostics, simple repairs, and emergency control of a vehicle in critical situations.

For the implementation of the abovementioned services within ICTS, computing components are leveraged as onboard vehicle equipment, specialized applications, transceivers, and electronic gadgets, including mobile gadgets [9].

On the side of the cloud control and monitoring there should be a fog network of transponders mounted in the road infrastructure for telemetry data collection and transmission to the local and/or global big data centers. Services of traffic control clouds perform data analysis on each vehicle and local traffic situations with subsequent sending of actuation signals to drivers and traffic lights.

The cyber physical infrastructure for exact transport monitoring and control includes each vehicle's cloud cabinet [10], interacting with the telematics module satellite navigation and communications, the driver, and the onboard systems for monitoring, data storage and transmission, and automatic vehicle identification.

An extremely important innovation is an anonymous delegate of the vehicle's positioning and routing to the cloud service for monitoring and traffic control. This makes it possible to optimize traffic flows on the local and global scales by issuing recommendations for moving through problem traffic areas when the route is booked in advance. The driver's knowledge of traffic conditions on the movement route would eliminate emergency situations on the basis of accurate predictions of traffic signals and potential collisions with cars emerging from obscured crossings. This vital information should be supplied to the vehicle screen from the cloud service in order to warn the driver of a possible collision, while maintaining the same motion parameters.

Cloud microservices of the transport enterprise include functionality: protection, storage, and analysis of data; vehicle identification; monitoring and diagnosis of technical conditions; prediction of the operational parameters of the vehicle components [11]; the operational control of the vehicles [12]; and monitoring of the impact of transport on the environment and decision making to prevent its contamination.

Thus, monitoring of vehicle positioning and movement in transport companies and their subsequent cloud management enable us to optimize the execution of orders for transportation of passengers and cargo, significantly reducing material and/or time consumption. It should be noted that the monitoring parameters of the technical condition of vehicles in the framework of cloud services enables transport companies to eliminate accidents in the operation of road transport, and thus to significantly increase revenues.

10.6 Formal Model of a Traffic System

The goal is to improve the quality and safety of traffic by creating a cyber physical system—an intelligent traffic control system (TCS)—which provides monitoring and control in real time [13–15], based on the use of vehicle mobile gadgets and cloud (virtual) traffic lights, making it possible to improve the quality of life of the driver, minimize the time and costs involved in the organization of traffic, and

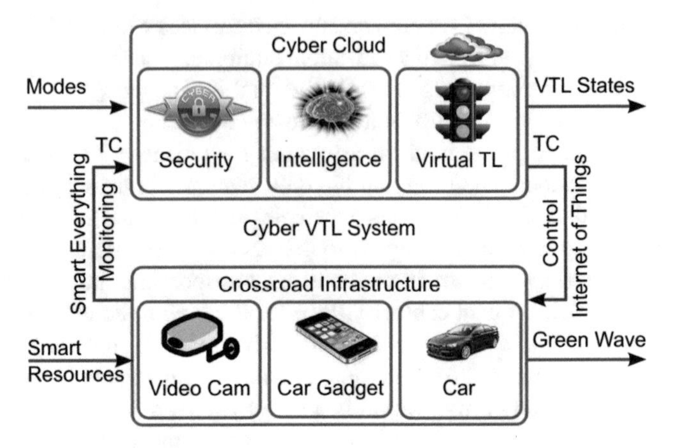

Fig. 10.3 Computer control of a crossroad

generate innovative solutions for social, human, economic, and environmental problems. The research object is cloud cyber system technology for monitoring and management of vehicles through use of virtual traffic lights and road signs, and mobile gadgets for identification, radar, and navigation of road users.

The subject of research includes traffic flows, the virtual infrastructure of traffic messages, and also software–hardware mobile systems for identification, monitoring, and management of traffic through the use of virtual traffic lights. The main point of the research is creation of a CPS in the form of smart cloud traffic control in real time on the basis of development of a cloud traffic infrastructure (Fig. 10.3), integrated with virtual traffic lights (VTLs) and road signs, and mobile tools for vehicle identification in order to improve the quality and safety of vehicle movement, minimizing time and costs when traveling on the specified routes.

Innovative Proposal: Smart cloud traffic control is aimed at introducing the time parameter into the digital map of the planet, as well as a gradual transfer of traffic signs and traffic lights to the cloud, to radically improve traffic infrastructure and create the potential for saving thousands of tons of metal currently needed for the manufacture of traffic lights and unnecessary vehicle license plates, millions of kilowatt hours of electricity, and millions of dollars for installation of traffic lights, road signs, and operating costs, as well as reducing the time needed for installation and updating of traffic lights in the virtual infrastructure of cities from a few days to a few minutes. The formal cyber system model is represented as two cloud components or engines: f is monitoring and management, and g is the executive infrastructural engines, which are interconnected by signals for monitoring, management, and initiation of both components for implementing the services. An analytical form [16] for describing the TCS and its structural equivalent is represented in Fig. 10.4.

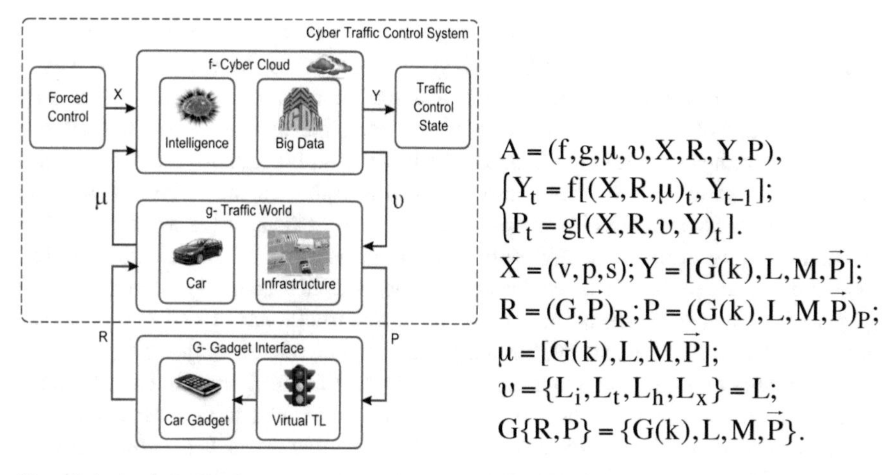

Fig. 10.4 Analytical and automaton form for describing the traffic control system

In the model there are the following elements $A=(f, g, \mu, \upsilon, X, R, Y, P)$—correspondingly, control and execution units, monitoring and control signals, inputs of control tasks and executive resources, and outputs for indicating the status of the algorithm of task execution and providing a service. Here also there are signals for external traffic management $X = (v, p, s)$ for regulating movement of government persons, police cars, and special purpose vehicles, respectively.

The transport control signals $\upsilon=(L_i, L_t, L_h, L_x)=L$ use a virtual traffic light operating in the following modes: (1) an intellectual one, functionally dependent on road conditions; (2) an automatic one with fixed switching periods; (3) a virtual manual mode based on digital monitoring of intersections on a police computer screen—its analog is air traffic control using the monitor of the flying control officer at the airport; and (4) an emergency stop $L_x \in \upsilon$ of a vehicle upon a digital request from the police, which is visualized on the screen of the car gadget. The following objects are subject to cloud monitoring: $\mu = \left[G(k), L, M, \vec{P}\right]$—all mobile gadgets in the cars with their coordinates, and the status of traffic lights tied to terrain map M, as well as execution for ordering traffic routes.

Transfer of the traffic light $L \in \{\upsilon, P\}$ from the real crossroad to the cloud one completes the creation of a virtual infrastructure on the planet, forming a closed loop of a monitoring and control system involving a single real component in the form of the mobile gadget $G\{R, P\} = \left\{G(k), L, M, \vec{P}\right\}$ of the road user (RU). The gadget realizes the interface function for communication with the cloud: R is entry to the cloud—the order of service $R = \left(G, \vec{P}\right)_R$ (delegation of the ID gadget and the movement path to the cloud); and P is exit from the cloud—obtaining service $P = \left\{M, G(k), \vec{P}, L\right\}$ (the map, the gadget coordinates, the best route, traffic lights).

The user receives a service for scrolling the map window and traffic lights in real time on the way $\vec{P}$ if he or she has delegated a gadget to the cloud. If the way is ordered, a user additionally receives a quasioptimal route of travel and priority traffic signals. Actually, in the view of the user, the system is created by two components: the cloud and the gadget. The novelty and originality of the proposed system is in providing the cloud service—traffic lights on the screen of the road user's gadget. Everything else (maps, routes) already exists and is working. Implementation of the proposed TCS will be through creation of virtual traffic lights, which duplicate real ones in synchronous mode, and then gradual elimination of all physical devices and signs of the ground road infrastructure as drivers purchase new technological culture in an evolutionary way. Moreover, all major cities already have virtually centralized computing (cloud) controlled traffic lights. Therefore, the transfer of traffic lights to the cloud will not be associated with substantial additional costs—rather, the opposite—the large operating costs for maintenance of traffic lights and signs in the urban infrastructure in working condition are transformed to zero. The mobile vehicle gadget G is the main control unit for the TCS, as well as being a major consumer of traffic light signals L of car motion control, displayed on the windshield:

$$L = \{L_i, L_t, L_h, L_X\} = F(L, G, V, T, D, \vec{P}).$$

where V is special control signals; T is a programmable cycle of autonomous control of traffic lights; D is accumulated intellectual statistics on traffic lights (in the street or district), including taking into account the time of year and day; and $\vec{P}$ is incoming orders on traffic routes. Creating a virtual system of traffic lights involves practically no financial, time, material, or energy costs for adding new traffic lights in the virtual space by programming, as well as for removing them from the cloud during modernization of the infrastructure. Visualization of traffic lights on the windshield (mobile monitor) and voice dubbing will improve the quality and safety of traffic, reducing emergency situations for the driver and the urban infrastructure in general. The cloud traffic light, as a digital signal—unlike the analog perception by the driver of real traffic lights—is a more reliable tool for managing the vehicle, including for subsequent introduction of an autopilot in traffic that perceives only deterministic control signals.

Road users are identified in the cloud by means of a gadget or cellphone, which are matched when he or she gets in the car. User status increases with passage through traffic lights if the route is ordered in advance. Other road users (pedestrians, motorcyclists, and cyclists) also have the right to order a route, raising their status for the use of traffic lights. Pedestrians are able to get a service for ordering a combined route, including all kinds of ground and underground transport (bus, metro).

For the vehicle, a control system forms the criterion for evaluating the quality of service, which depends on the following variables (time, route length, and route quality): $Q = \min f(T, \vec{P}, K)$. For traffic lights, a control system uses the formula, minimizing the total downtime of vehicles during the day (Z is the switching cycle of the traffic light):

$$Q = \min \frac{1}{n} \sum_{i=1}^{n} \left[\frac{T_i(\vec{P}_i, V_i, L_i, J_i)}{Z(\vec{P}_i, V_i, J_i)} \right]^{-1}.$$

where the numerator and the denominator show the functional dependence of downtime and the cycle on the parameters is shown in parentheses. If the result is a quality score close to unity, $Q = 1$, then the intersection operates normally. Otherwise, it is necessary to modify the streetlight switching cycle or reconstruct the intersection.

For the infrastructure of a city or a closed area, the traffic cloud control system forms a function that optimizes the criterion of the service quality for vehicles over a time period (hour or day). The criterion depends on the total time of cars driving along the ordered routes and the recommended speed, and the idling time of cars at traffic lights and in traffic jams related to the ideally traveled routes at the permitted speed without delays at traffic lights and in traffic jams:

$$Q = \min \frac{1}{n} \sum_{i=1}^{n} \left[\frac{T_i(\vec{P}_i, V_i, L_i, J_i)}{T_i(\vec{P}_i, V_i)} \right]^{-1}.$$

For the city's streets, the traffic cloud control system forms a function that optimizes the criterion of the total driving time from the beginning to the end of the street for the time interval (hour or day):

$$Q = \min \frac{1}{n} \sum_{i=1}^{n} \left[\frac{T_i(V_i, L_i, J_i)}{T_i(V_i)} \right]^{-1}.$$

The quality of the road infrastructure is determined by the integral criterion of the effectiveness of the cloud monitoring and TCS, consisting of the following conflicting parameters: L is the level of erroneous decisions leading to a traffic collision; Y is the quality of service that reduces to zero the time of forced downtimes of the vehicles T. The criterion also takes into account the cost level of infrastructural complexity H^a and operational (driving cost) complexity H^s, creating mechanisms for monitoring and control traffic flows [1]:

$$\begin{aligned} E &= F(L, T, H) = \min \left[\frac{1}{3}(L + T + H) \right], \\ Y &= (1 - P)^n; \\ L &= 1 - Y^{(1-k)} = 1 - (1 - P)^{n(1-k)}; \\ T &= \frac{(1-k) \times H^s}{H^s + H^a}; H = \frac{H^a}{H^s + H^a}. \end{aligned}$$

Here, k is the passability (controllability + observability) [17] of the infrastructure, and P is the probability of collisions; n is the number of hidden infrastructure errors. The time required for the passage of the transport stream through the infrastructure fragment is determined by k, multiplied by the basic expenses of the road operation related to the total sum $H^s + H^a$, where the parameters are called basic (when driving) and additional (infrastructural) expenses. Parameter H represents the cost of infrastructure in the total transportation system expenses. So, the optimal smart road infrastructure ensures nonstop car driving on the ordered path at the permitted speed.

10.7 Technology of GNSS Navigation

Navigation is defining the location of a moving object in a given coordinate system, as well as the velocity vector and angular orientation, with accurate reference of time to the navigation parameters.

Since the functions of the cyber system involve not only monitoring of vehicles but also management of their movement, the requirements for the accuracy and reliability of navigation operations must meet the current and future requirements of public navigation plans, as well as the requirements for critical control of objects in typical and difficult road conditions. In this case, the signals of the global navigation satellite systems (GNSS) are used for control—GPS (USA), GLONASS (Russia), BeiDou/Compass (China), and Galileo (EU)—and also their regional functional complements (XPD)—WAAS (USA), EGNOS (European Union), MSAS (Japan), QZSS (Japan), and GAGAN (India) [18]. As an analysis of the work shows, the main trends in the navigation market in the world are determined by use of satellite navigation technologies in transport, especially on roads.

A good example of the use of satellite technology is the project for emergency response on roads known as ERA-GLONASS (Russian Federation). The prospects for satellite navigation technologies lie in high-precision transport navigation, which will be available in 5 years, due to the differential technologies RTK (Real Time Kinematic) and PPP-RTK for determining the trajectory of the object's movement with centimeter (millimeter) accuracy. Methods based on wide area augmentation—WAAS (USA), EGNOS (European Union), MSAS (Japan), QZSS (Japan), and GAGAN (India)—allow creation of wide area differential systems (WADGNSS). Broadcasting of differential corrections from satellites improves the accuracy of determining the plane coordinates up to 1.0 m. The classical differential method for correcting errors of GNSS measurements by using differential corrections from the individual GNSS stations allows an increase in the accuracy of determining the plane coordinates to the level of 0.5–0.8 meters with less reliable navigation definitions compared with the method of wide area navigation.

The navigation equipment of the vehicle (the radio navigation receiver and radio modem) should provide (1) determination of the object coordinates with an accuracy of 0.5–1.0 m on the plane and 1.0–2.0 m in height; (2) determination of the velocity vector components with an accuracy of 0.05–0.10 m/s; (3) determination of the angular orientation parameters of the object with an accuracy of 0.15–0.30 angular degrees or 0.5–1.0 angular degrees of pitch and/or roll; and (4) definition of drift of the onboard time scale with an accuracy of 50–100 ns. When controlling moving objects, the onboard navigation equipment has to provide definition of the coordinates of dynamic objects with an accuracy of 0.05 m (on the plane) and 0.1 m (in height). The above values of the accuracy characteristics (tolerance levels) of navigation definitions correspond to 95% probabilistic confidence intervals. The requirements for the reliability of navigation definitions are (1) accessibility or availability with a probability of 0.999; (2) integrity including an alert limit of 2 m and a time interval up to a message about characteristics out of tolerance (time to alarm) of 5 s; and (3) continuity with a probability of 0.999/h.

To solve these problems with the specified requirements for this project, it is necessary to create a prototype of an integrated onboard high-precision navigation system (GNSS plus an inertial navigation system (INS)) using RTK/WADGNSS/DGNSS network technologies. If we need additional information about the parameters of the angular orientation of objects, it is possible to use GNSS technology for determining the angular parameters in conjunction with the inertial subsystem. For example, one of the world leaders—NovAtel implementation of RTK technology—requires use of telecommunication tools (mobile Internet, GSM/GPRS) for obtaining differential corrections from a network of reference stations. Such a network already exists in Ukraine and provides almost complete coverage of its territory.

The navigation technology WADGNSS receives corrections from consumers via both geostationary relay satellites on the L1 GPS frequency and the Internet in terms of urban developments and high-rise buildings with masking signals from the relay satellite corrections. Integration of GNSS + INS allows exclusion of loss of navigational information in areas where there are no tracking signals of navigation satellites—under bridges, in tunnels, in urban "canyons"—and implements instant tracking recovery of GNSS navigation signals. GNSS sensors provide the initial conditions (the current position and velocity components of the object) for the inertial subsystems. If RTK definitions are supported, the integration of GNSS + INS allows provision of the required characteristics for the reliability of navigation definitions (availability, integrity, and continuity). At the present stage, RTK technology does not separately provide the necessary reliability of navigation definitions for cyber management problems. An INS device may include a system of sensors–accelerometers and sensors–gyroscopes in various configurations to determine location and angular orientation parameters. In recent years, for transport applications, inexpensive inertial microelectromechanical systems (MEMS) sensors have been used, which are integrated with GNSS OEM modules.

Thus, the best variant for building an onboard subsystem of navigation definitions for transport monitoring and control is the following: (1) integration of GNSS + INS based on MEMS technology (coordinate and, if necessary, angular definitions); (2) a multisystem approach, simultaneously receiving signals from all four GNSS + WADGNSS correctional signals; (3) RTK/WADGNSS/DGNSS technology for precise positioning/navigation; and (4) executability of angular definitions by GNSS signals with INS integration. There exists equipment for implementing these functions with a given quality; its current price is US $15,000–20,000. Within 5 years we can expect a reduction in price to ~US$1000 with mass production.

10.8 Conclusion

The obtained scientific and practical results provide an opportunity to significantly improve the comfort of movement for drivers and passengers, save billions of dollars in materials and time, and also preserve the ecology of the planet for future generations. The advantages of precise digital monitoring and cloud traffic control are global and local in nature, and are as follows:

1. Preservation of global and regional ecology by reducing pollution, improving the quality of life of drivers, saving fuel and reducing travel time by choosing the optimal route, and reducing the volume and complexity of traffic by using intelligent infrastructure and smart traffic lights, with combinations of technologies related to satellite navigation of cars, cloud monitoring and control services, and intelligent traffic lights provide new opportunities to create a clean planet (Fig. 10.5)
2. Complete elimination of traffic police through identification of vehicles that violate traffic rules, and also precise monitoring of vehicle positioning in time and space to avoid theft, collisions, and travel by unauthorized routes; remote control, via cloud emergency services, to turn off the car engine when it creates a real danger to other road users; significant reductions in accidents through monitoring and definition of safe maneuvers, reducing the consequences of road traffic accidents and increasing the safety and comfort of road users
3. Monitoring of the positioning and movement of vehicles in the transportation companies; optimal execution of orders for carriage of passengers and goods in order to minimize material and/or time costs
4. Provision of services to the driver associated with online generation of the best routes and the timetable, taking into account the negative factors on the road in the existing infrastructure to minimize material and time costs; significant reductions in accidents through digital monitoring of parts of the road with a limited view and definition of safe maneuvers; cloud and smart physical infrastructure interactions to optimize decision making, as shown in Fig. 10.6
5. Provision of services to passengers for monitoring of the positioning and movement of vehicles through stop points or transport terminals through the use of stationary monitors and mobile gadgets associated with corresponding cloud services; visualization of critical points on the route on car monitors in real time through use of video cameras; creation of a hybrid destination route through use of various kind of vehicles

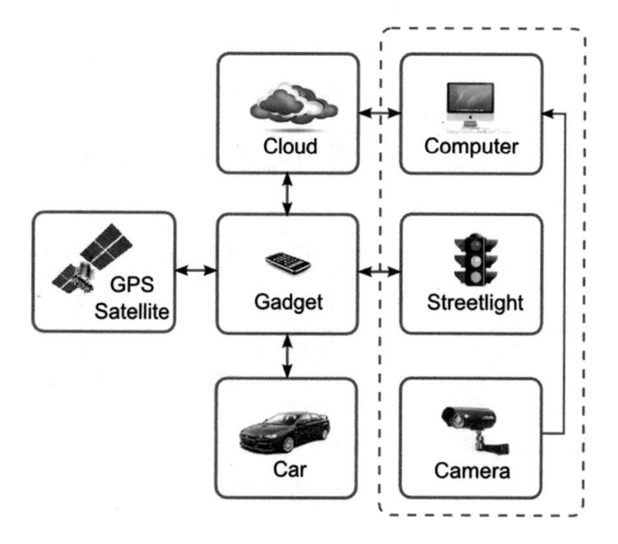

Fig. 10.5 Cyber and physical technology integration

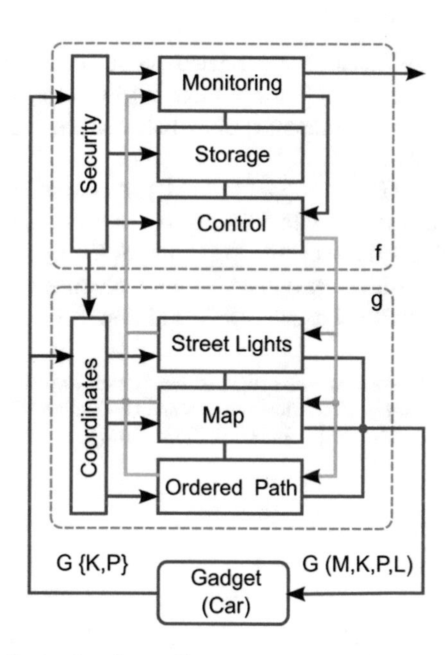

Fig. 10.6 Cloud and infrastructure interactions

6. Automatic and accurate digital registration in time and space of all traffic violations and traffic accidents, resulting in infringement fines being directly debited to the driver's personal account and also forming the driver's history for

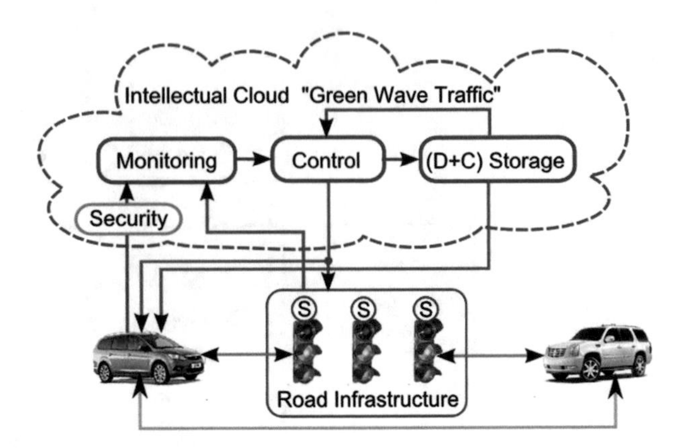

Fig. 10.7 Cloud traffic control system

the insurance company, making it possible to eliminate police participation in the procedures of road accident examination; cloud control, car computers, and an infrastructure of streetlights creating a system for robust and reliable transportation around the world (Fig. 10.7)

Directions for future research are related to the creation of minimal traffic infrastructure, duplication of tools for positioning vehicles, and creation of cloud services for design and telecommunications between the car, the cloud, and the infrastructure components.

The main drawback of the proposed cloud technology is the lengthy process required for total education of humanity in the new driving culture. It involves rejection of the existing system of road signs and traffic lights, the maintenance of which currently costs billions of dollars. The development cost for global cloud traffic control services will not exceed US$10 billion.

Acknowledgments The results of this investigation were obtained with the financial support of a State Grant from the Ministry of Education and Science of Ukraine "Cyber Physical System—Smart Cloud Traffic Control" No. 0115 U-000712 (2015–2017).

References

1. Bondarenko, M. F., Hahanov, V. I., & Litvinova, E. I. (2012). Logical associative multiprocessor structure. *Automation and Remote Control, 73*, 1648–1666.
2. Hahanov, V. I., Gharibi, W., Litvinova, E. I., & Shkil, A. S. (2015). Qubit data structure of computing devices. *Electronic Modeling, 1*, 76–99.
3. Hahanov, V. I., Amer, T. B., Chumachenko, S. V., & Litvinova, E. I. (2015). Qubit technology analysis and diagnosis of digital devices. *Electronic Modeling, 37*(3), 17–40.

4. Hahanov, V., Litvinova, E., Gharibi, W., & Chumachenko, S. (2015). Big data driven cyber analytic system. In *2015 I.E. international congress on big data*, New York.
5. Marr, B. *Forbes. 9 Attractors in the Computer Market. 17 'Internet Of Things' Facts Everyone Should Read.* http://www.forbes.com/sites/bernardmarr/2015/10/27/17-mind-blowing-internet-of-things-facts-everyone-should-read/
6. Galer, S. *Forbes. IDC Releases Top Ten 2016 IT Market Predictions.* http://www.forbes.com/sites/sap/2015/11/05/idc-releases-top-ten-2016-it-market-predictions/
7. Dimitrakopoulos G., Bravos G. (2016) Embedded intelligence in smart cities through urban sustainable mobility-as-a-service: Research achievements and challenges. In *Proceedings of ICOMP'16*
8. Kakurin, N.Ya., Hahanov, V., Loboda, V. G., & Kakurina, A. N. (1988). Patent 1439682 USSR. Shift register. № 4251904/24–24, 07.04.87; published 23.11.88, Bul. № 43.
9. Vlasov, V. M., Nikolaev, A. B., Postolit, A. V., & Prihodko, V. M. (2006). *Information technologies in road transport.* Moscow: Nauka.
10. Osipkov, V., Ksenevich, T., Belousov, B., & Karasev, O. (2016). Intelligent transport systems: Revolutionary threats and evolutionary solutions. SAE technical paper 2016-01-0157.
11. Volkov, V. P., Gritsuk, I. V. et al. (2016). Technical regulations and results of information software system "MonDiaFor HADI-15: Monitoring, diagnosis, forecasting technical condition of the vehicle under ITS". Certificate of copyright number registration 64765, 04.04.2016. Application of 10.02.2016№65240.
12. Volkov, V.P., & Komov, P.B., et al. (2012). Technical regulations software "virtual operator" NADI-12 during normal operation. Certificate of copyright number registration 47230, 01.15.2013. Application from 15.11.2012 №47522.
13. Bazzi A., Zanella A., Masini B.M., Pasolini G. (2014) A distributed algorithm for virtual traffic lights with IEEE 802.11p. In *IEEE European conference networks and communications (EuCNC).*
14. Ferreira, M., & d'Orey, P. M. (2012). On the impact of virtual traffic lights on carbon emissions mitigation. *IEEE Transactions on Intelligent Transportation Systems, 13*(1), 284–295.
15. Conceicao, H., Ferreira, M., & Steenkiste, P. (2013). Virtual traffic lights in partial deployment scenarios. In *IEEE intelligent vehicles symposium (IV).*
16. Hahanov, V. I., Melikyan, V. S., Saatchyan, A. G., & Shakhov, D. V. (2013). "Green wave"—A cloud for traffic monitoring and management. *Armenia, Bulletin Information Technology, Electronics, Radio Engineering, 16*(1), 53–60.
17. Ngene, C. U., & Hahanov, V. (2011). A diagnostic model for detecting functional violation in HDL-code of SoC. In *Proceedings of IEEE east–west design and test symposium.*
18. Diggelen V., & Tan, K. F. (2014). Interchangeability accomplished. Tri-band multi-constellation GNSS in smartphones and tablets. In *GPS World.*

Chapter 11
Cosmological Computing and the Genome of the Universe

Vladimir Hahanov, Eugenia Litvinova, and Svetlana Chumachenko

11.1 Introduction

The most significant discoveries and innovations are currently being created by the superposition of interdisciplinary research, as well as through the introduction of technological achievements from one culture into other branches of knowledge. Computing models are increasingly recognized to explain physical and cosmological processes, and their use in monitoring and managing phenomena in all areas of human activity and nature [1–4]. The use of an automaton model of computing to describe brain processes leads to accurate monitoring of violations and subsequent efficient recovery of functionality by applying actuation signals [5]. Scientists are making successful efforts in the recognition of computer structures for natural and cosmological solutions [6], which will certainly lead to new knowledge of the microcosm and macrocosm. An interesting association of deep neural networks has been traced to describe cosmological processes, starting from the interaction of elementary particles of atoms [7].

The concept of computing evolves from a classical automaton model that combines control and execution mechanisms, monitoring and actuation signals, inputs for instructions and data, and outputs of the system status and results. Computing is the process of achieving defined goals through leverage of control execution mechanisms in a feedback cycle system with given inputs and outputs, and signal monitoring and actuation. This definition of the computing model is then used to form a coherent cosmological structure of interactive actions between entities and forms of the universe.

V. Hahanov (✉) • E. Litvinova • S. Chumachenko
Design Automation Department, Kharkov National University of Radio Electronics, Ukraine, Nauka Avenue 14, Kharkov 61166, Ukraine
e-mail: hahanov@icloud.com; litvinova_eugenia@icloud.com; svetachumachenko@icloud.com

V. Hahanov, *Cyber Physical Computing for IoT-driven Services*,
DOI 10.1007/978-3-319-54825-8_11

11.2 Cosmological Computing

The goal is the creation of a cosmological computing model of the universe, which explains its development and functioning on the basis of the proposed lower harmonic genome of the interconnected synchronous change of energy, matter, space, and time.

The objectives are the following: (1) to create a harmonic cyclic cosmological model of the universe's computing; (2) to develop a genome algorithm of the universe's functional cycle as a superposition of interrelated harmonic functions; and (3) to determine the structural components of the universe in a synchronous cyclic variation of forms and substances.

The definitions are as follows. The universe U is a strongly connected system of entities and forms, aimed at cyclic synchronous and harmonious change of energy E, matter m, space S, and cosmological time t. The wordings of these four equivalent components of the universe use the addition principle of the universe: $E = m \oplus S \oplus t, m = E \oplus S \oplus t, t = m \oplus E \oplus S, S = m \oplus E \oplus t$. Here are $U = \{V,W\}$, $V = \{m,E\}$, $W = \{S,t\}$. All primitives of the universe m и E, S и t are changed in pairs from 0 (Ø) up to 1 ($U = \{V,W\}$), but their intersections equal the empty set: $m \cap E = \emptyset, S \cap t = \emptyset$, and the union is the relevant universe, equal to 1: $m \cup E = V(1)$, $S \cup t = W(1)$. To set the connected components of the universe changes uses an ($\oplus$) *xor* operation, which serves as the symmetric difference of primitives $\{a, b\}$ in set theory: $a \oplus b = U(1), b \oplus a = U(1), a \oplus a = \emptyset(0), b \oplus b = \emptyset(0)$. It is obvious that this operation for different values of the universe turns into a primitive union operation. The unity and the preservation of the entities V and forms W of the universe in each cycle phase $\{(Ph), (p)\}$ are defined as follows: $E(p) \oplus m(p) = 1$ (V) и $S(p) \oplus t(p) = 1(W)$.

This means energy and matter, time and space, transforming into each other harmonically and synchronously in pairs while maintaining, in every phase of the cycle, the overall balance of zero $[E(p) \oplus m(p)] \oplus S[(p) \oplus t(p)] = 0$. The material and energy nature of V in the space–time form W is strongly interconnected to synchronous change in the cycle.

Energy is the system's ability to change matter, space, and time. Energy is a unstable form of matter existence. Clean energy is the ultimate form of matter existence, laid out in elementary particles (photons) moving inside a singular point at light speed.

Matter is a physical substance that exists in varying forms of energy, space, and time. It is a stable substance storing potential energy. Pure matter is the ultimate form of existence of zero-point energy in maximum space in the absence of time. Energy states of matter in the cycle of the universe are singular: plasma, gas, liquid, and solid.

Time is a cyclic form of matter changes from clean energy to clean space, and vice versa. It is a form of consistent change of the entity. Pure time is a cycle phase point of the universe, corresponding to a pure energy state when the elementary particles of matter move inside a singular point at light speed.

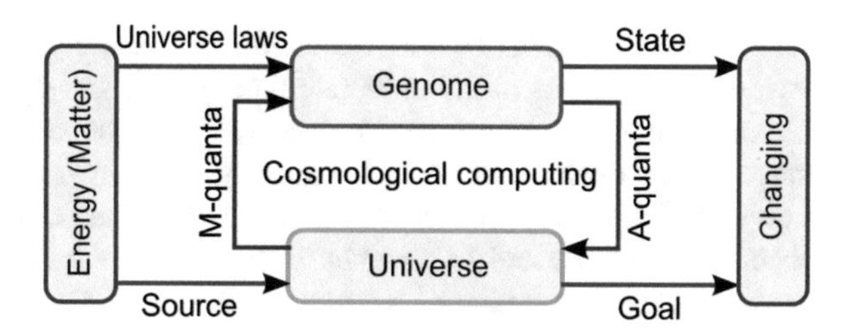

Fig. 11.1 Cosmological computing

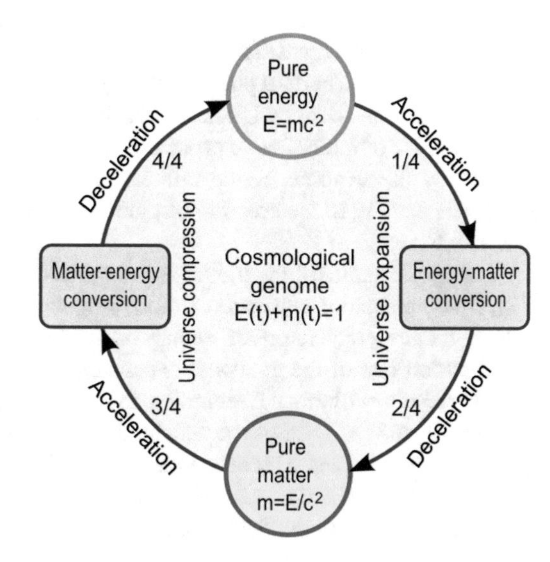

Fig. 11.2 Genome algorithm of the cycle of the universe's change

Space is a form of change in the structure of matter, time, and energy. Space is the form of the parallel changes of nature. Pure space is the maximum amount of matter structure with zero entropy in the complete absence of energy and time [8].

The nature of the universe is the matter–energy substance that synchronously and cyclically changes from pure energy to pure matter.

The form of the universe is the space–time continuum, which synchronously and cyclically changes from pure time to pure space.

The universe is the matter–energy substance in the form of space–time, which synchronizes with cycle phases of its harmonic changes $U = \{E,m,S,t\} = f(p)$.

The aim of cosmological computing is change. The source (Fig. 11.1) of the changes in the cosmological system is presented with energy and matter, while a result of its actions (goal) is a change in the material and energy substance.

The nature of the universe is cyclical transformation of energy into matter and vice versa (see Fig. 11.2). The universe is a matter–energy substance that has a life cycle of transformation from energy into matter (the expansion of galaxies with acceleration and deceleration), and vice versa from matter into energy (compression of galaxies with acceleration and deceleration). The cycle contains two boundary states of the universe: pure energy and pure matter.

Matter and energy are able to maintain the genomes of the universe at every stage of the cycle. Energy and matter overlap each other in the genome of the universe. Interesting cosmological conclusions can be drawn, based on Einstein's theory of relativity and its formula for the equivalence of matter and energy with a factor equal to the square of light speed: $E = mc^2$. Cosmic radiations (photons) on all bands of the frequency spectrum not only act as sensors for monitoring the state of matter in space objects, but also are actuators of matter development in time and structure of the universe. The conclusion is that each "cell" of the universe knows what it will be in a certain phase of the cycle change. This means that the Earth did not appear by accident and, naturally, it has a beginning and an end. Naturally, the random noise of intergalactic processes influences the determinism of the universe. Therefore, the genome of the universe must have a multiverse of intergalactic decisions leading to the creation of planets suitable for the development of biological life forms.

The cyclic nature of the universe, recognized by a certain group of scientists [9–12], involves matter expansion from a singular point of energy by the Big Bang, which means conversion of energy into matter, which is expanding with the acceleration rate of the galaxies' expansion. It is the current phase of the cycle of the universe's evolution. Eventually, the running speed of the galaxies reaches its maximum, and then begins the deceleration phase of the galaxies' expansion, to the point where the speed becomes equal to zero. The force received by cosmic bodies because of the Big Bang eventually will be equal to the force of gravity. The energy of the universe at a boundary point of the cycle momentarily becomes zero. After that there is a reverse phase called compression of galaxies with acceleration and deceleration of cosmic bodies to the center of the universe due to centripetal forces of gravity that will end with the transformation of cosmic matter into a singular bunch of pure energy, which will create a new Big Bang.

The timelines of energy transformation into matter and vice versa have a harmonic character and are phase shifted by 180 degrees (Fig. 11.3). This means that the matter–energy impulse $m(t) + E(t) = 1$ of the universe at any given time is a constant. The universe is the "engine" that converts energy cyclically into matter and vice versa. The genome universe program sets its development in time and space according to the two harmonics shown below.

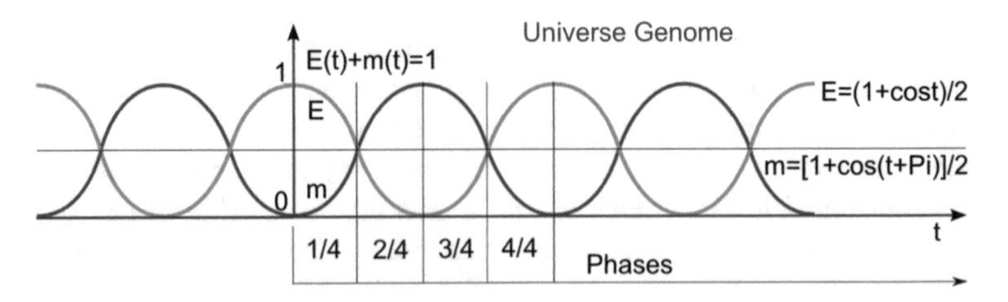

Fig. 11.3 Harmonic genome of the universe's essence in time

The nature of cosmological computing, as mentioned above, is to change anything in an arbitrary phase slot: time, space, energy, matter, strength, speed, density, and gravity. Naturally, the execution of this functionality can be programmed with only the most primitive and the most varying function known in nature. The genome of the universe has to be the easiest to understand and the most substantially changing. What satisfies these requirements? Only the harmonic function, whose derivative practically at all points equals not zero. It is ideal for describing the genome of the universe. The essence of this function and the universe is change. Moreover, the energy function $E = [1 + \cos(t + \pi)]/2$ is here shifted on the x axis by a half period that duplicates the function of matter in the universe's genome. In fact, the universe takes care of genome duplication by its writing in energy and matter in order to avoid mistakes in the universe's evolution, which is read from the energy and matter carriers.

11.3 3D Model of the Universe's Genome

The computing model of the universe's changes presented in Fig. 11.1 needs to be supplemented, if it is assumed that space and time are equivalent components. In this case the automaton model will have four equal parts, which are functionally dependent on the phase of the universe's evolution $\{m, E, S, t\} = f(p)$, which is shown in Fig. 11.4.

Changes in the four components $U = \{E, m, S, t\} = f(p)$ define the essence (energy and matter) and forms (space and time) of the universe's cycle phase. Figure 11.5 explains the interaction of forms and essences, which are synchronized with phases of the universe's change cycle. The harmonics do not have a time axis, because time is the harmonic function here. The time varies depending on the phase p of cyclic period $t = [1 + \cos(p + \pi)]/2$ and within the discrete entropy interval (1–0) of the universe. Time, almost a Lorentz factor $t = 1 - \left(1 - \frac{V^2}{C^2}\right)^{\frac{1}{2}}$, equal to $t = 1$, characterizes the phase point when the entropy of the universe has a maximum value of 1, meaning the light speed of change in all processes related to pure energy, and the absence of matter and space. It means the phase point of the universe's birth. Human time (T) here, according to Einstein, as the reciprocal of light speed at this phase point, is zero:

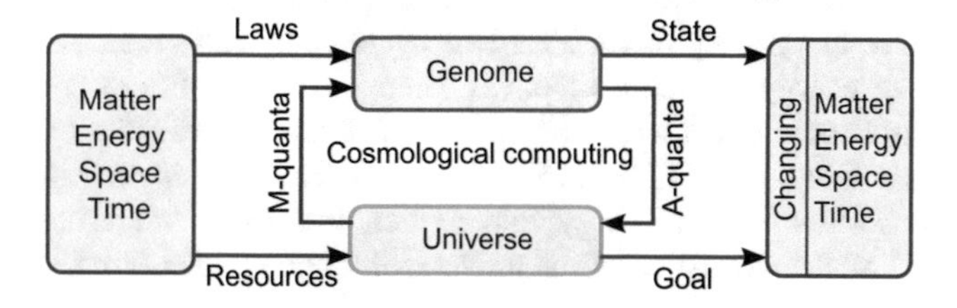

Fig. 11.4 Four-component computing model of the universe

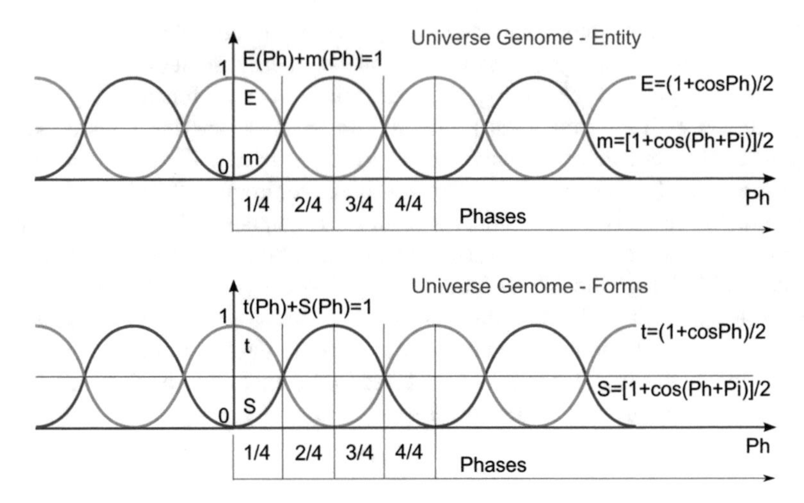

Fig. 11.5 Harmonics of the universe's genome in the cycle phases

$$T = \left(1 - \frac{V^2}{C^2}\right)^{\frac{1}{2}}, \quad V = C$$

Time, equal to $t = 0$, characterizes the phase point when the entropy of the universe has a minimum value of 0, which means a zero rate of change of all processes associated with pure matter and maximum space, with lack of energy and time. It means the phase point of the universe's death. Human time here, according to Einstein, as the reciprocal of light speed at this phase point, is infinite:

$$T = \left(1 - \frac{V^2}{C^2}\right)^{\frac{1}{2}}, V = 0.$$

The diagram of the universe's cycle has two unstable states and two long transitions (see Fig. 11.5), initiated with the contest of two forces: the Big Bang F and gravity G. The states mark a boundary point of the matter–energy and space–time substance of the universe. (1) A singular bunch of pure energy and pure time, lack of space and matter, is followed by detonations, when $E = 1$, $m = 0$, $t = 1$, $S = 0$, $F > G$. This state is preceded by a half cycle of transformation of matter into energy, and space into time, which is accompanied by a compression of the universe under the gravity force. This half cycle ends with receipt of a singular bunch of pure energy and pure time. (2) There is a state of zero entropy in the maximum space of pure matter, when the energy of the universe is zero and there is no time, when $E = 0$, $m = 1$, $t = 0$, $S = 1$, $F = G$. This state is preceded by a half cycle of transformation of energy into matter, and time into space, which is accompanied by the expansion of the universe through the Big Bang.

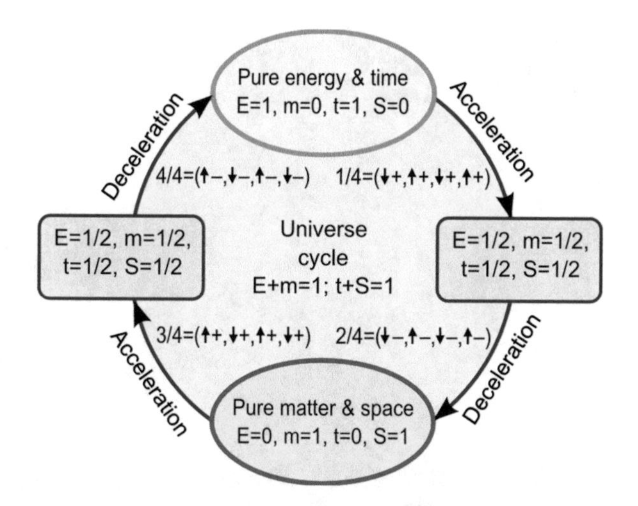

Fig. 11.6 Change cycle of the universe's substances

The cycle of the universe can be divided into four detailed phases, considering that each transition consists of acceleration (+) and deceleration (−) of the cyclic transformation process of energy into matter and time into space. So we can get a phase portrait of the universe's life cycle in the changes metric (↓↑ represent a decrease and an increase in parameters, respectively (*E, m, t, S*)) (Fig. 11.6).

(1) Phase 1/4 [(1, ↓+, 1/2),(0, ↑+, 1/2),(1, ↓+, 1/2),(0, ↑+, 1/2)] is the acceleration of structural matter expansion after the Big Bang, where the maximum speed rate is fixed at the end of this phase. (2) Phase 2/4 [(1/2, ↓−, 0),(1/2, ↑−, 1),(1/2, ↓−, 0),(1/2, ↑−, 1)] is the deceleration of structural matter expansion after the Big Bang, where the minimum speed rate, equal to 0, is fixed at the end of this phase. (3) Phase 3/4 [(0, ↑+, 1/2),(1,↓+, 1/2),(0, ↑+, 1/2),(1, ↓+, 1/2)] is the acceleration of structural matter compression after the universe's death point, where the maximum speed rate is fixed at the end of this phase. (4) Phase 4/4 [(1/2,↑−, 1), (1/2, ↓−, 0), (1/2,↑−, 1), (1/2, ↓−, 0)] is the deceleration of structural matter compression after the universe's death, where the end of this phase is followed by a singular bunch of pure energy and pure time and, after that, a new Big Bang. So, the phase portrait of the universe components' cycling indicates acceleration and deceleration synchronicity of the substances and forms: energy and matter, time and space, which leverage the pairwise variation in opposite phases (*E, m, t, S*).

Assuming that the proposed cosmological cycle indeed exists, and taking into account the fact of the real expansion of structural matter or galaxies, our universe now is in the first phase of the 1/4 cycle. Because 14–15 billion years have passed since the moment of the Big Bang [9, 12], the duration of the universe's cycle is equal to not less than 60 billion years.

But it will be clear only after humanity defines the starting point of the deceleration process of the galaxies' expansion. The harmonic universe genome determines that the substances of space, time, matter, and energy exist in change and have a beginning and an end, which generates a new beginning after a new Big

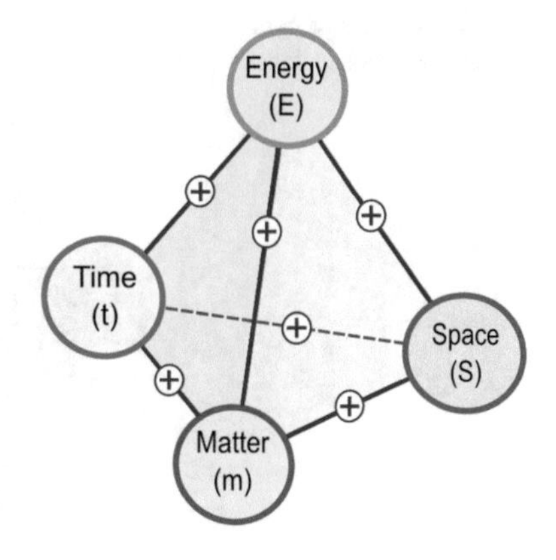

Fig. 11.7 Strong interaction between entities and forms of the universe

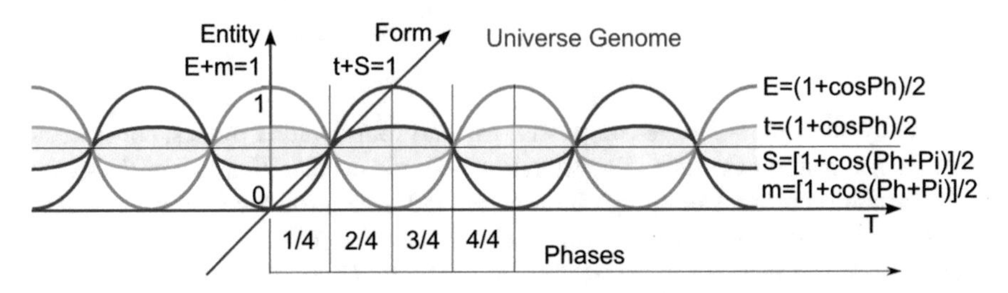

Fig. 11.8 3D genome of the universe

Bang. Time has scalable cycles: 60 s, 60 min, 24 h, 7 days, 12 months, 12 years ... 60 billion years. New cycles occur after the Big Bang. Time is subject to acceleration and deceleration (according to Einstein), like the other components of the computing cosmological model (energy, matter, and space). Time proceeds only with a change in the relative range (0–1) (deceleration, acceleration). The system interaction of the four structural components of the universe is determined by the tetrahedron constraints shown in Fig. 11.7: $[E(p) \oplus m(p)] \oplus S[(p) \oplus t(p)] = 0$. It means that any change in one component leads to changes in the others.

The cyclic system model of the universe can be described by the equation of the 3D genome, where every phase point (p) satisfies the equation $E(p) + m(p) \oplus S(p) + t(p) = 0$, because $E(p) + m(p) = 1$ and $S(p) + t(p) = 1$.

The genome regulates the interaction of all four components, not in time but in the phase $p = Ph$ of the two entities $E(p) + m(p) = 1$ and the two forms $t(p) + S(p) = 1$ of the universe. Those equations can be associated with two polarized harmonics: $E(p) + m(p) = 1$, $t(p) + S(p) = 1$, shown in Fig. 11.5. They are 90° rotated relative to each other in space. Together, the two harmonics of the entities' and forms' interconnections create the 3D genome of the four components as a chain of connected sphere-like structures, represented in Fig. 11.8. This 3D genome corresponds to the universe's cycle (see. Fig. 11.4), which contains four phases of harmonic interaction between the four components.

Einstein's formula for the equivalence of matter and energy ($E = mc^2$) applies to a similar transformation of space and time: $S = TC^2$, where the human time interval ($T = 1/t$) corresponds to a fragment of space. The constant C^2 is the coefficient of interaction between matter and energy, and space and time. There is only one constant in the computing of the universe, which defines the amount of the changed entity $E + m = 1$ and forms $t + S = 1$ in a closed system. Larger space S means smaller time t, because $t = 1 - S$. If there is no space S, it means there is no human time $T = S/C^2$. The Big Bang generates a space that creates time T in human understanding. As a form of the universe, time is designed to identify the changing status of the universe's space. Cosmological time can be defined as the matter (energy and space) change rate, related to light speed: $t = V/C$. Human time T also depends on the speed of processes and events, but as an addition to 1: $T = \sqrt{1 - V^2/C^2}$. Time is a way of measuring space. Space is a form of time's existence. The happiness of the universe is a moment when all the matter $m = 0$ become pure energy $E = 1$ of elementary particles moving along nonintersected closed orbits in the same direction at light speed, when human time $T = 0$ and structured space $S = 0$ are absent. Another opinion states, "Singularity is a condition where everything: matter, energy, space and time disappear into nothing" [13]. Pure energy differs with a change in speed from pure matter in accordance with the formula $E = mC^2$. Clean energy is the fifth form of matter existence at the singular point. The volume of the universe is determined by the distance, depending on the evolution cycle, multiplied by the square of light speed: $S = tC^2 = 60$ billion years multiplied by $300{,}000^2$. The space formed by the formula of the volume sphere is $S = 4/3\pi R^3 = 4/3\pi(tC^2)^3$. Setting one component of the universe by means of the other three provides the ability to change the relationship of humanity to the universe. (1) Time can be controlled by changes in energy, matter, and space. (2) Space depends on changing the three remaining components: energy, matter, and time. (3) Energy depends on changing matter, time, and space. (4) Matter can be derived from energy, space, and time. Thus, the equation of the universe provides a methodology for transforming or producing one of the universe's components by changing the others.

The universe $U = \{V, W\}$ consists of the essence $V = E(p) + m(p)$, represented by two primitives—energy and matter—and the form $W = S(p) + t(p)$ given by two components—space and time. Essence and form are synchronized in phases (p) of universe changes with zero symmetric difference of their scalar values: $V(p) \oplus W(p) = 0$. Energy and matter contain an integral changing nature of the universe, called the matter–energy universal set: $E(p) + m(p) = 1$, $E(p) + m(p) = V$. Energy is a changeable essence, in addition to matter in the matter–energy universal set: $E(p) = 1 - m(p)$, $E(p) = V\setminus m(p)$. Matter is a changeable essence, in addition to energy in the matter–energy universal set: $m(p) = 1 - E(p)$, $m(p) = V\setminus E(p)$. Space and time contain an integral changing form of the universe's nature, called the space–time universal set: $S(p) + t(p) = 1$, $S(p) + t(p) = W$. Time is a changeable matter and energy form, in addition to space in the space–time universal set: $t(p) = 1 - S(p)$, $t(p) = W\setminus S(p)$. Space is a changeable matter and energy form,

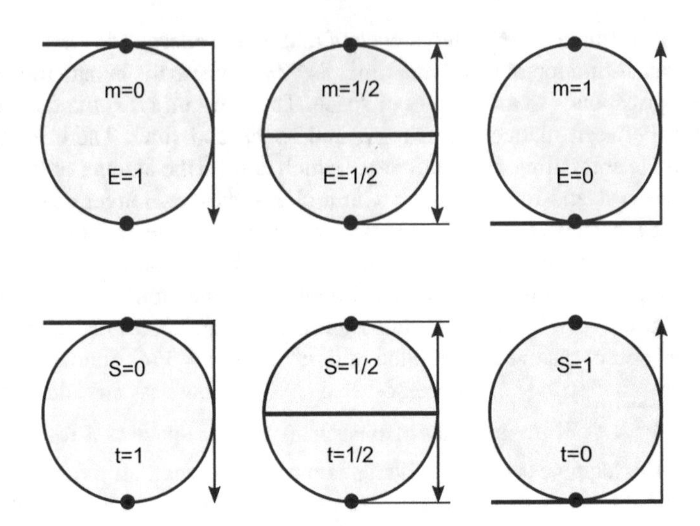

Fig. 11.9 Universal sets of the substance and form of the universe

in addition to time in the space–time universal set: $S(p) = 1 - t(p)$, $S(p) = W \backslash t(p)$. The area of a circle, which is equal to 1, corresponds to the universal set $U = \{V, W\}$, divided into two parts with a diameter, which moves from top to bottom, and vice versa, changing the ratio between the two pairs of primitives: space and time, matter and energy (Fig. 11.9).

11.4 Conclusion

(1) An automaton model of computing formulates and explains technology monitoring and control of processes and phenomena in cosmological space. (2) Verbal and structural determination of the components of cosmological computing, based on current trends in cyber ecosystem evolution of the planet, are presented. (3) The harmonic genome of the universe as a computing system for evolutional changing of: space and time, matter and energy, infinite and mutually transformed into each other, are defined. (4) A methodology for predicting cosmological processes affecting the solar system, on the basis of the proposed genome of the universe, in order to plan the future of humanity, is suggested.

The speed of light is the single constant, the square of which forms the interaction of the harmonic changing space–time form and the matter–energy substance of the universe. The relationship of all four components of the universe is functionally dependent on the speed of their transformation, which models changes in the cycle from 0 to 1 (1 is the upper limit of the speed of light model). Figure 11.9 shows the

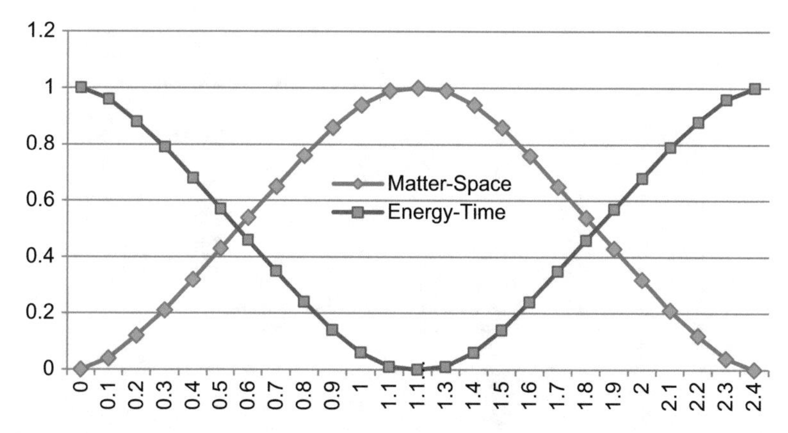

Fig. 11.10 Harmonic cycling of the substance and form of the universe

harmonic evolution model of the universe's component relationship. The consecutive movement of the chord through the matter–energy and space–time circle–universe represents the relationship model. The change in the areas above the chord and under it corresponds to a harmonic transition of energy into matter, time into space, and vice versa. It is interesting that the area of the circle $S = \pi R^2$ and the Einstein equation $E = mC^2$ are similar precisely in the form of the interaction between the components of the universe, which reduces to the area circle formula, where $m = \pi$. The circular discrete model indicates the unity of the form and substance of the universe, which together give the area of the circle at each moment of its harmonic evolution. The spherical image of the universe, where chords of cyclic change move perpendicular to each other—matter–energy, space–time—correspond to the harmonic genome of evolution.

Figure 11.10 illustrates the model phases of the change in two parts of the circle area, divided by the reciprocation motion of the chord $h = Ph$. The changes correspond to the evolution of the matter–energy substance and the space–time form of the universe $Q = \pi R^2 = (m \backslash E) \vee (S \backslash t)$. At a radius $R = 0.5643$, the area of the circle is equal to $Q = 1$, which corresponds to the total model value of the substances $m + E = 1$ and the forms $S + t = 1$ at each point of the evolution phase. Two genome harmonics of matter–space and energy–time illustrate the constancy or infinity of changes in all four components of the universe. Harmonic change is the basis of God's idea of the cyclical evolution of the matter–energy substance and the space–time form.

Naturally, the Big Bang theory, when all four components are zeroed out at a singular point, is strange at least from the standpoints of both materialism and God's governance of the universe. To create something, one needs to have something. The thermal death of the universe is absolutely excluded in its harmonic evolution, when in the first half cycle, energy (time) is transformed into matter (space), and in the second half, matter (space) is transformed into energy (time). The universe, as an ideal system, does not interact with the outside world, which

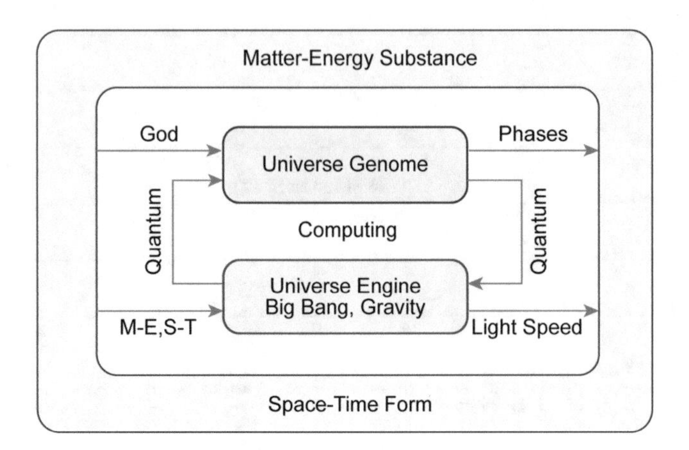

Fig. 11.11 Computing of the universe

does not exist. Here the law of matter and energy, and space and time conservation is fulfilled, the cosmological efficiency is equal to 100%, and there are no losses. As each cell of the organism knows its own genome (algorithm) of evolution, so each material–energy particle of the universe contains information about its own genesis. The matter–energy cells of the universe interact with each other by means of information electromagnetic transactions. The little-modified automaton model of cosmological computing (Fig. 11.11) includes the following. (1) The control mechanism is God or the genome of the universe. (2) The mechanism of execution is the forces of the Big Bang and gravity. (3) The monitoring signals and actuator effects are electromagnetic fields or radiation. (4) The resources are the matter–energy substance and the space–time form. (5) The goal is to achieve the speed of light by changing all components. (6) States are phases in the universe's evolution. (7) The initiating signal is God's will or the super-idea.

Time creates space in the first phase of evolution (expansion) of the universe. The space in the second phase of evolution (convolution) is transformed into time. Otherwise, a decrease in the space of the universe leads to an acceleration of the stream of time or the speed of change in the matter–energy substance. A similar effect is observed when the matter–energy substance of the universe accelerates. Thus, it can travel in time, due both to near-light speed and the concentration of the matter–energy substance in a fixed volume of the universe. Human time is the difference (derivative) between the speed of light and the movement of the matter–energy substance. The closer the derivative is to zero, the more the existence of bioenergy objects turns into "eternity." Similarly, it should be assumed that human time turns into "eternity" if the derivative between the density of the matter–energy substance at the point of the Big Bang and the density of a particular section of the universe tends toward zero. Otherwise, one can achieve a change in time toward human "eternity" in a local part of the universe, which today is identified with a black hole. However, the substance of the time acceleration to a model value 1 is one—an increase in the movement speed of a matter–energy substance in a black hole or outside this section of the universe.

Thus, the black hole is characterized by the light speed of motion of the matter–energy substance with the maximum 1 model density in circular orbits. The light speed constant creates a relative model time in the interval between zero and one. The ideal model time, equal to 1, is formed in processes where the matter–energy substance, for which time is determined, moves at the speed of light. In this case, the difference between the speed of light and the substance moving will determine the relative (human) time or the time of the process flow of the substance. If the Earth moves in space at the speed of light, the land time will stop: $C - T = 0$; the movement of all land particles relative to each other will be equal to zero. This means that all particles move at the speed of light, without overtaking and not slowing down; they will forever fix each other in the "happy eternity" of the given relations. Time is the change speed of processes, reduced to the speed of light, which is the single constant value on the scale of the universe. The speed of light is a single metric standard for measuring the matter and energy, and the space and time of the universe. Naturally, all four components of the universe tend to achieve the moment of truth in their evolution—to reach the speed of light in order to explode again and begin a new cycle of universe changes. "It is almost immobility anguish, rushing somewhere at the speed of sound, knowing perfectly that there is already somewhere, someone who is flying at the speed of light" (Leonid Martynov).

References

1. Bondarenko, M. F., Hahanov, V. I., & Litvinova, E. I. (2012). Logical associative multiprocessor structure. *Automation and Remote Control, 73*, 1648–1666.
2. Hahanov, V. I., Gharibi, W., Litvinova, E. I., & Shkil, A. S. (2015). Qubit data structure of computing devices. *Electronic Modeling, 1*, 76–99.
3. Hahanov, V. I., Tamer, B. A., Chumachenko, S. V., & Litvinova, E. I. (2015). Qubit technology for analysis and diagnosis of digital devices. *Electronic Modeling, 37*(3), 17–40.
4. Hahanov, V., Litvinova, E., Gharibi, W., & Chumachenko, S. (2015). Big data driven cyber analytic system. In *IEEE international congress on big data*, New York City.
5. Dehorter, N., Ciceri, G., Bartolini, G., Lim, L., del Pino, I., & Marín, O. (2015). Tuning of fast-spiking interneuron properties by an activity-dependent transcriptional switch. *Science, 349* (6253), 1216–1220.
6. Merolla, P. A., et al. (2014). A million spiking-neuron integrated circuit with a scalable communication network and interface. *Science, 345*(6197), 668–673.
7. Lin, H. W., & Tegmark, M. (2016). Why does deep and cheap learning work so well? http://arxiv.org/pdf/1608.08225v1.pdf
8. Lloyd, S. (2014). *Programming the universe: Quantum computer science and the future*. Moscow: Alpina Non-fiction.
9. Morison, I. (2013). *Introduction to astronomy and cosmology*. New York: Wiley.
10. Peterka, T., et al. (2012). Meshing the universe: Integrating analysis in cosmological simulations. In *High performance computing, networking, storage and analysis (SCC)*. SC Companion, Salt Lake City.

11. Habib, S. (2015). Cosmology and computers: HACCing the universe. In *2015 international conference on parallel architecture and compilation (PACT)*, San Francisco.
12. Sadokhin, A.A. *Concepts of modern science, cosmology and cosmogony*. http://www.gumer.info/bibliotek_Buks/Science/sadoh/06.php
13. http://ru-universe.livejournal.com/712762.html?page=1

Chapter 12
Cyber Social Computing

Vladimir Hahanov, Tetiiana Soklakova, Anastasia Hahanova, and Svetlana Chumachenko

12.1 Background of Cyber Social Computing

The motivation for the proposed investigation is determined by the desire of many constructive citizens to make the state morally civilized, technologically cybercultural, investment attractive, and cost effective. For this, we consider the cyber system components, which are essential for the evolutionary creation of attractive statehood [1–5]: (1) the structure of cyber social computing; (2) the metric of social significance; (3) statehood and corruption as two sides of the same coin; (4) unification and intersection of interests in the management of society—a metric of the social group intellect; (5) the harmonic genome of community development; and (6) cyber social computing—the moral future of mankind.

The purpose of this investigation is to show new technologies for improving cyber community management based on the use of cloud services and accurate digital monitoring of each citizen's opinion.

The objectives are (1) the state of social group management in developing countries; (2) technologies, models, and methods of cyber social computing; (3) cyberculture, measuring the intelligence of a social group and cyber social management; and (4) metric cyber relationships and the future, such as cyber statehood.

The structure of cyber social computing is considered below. Modern cyberculture imposes strict requirements on the structure of cyber social computing, which should include the following components in descending order of

V. Hahanov (✉) • T. Soklakova • A. Hahanova • S. Chumachenko
Design Automation Department, Kharkov National University of Radio Electronics, Ukraine, Nauka Avenue 14, Kharkov 61166, Ukraine
e-mail: hahanov@icloud.com; tetiana.soklakova@gmail.com; hahanova@icloud.com; svetachumachenko@icloud.com

V. Hahanov, *Cyber Physical Computing for IoT-driven Services*,
DOI 10.1007/978-3-319-54825-8_12

importance: (1) digitized horizontal and vertical relationships in a social group determined by legislation, culture, history, and traditions; (2) cloud management of processes and phenomena based on accurate digital monitoring; (3) personnel who, through fair relations and competent management, become a productive force that creates products, services, and financial success; (4) electronic infrastructure of a social group, providing comfortable conditions for creative, productive work and leisure of the team on a 24–7 timing format; (5) a road map that forms strategic market-focused goals and objectives related to product quality, employees' quality of life, and saving the ecosystem; (6) financial resources and metric cyber management creating a globally successful company or rapid bankruptcy in the market segment; and (7) manufacturing processes as the basis of cyber social computing, aimed at creating quality products and services for sale in the market.

Seven points form the quality of a social system, enterprise, or state. Ernst & Young conducted an international study of the level of corruption in business in 2017. Ukraine took the "honorable" first place with a result of 88% [6]. What is lacking in that country to become successful and to defeat corruption? Systematically, in the state, there are not only two first points—moral relations and competent management. The last two components have to be created by the political elite (comprising 500 officials), who should have knowledge about relationships and management in highly developed countries, such as the USA, Germany, Japan, and the UK.

A metric of social significance is described further. Technological perfection should translate into the social significance of obtaining a well-deserved moral and material reward: (1) morality of actions and deeds; (2) respect for cultural, linguistic, and historical values and traditions of all peoples; (3) education, competence, and knowledge of special technologies; (4) outstanding results in the performance of official duties; and (5) merit in society for voluntary contributions to socially useful activity.

Statehood and corruption are two sides of the same coin. Statehood introduces taxes, collecting some of the citizens' funds for subsequent provision of infrastructure services, health protection, and protection of private property, honor, and the citizens' dignity. The taxes of citizens do not reach the providers of these services. They are melted away in the offices of government officials. In this case, the producers of services for the population create private companies—police, troops, clinics, courts, and prosecutors—who sell the services necessary for human life in exchange for a secondary levy on the population (the first was in the form of taxes). The population has, in fact, double taxation. The way out is simple: to create cloud-based cyber services to distribute budget funds, bypassing government officials. Another way out is to reduce the taxation of citizens and the number of state structures to zero by creating private commercial organizations. Corruption exists while taxes from citizens are accumulated in the accounts of a privileged corporation, which is called the state. At the same time, there are always buyers and officials who are ready to sell goods that do not belong to them—state finances and resources.

Uniting and intersection of interests in the management of society are defined by two primitive logical operations. The first one unites people by taking into account their interests, cultures, histories, languages, and traditions, which are put on an equal footing, regardless of the number of carriers of these concepts. The state, governed by the political elite and based on the unification doctrine, becomes attractive and competitive in the market. At every corner and in the mind of every US citizen, it is written, “United we stand.” The result of such a positive influence is the social, economic, and spiritual prosperity of the country. The second operation divides the community into separate and creatively weak social groups, hating each other due to the propaganda of unequal relations of power in terms of languages, histories, cultures, and traditions. Such management by the political elite distracts the population from criticism of ignorant and incompetent leaders but leads to destruction of citizens and the state. “A great civilization is not conquered from without until it has destroyed itself from within” (W. Durant). The law of intelligence additivity is represented below. The metric distance between all members of the community determines the intellect of a nation or social group:

$$I = \bigoplus_{i=1}^{n} P_i.$$

The function of the symmetric difference between participants (components) of the society is performed. For Cantor’s set theoretic alphabet $A = \{0,1,X = \{0,1\}, \varnothing\}$ this operation is represented by a truth table illustrating the interaction of the intellects of two people:

$\oplus$	0	1	X	∅
0	∅	X	1	0
1	X	∅	0	1
X	1	0	∅	X
∅	0	1	X	∅

The exceptional advantage of this function is that it additively combines all the differences and turns all identities into a void. Paradoxically, the level of intelligence of 500 like-minded people turns into an empty set. Experts differing in competencies additively create intelligent power equal to 500. Otherwise, the nonadditivity of this law with respect to two identical individuals asserts that $1 + 1 = 0$. In accordance with the reflexivity axiom, the distance between oneself is just equal to an empty set or zero. If two persons are completely different, their intellects are summed up: $1 + 1 = 2$.

Thus, several different experts are more important for a social system than 100 equally minded specialists. Even without knowing this law, experienced experts–managers of successful companies make the right decisions when hiring specialists who do not intersect by competencies. Leading universities in the world do not collect expert boards consisting of 24 scientists to assess the quality of scientific research; three leading experts in a particular field are quite sufficient.

Three people create a reliable system, which is sustainable in the event of unforeseen technical failure of one of them and nothing more. As a rule, in a region or country there are no more than three experts in a particular research object. Theoretically, the quality of the expertise of 24 scientists is much worse than that of three experts, because of possible democratic dominance of the same incompetent majority when making a decision based on secret ballots. This is particularly negative in the election of managers at all levels. The incompetent majority, as a rule, makes a choice in favor of a corrupt candidate. When making a socially significant decision, it is necessary to follow the axiom that one expert with argumenta proofs is more right than dozens of incompetent persons. The transition from an always ignorant democracy is a way to the quality of social solutions provided by the experts. A positive example of the use of this law is the 200-year-old political culture of the USA, expressed in the doctrine: "All roads of talented people lead to the United States." From the proposed law of additivity of the intelligence follows an important, mathematically proven conclusion for any country that wants prosperity: "The power of the state is in a tolerant union of talented people, different nationalities, languages, cultures, histories, religions, and traditions." How many billions of dollars do you need to become Germany? Zero. It is only necessary to change the constitution, according to the last thesis. The rest of the people will do it themselves—if the officials do not interfere.

The harmonic genome describes social group development. The harmonic curve is an ideal and actual model of society development in time: always with change, depression and flowering, recessions and booms. The natural wisdom of such a model is the potential opportunity to transform social relations into more progressive ones by searching and comparing the existing cultures in the development of mankind in order to adopt a more correct form of social relations. The backwardness of a country can give way to optimism through its transformation into a foremost country by assembling a team of moral and competent top managers from highly developed countries. It is the quickest way. The ignorance of the community is manifested by nonadmission of other managers so they cannot intrude into the corruptive sanctum with their charter of honest relations. It will happen even if they are Nobel laureates in economics. Peter the Great used the experience of foreign talent and won lots of time for the development of a new technology. The USA has been collecting the best for 200 years by creating ideal conditions for creativity, and now it is living the dream.

Computing is a branch of knowledge that develops the theory and practice of reliable metric management of virtual, physical, and social processes and phenomena based on the use of computer centers and networks, big data, and digital monitoring of cyber physical space using intelligent retrieval and analytical services, personal gadgets, and smart sensors.

The Internet of Things (IoT) is a global and scalable technological structure of cloud metric human-free management of virtual, physical, and social processes and phenomena based on the use of computer service platforms, big data centers, and

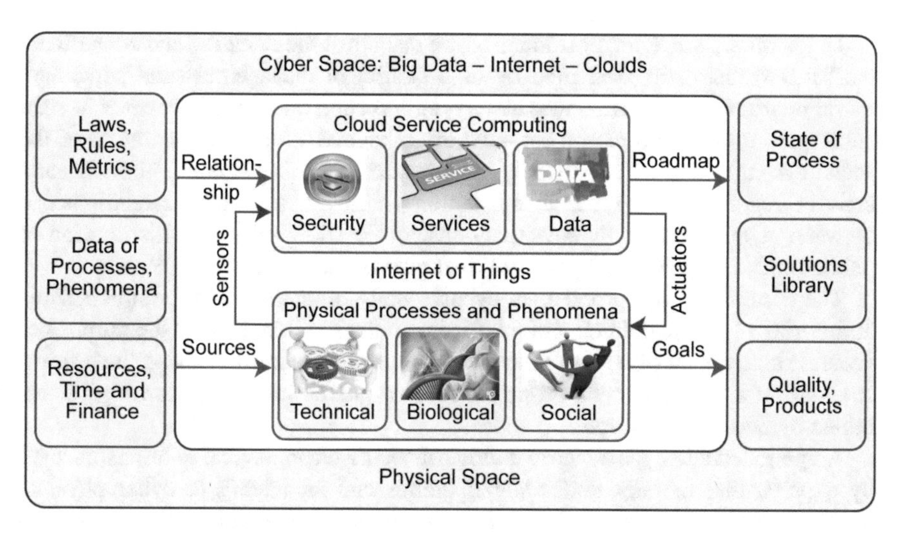

Fig. 12.1 IoT computing

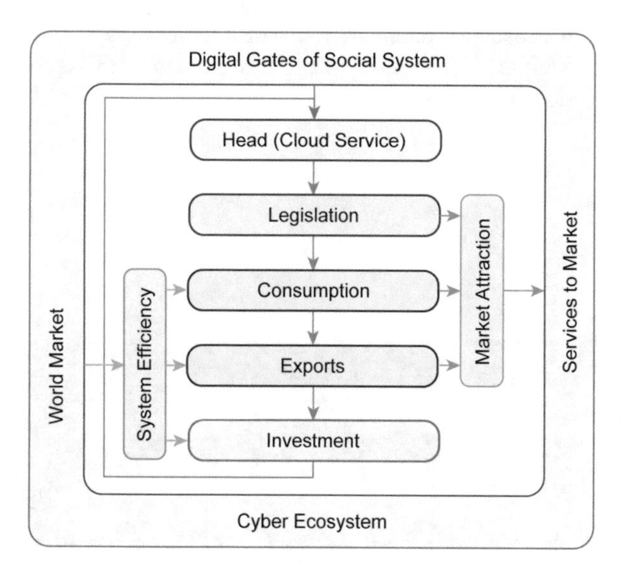

Fig. 12.2 Efficiency of the social system

electronic infrastructures for digital monitoring of cyber physical space by using smart retrieval and analytical services, personal gadgets, and smart sensors in order to ensure quality of life and preserve the ecology of the planet (Fig. 12.1).

The effectiveness of the social system (for instance, a university) is determined by three purely economic estimates of the following levels: consumption, exports, and investment (Fig. 12.2).

Universities, which are positioned as the leaders of the scientific and educational market, are metrically evaluated by high quality of management and personnel, moral relations in the team, laws, history, culture, and traditions. Investments are a consequence of—not a reason for—the effective and sustainable operation of the social system; money likes silence and sustainability of economic, political, and creative relations. The roots of corruption lie not in staff but in relationships between people, defined by laws that consciously allow subjective distribution of public funds and positions by the political elite of the state.

The solution to the problem is cyber social computing, with comprehensive monitoring of public opinion and intelligent analysis of big data for the purpose of accurate cloud management of financial and human resources by using the e-infrastructure of a scalable social group based on digitized metric relationships determined by current legislation.

A cyber democracy is a metric culture of social–technological relations, formed by experts, that morally unites social groups and an intelligent cyber physical infrastructure for exhaustive digital monitoring of public opinion and cloud management of social processes and phenomena in order to preserve the planet's ecology and achieve a high quality of life. More popularly, cyber democracy is the integration of exhaustive democratic debate and accurate expert monitoring of social problems with the cloud metric of cyber management of the society based on digitized legislation (Fig. 12.3).

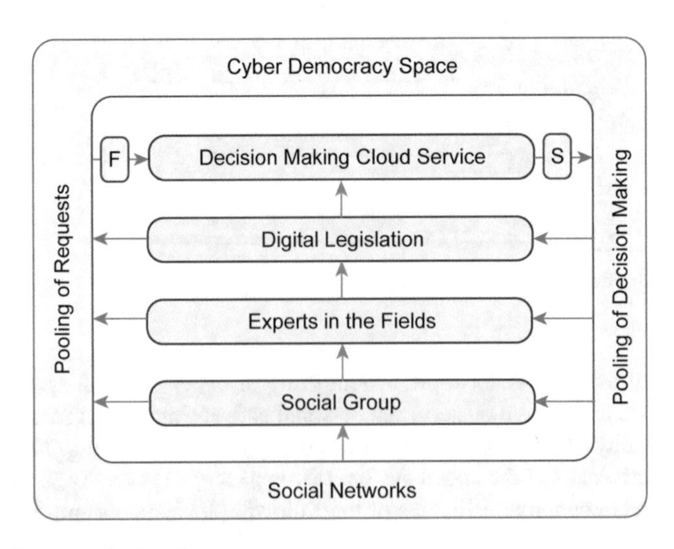

Fig. 12.3 Structure of cyber democracy

12.2 Cyber Social Governance

Of particular importance in the twenty-first century is cyber social governance, capable of destroying revolutions, wars, social cataclysms, terrorism, crime, corruption, immorality, and injustice on the part of the political elite. One of the main reasons for all conflicts is the existence of a critical mass of incompetent leaders and an uneducated political elite in local areas.

The purpose of such people is to buy the "love" of the electorate, which is incompetent in matters of jurisprudence and economics, for their subsequent legitimate robbery. At the same time, the elite officials do not think about the people's quality of life and the market attractiveness of the state system. Their mission's legacy is a negative or even bloody trail in history.

For the people, this track often results in decreased living standards according to the anecdotal axiom: "Each succeeding ruler is worse than the previous ones." Therefore, it is necessary to select a competent leader metrically by using an expert group and to limit its possible negative impact on social processes. For this, ten commitments of cyber social management are offered: (1) digitization of all social relations, constitutions, legislation, and bylaws; (2) electronic document circulation (EDC) for public institutions and private companies on the basis of primary authentication, with introduction of an electronic money system based on e-infrastructure; (3) creation of an e-infrastructure for state institutions and private companies; (4) cloud services for smart monitoring and accurate management of all social processes and phenomena; (5) implementation of cloud services for statehood: cyber democracy, cyber management, and a cyber parliament; (6) legalization of bioauthentication based on primary signs (fingerprints, iris, DNA); (7) implementation of cloud services for electronic voting and elections for managers at all levels of governance; (8) metric digital assessment of physical and social processes and phenomena in decision making; (9) distribution of finances, resources, and personnel appointments on the basis of competitive evaluation of applicants; and (10) online access to cloud services for public administration of the country, city, and district on a 24–7 timing format.

The downsides of democracy have hindered humanity's development in the past, and still do, but will not exist in a more educated cultural community. Democracy in developing countries is an immoral dictatorship of an always ignorant majority over an educated and competent minority. The Pareto principle declares that in a social group, 20% of people create 80% of the production. Another version is also true: the remaining 80% of citizens create 20% of the production. It is only natural that the best 20% should receive 80% of the moral and financial awards. However, the worst 80% have a powerful weapon—democracy—which, at best, bloodlessly defeats the constructive minority and destroys the state enterprise's efficiency. Immoral democracy also initiates revolutions, such as the Maidan, uprising in Ukraine, in order to physically destroy the best part of the social group. As a result of destruction of creative people, society has regressed decades into the past. In order to replace the destroyed scientists, it is necessary to bring up a new generation of creative people over the next 30–40 years. It should be borne in mind that the dictatorship of immoral democracy always destroys the strongest and genetically best people, and then the new generation of the 80% genetically deficient part of the society will multiply. Revolutions, dictatorial regimes, and the First and Second World Wars took tens of millions of constructive citizens with better genes, raising the concentration of noncreative people.

The authoritarian rule of a moral, competent, and creative leader in this situation is the perfect form, rather than the democracy of the uneducated majority. Such a positive dictator always refers to the society as his own system, which must be competitive in the external environment. Therefore, it creates a system of a capitalist relationship, where the best 20% receive 80% of all benefits, which ensures world recognition of the social group and a high standard of living for creative citizens. Naturally, in a cost-effective system, a social protection mechanism is created for the noncreative 80% of the population, which provides a powerful potential for further development of the social group. The Pareto principle need not always have a separation of 20%/80%. There are powerful organizations (e.g., Synopsis, Apple, NASA, Cambridge University, Harvard University, Stanford University) where these asset and liability proportions swap places, i.e., 80%/20%. What prevents the social system from being successful with the 80%/20% metric? The answer to this question lies in its history. In the post-Soviet countries, socialism corrupted citizens by equalizing their incomes so that all people—active and passive—received the same salary. The best people and systems suffered, but the passive part (80%) of society was guaranteed to enjoy a miserable existence. Today, nothing has changed. We pretend that we are working; top officials pretend that they pay us. The most terrible thing is that from this historical impasse of the development of society, there is no legitimate democratic way out.

This is illustrated by the example of a university, with initial data of 20% of staff working actively and 80% working passively. There are two candidates for the election of the rector: the first proposes to leave in place socialist relations, with equality of salaries in accordance with posts, regardless of the results of the scientists' activity. This leads to a guaranteed degradation of the university. The second candidate proposes to introduce a moral relationship between employees, with material rewards according to metric assessment of each scientist's or employee's achievements. This leads to the creation of an economically and scientifically wealthy institution for higher education. The voting results are that in five election campaigns occurring within 3 years, the first candidate for rector wins. The conclusion is that the university employees want to have a leader who will be loyal to their shortcomings, incompetence, bribery, and passive activity. Most of the staff and the rector are not focused on creating an economically efficient system, if only because for decades the university has been guaranteed to receive its state budget without any responsibility for targeted expenditure and quality scientific and educational products. There is no science, there is no education, but there is a fake university.

The way out is when the president or minister appoints the rector by leveraging metric parameters to assess each candidate's competence as a manager, businessman, professor, scientist, and public figure. The scalability of such an election campaign extends to all levels of social importance and responsibility, up to the election of the country's president. Cyber democracy offers a metric assessment of the candidates' competencies, and then an election takes between the two who scored the most points, through voting by a limited number of management experts. The closest to the proposed metric of cyber democracy is the US presidential election system. How can the influence of a negative leader be eliminated? By creating cyber governance, designed for human-free resource distribution based on metric performance of people and organizations. In this case, cyber democracy oriented toward human-free moral and metric monitoring and governance of the social group becomes relevant. Ironically, the effectiveness of the state is also assessed by the Pareto principle. If the ratio of private property to state property is 20%/80%, then it is a developing country. The closer the level of private property is to 100%, the greater the country's economy is, and the better the quality of a citizen's life is, with a level of corruption that tends toward zero. No man in the world will steal from himself.

The future of humanity is determined by cyber social, moral power: cyber democracy, cyber governance, and a cyber parliament, which, in the twenty-first century, will implement fair management of cyber services, erasing the borders of states from the map of the planet for the benefit of every citizen of the Earth.

There are innovative parallels of cyber (physical–social) computing based on MAT (memory–address–transaction) computing. The computer (quantum) of the future is memory and address transactions. Memory is organized into any form of matter existence. Execution and control mechanisms are implemented in memory. Address transactions create all processes and algorithms. Delivering one instruction from the control unit to memory is more efficient than transferring huge amounts of data to the ALU and back. That is the computing bottleneck, which has existed for 70 years. The data (big data) should be processed where it exists, and that is the near future of cyber physical computing and cyber social computing. As an analogy, megacities currently absorb millions of people in the morning through a bottleneck of terrible traffic for their utilization in offices, and in the evening push them back out again to villages near and far.

There are huge material, temporal, and financial costs for gasoline, rental of offices, movements in real space, stresses in traffic, and pollution of the atmosphere. The solution to the problem is cyber physical monitoring and control. Deliver instructional tasks through gadgets to employees in their personal virtual offices, due to legitimized online management, where the goal is to create maximum comfort for working employees regardless of their geographical positioning (at home, in a hotel, during travel, at leisure). The condition for remuneration is timely and quality performance of the day's assignment. Education and science must be performed online. Onsite meetings, lectures, and seminars are seen as a luxury of teacher–student communication. Today, there are all of the necessary components of the e-infrastructure for creating cyber social computing: cloud management services, edge gadgets of users, and smart things for implementation of the interaction. The only thing that does not exist is legislation to open the doors to this multibillion-dollar innovation on the scale of countries and the planet.

12.3 Cyberculture and the State

The state's success in the world is determined by the quality of its legislation, economy, ability to create products and services for export, citizens' standard of living, and government competence. A country's economic and scientific–educational backwardness is an effect of lack of morality and competence of the political elite in generating inhumane legislation, which leads to the degradation and disintegration of the state. The goals of post-Soviet officials are not related to the well-being and unity of the people. Their problem is to preserve power and steal more money for a limited term of government. What is the technology for retaining power that is used by incompetent managers of the economy? There are at least two win–win methods for fooling their peoples who are living in poverty, compared to the surrounding countries.

1. Create an image of an external enemy. The power usually says: "Our neighbors create our troubles." They need to be destroyed, and then we will be rich and happy. But if the neighbors are stronger, than our infantilism is to complain to the world community, recognizing our incompetence and ineffectiveness in political decision making. The trump card of patriotism (volunteering) never dies. Patriotism is addressed in all cases when an incompetent manager has no money to adequately pay for the work of his citizens. In 99% of cases, the "real" patriots are militant idlers, incapable of productive work.
2. Create an image of an internal enemy. One can constitutionally introduce discriminatory relations between social groups existing within the territory of the country. The purpose of such policy is to divide peoples according to the metrics of unequal relations of official authority on the basis of language, culture, history, religion, and nationality. Divide and rule. Constitutionally, the dominance of the titular nation is declared; all the rest are background citizens. The result is pathological hatred passing through peoples, social groups, and families. It distracts people from their constitutional right to dismiss the government and to eradicate poverty, which is a direct effect of the ruling elite's ignorance. The low living standard of 80% of citizens leads to an increase in crime and the destruction of infrastructure, the economy, the political system, and the state.

Alas, the country's economy will not improve due to the introduction into the mind of society of destructive doctrines regarding the presence of internal and external enemies. Such a policy stimulates the immoral and animal instincts of envy, destruction, and murder, and paralyzes the will of citizens to do creative work, making them unable to even bake a loaf of bread. How are we to solve the problem of the country's backwardness?

1. Unite peoples by creating a moral relationship (constitution) with equality of linguistic, historical, religious, and national cultures.
2. Metrically elect a government of competent leaders and parliamentarians.
3. Create friendly cooperation with the external environment of the country to contribute to economic effectiveness and the quality of citizens' lives.
4. Do not expel talented people from the country who are smarter than the officials, but create favorable conditions for creativity and a mass invitation to experienced persons from outside.
5. Create politically and economically stable conditions that are invariant despite changes in governments, presidents, and parliamentarians, to attract foreign and domestic investments in the form of capital and competent personnel.

What are the role and methods of emerging cyberculture in the formation of an economically attractive state with total elimination of corruption? There are five simple theses for ease of understanding by the public:

1. Digitization of all objects and phenomena in the country's territory, including citizens, the constitution, laws, money, and documentation.
2. Metric estimation of all processes and phenomena for accurate monitoring and adequate management of personnel and resources. A metric is a way of measuring distance in cyberspace between processes or phenomena by comparing their parameters.
3. Creation of an electronic infrastructure for digital monitoring and cloud management of an intelligent society, country, city, home, organization, transport, finance, science, and education (Fig. 12.4). "Intelligent" ("smart") is the definition of a process or phenomenon associated with the network interaction of addressable system components in time and space between themselves and the

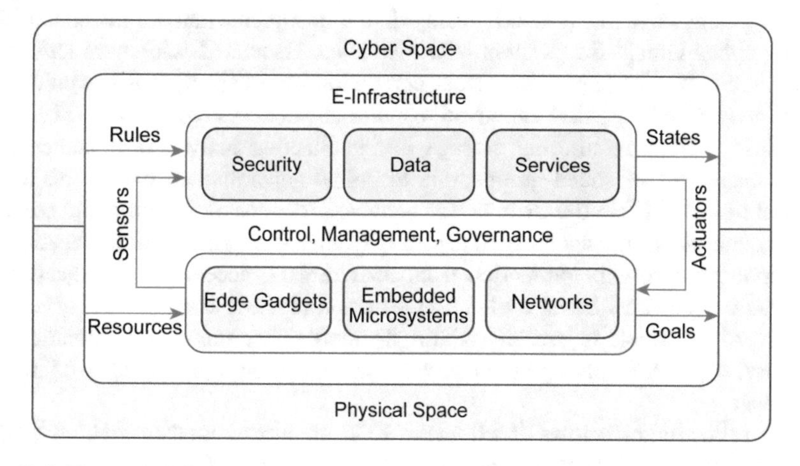

Fig. 12.4 Electronic infrastructure

environment, based on self-learning technologies to achieve their goals. E-infrastructure is an interconnection of cloud services, large data centers, computer devices, systems, and networks, as well as laws, standards, and means of authentication, cybersecurity, telecommunications, testing, and repair to provide scalable IoT computing for reliable monitoring and sustainable management of processes and phenomena in order to improve the quality of human life and preserve the ecology of the planet.

4. Implementation of EDC based on authentication of the individual and digital signatures in all spheres of human activity.
5. Introduction of electronic elections and electronic voting in all matters of competence assessment and decision making.

All five points are fairly easy to implement on a national scale, given political will and a minimum level of cyberculture among the power elite. Of particular importance for citizen education in cyberculture is EDC, which should not be understood only as a service of automated production, reception, transmission, storage, and disposal of documents.

EDC is a cyber physical computer-aided system for accurate digital monitoring and cyber social group management by using an e-infrastructure of individual authentication, edge gadgets, and cloud services for online solving of social problems, in a paperless system, to increase the quality of citizens' lives and to save the planet's ecosystem.

The basis of EDC is legitimate intellectual transactions of digitized document flows (sensory signals and regulatory influences) in a smart, logically distributed data network designed to implement paperless relations with the outside world, with direct monitoring and management of scientific and educational processes and departments in the university. Digitized documents (available for understanding by computers and humans) serve as digital sensors and actuators in a cyber system (for example, a smart cyber university). This means the ability to generate digital reports and manage the system by using digital documents that are understandable to the cyber human-free system. EDC is often associated with transactions of electronic copies of papers for visual perception by a person, which would have been innovative in 1990. An innovative cyber physical system of EDC (Fig. 12.5) contains (1) cloud monitoring, storage, and intellectual analysis of documents for the management of social groups, due to online generation of documents at the request of users; (2) end-user interface gadgets for connection with cloud services that provide electronic documents; (3) complex means of primary authentication for the cloud system, with protection from unauthorized access and falsification of electronic documents based on the user's trusted identification; and (4) a scalable social group, which is created within the enterprise, university, organization, territory, or state, receiving cloud services and ensuring a high quality of life for each user.

The following principles or axioms of EDC are aimed at ensuring high-quality services, quality of human life, and forest preservation for future generations:

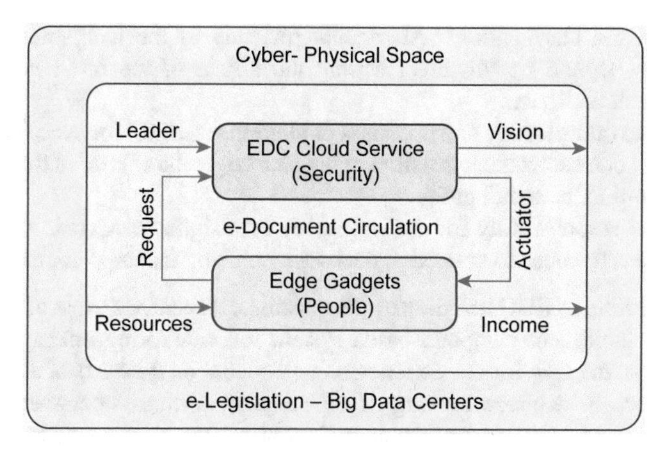

Fig. 12.5 Structure of electronic document circulation

1. Online digital monitoring, metric estimation, and forecasting of the creative activities of each citizen or social group for generating adequate control actuators, based on existing electronic documentation.
2. Online creation of a constructive control action in the form of a digital document, as a result of metric evaluation of a process or phenomenon at the user's request.
3. Creation of an electronic infrastructure for comprehensive protection of EDC against unauthorized access to document data based on primary authentication: an electronic digital signature, fingerprint scanning, DNA analysis, or a manual signature on a touch screen.
4. Exclusion of all secondary material carriers of authentication—electronic maps, passports, passes, driver's licenses, certificates, diplomas—which complicate a person's life and pollute the planet's ecosystem.
5. Ensuring the reliability of an electronic document, due to duplication of data and its virtualization in cyberspace, which is ten times higher than that of the material document, which can be stolen, destroyed, or lost.
6. Online accessibility of an electronic document or cloud service 24–7 from anywhere on the globe, making EDC especially attractive for all the inhabitants of the planet.
7. Total exclusion of physical circulation of metal and paper currency, which have monstrous shortcomings associated with cutting down of the best forests; pollution of the earth's biosphere; enormous costs for manufacturing, storage, and transportation of millions of tons of bank notes; and, most importantly, its involvement in the spread of infectious diseases throughout the planet. In return, we propose pure electronic money and billions of dollars in savings. This is the number one task for the financial and political elite of humanity.
8. Transforming the constitutions and legislations of states to electronic digital documents for direct monitoring and execution in the format of "fact–evaluation–action" by refusing declarative documents requiring hundreds of

bylaws and clarifications. All documents, due to the availability of cloud services, should be for direct action: the article of the law determines the activity of a citizen.

9. Absolute transparency and openness of all kinds of EDC, for every person, that does not contradict the legislation of the country or the charter of the enterprise, organization, or social group.
10. Personal responsibility for the legitimacy of a signed electronic document to exclude references to collective decisions made by incompetent experts.

The evolution of EDC will destroy the existing declarative format of documents. Instead, an intelligent computer-based system for monitoring, management, and forecasting of the activities of each person will appear on the basis of accumulation of experience in decision making. Each program-oriented document will be a digitized triad of "fact–evaluation–action", understandable for generation of adequate actuation effects through computing and considered as a response to the user's request.

12.4 Metric Cyber Relationships as the Basis of Management

No measurement—no governance—no science and education—no economy—no state. Metric estimation of processes and phenomena for adequate distribution of moral and material resources is the basis of fair social relations in a company, organization, or country. The public does not positively perceive this thesis, since it differentiates people into the 20% of creative workers who create 80% of production and the remaining 80% who qualitatively perform functional duties.

Naturally, moral and material incentives between the two specified groups of people in developed countries and the best companies on the planet with a high level of technological culture are distributed in the reverse proportion of about 80%/20%. But in this case, each of the employees, who are within that 80%, has a high level of income. A cost-effective system can provide decent salaries. However, passive majority democracy is unable to adopt a constructive metric for assessing people's activities. Therefore, unpopular and hard regulatory actions by the management who wield the cyberculture of system management of the company, organization, or country are needed. Thus, only creative metric relationships in the society are able to create a competitive structure in the market that can be scaled up to a country, company, organization, or university. The 26-year experience of higher education management by the ministry is interesting here: year after year it created metric relationships for evaluating scientists, units, and universities. The result surprises with its inconsistency associated with creating a redundant metric containing 107 virtual points; in their muddy waters, real–effective indicators have drowned. Suffice it to say that for the formation of summary tables of metric evaluation, a significant part of the university staff "creatively" works for a whole month without producing any socially useful products. On the scale of the country, it represents millions of dollars lost. But that is not all. Skilled university officials and department heads have learned to write fake paper reports based on virtual indicators that defeat the real scientific achievements of constructive scientists. Also, for many years, there have been strange penalties for employees who have zero performance results for certain indicators in the abovementioned 107-point metric.

Their goal is to punish the scientist (department) for noncompliance with certain metric items, instead of rewarding him for achievements in other types of activities. For example, if a scientist is awarded a Nobel Prize but on many other indices he has zeros, the penalties will make him an outsider in the rating evaluation, with subsequent moral and material penalties for failure to comply with their obligations. How should the existing system of rating evaluation, which destroys science in higher education, be corrected? It is necessary to reduce the metric of indicators to a dozen ones, leaving the criteria of scientific, educational, and volunteer extra activity of a scientist or collective to be recognized on the international market: (1) R&D grants and commercial, government, educational, and production projects; (2) scientometrics: the Hirsch index, citations, number of publications; (3) monographs, tutorials, textbooks and manuals for laboratory works, massive open online courses (MOOCs); (4) dissertations, researched under scientific supervision; (5) patents, market research and educational products and services; (6) participation of students in all types of extra activities; (7) awards, titles, prizes, and diplomas for social recognition of merit; (8) organization and holding of conferences, exhibitions, and seminars; (9) publishing in journals, films, and proceedings of conferences; and (10) agreements with enterprises and universities.

The primary criteria forming the university's image are only the first seven ones, which significantly affect the international rating, image, financial state, and economic state of the university and the quality of life of its employees. Only they need to be taken to real execution in order for the whole community to live better and make life more fun.

Elimination of Corruption: The structure of corruption, as a systemic state attribute, contains three interrelated components: officials, citizens (companies), and resources. Resources can be any state (hence non-owned) commodity that can be sold, including finances, natural resources, services, and positions. Officials are civil servants who trade goods that do not belong to them. Eliminating any component from the triad eliminates corruption as a state phenomenon. The most costly and hopeless way is to eliminate all citizens of the state, and corruption will disappear, because there will be no buyers and no state. The second solution to the problem is to eliminate officials by replacing their functions with incorruptible cloud services. The third way to eliminate corruption is not to accumulate state resources for subsequent distribution among the citizens. Cash flows from citizens should directly go to private companies that provide "public" goods and services. This variant also does not provide for the presence of officials as civil servants. They are transferred to private companies, which create and support the country's infrastructure, health, education, and legal protection of citizens' rights and

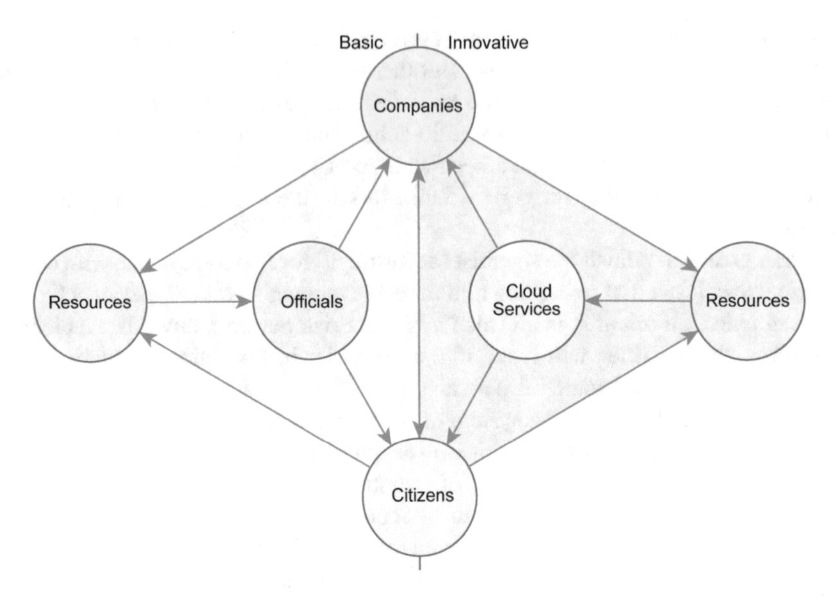

Fig. 12.6 Transformation of statehood into a corruption-free society

property. Thus, the structure of financial flows and services in a corrupt state (Fig. 12.6) is gently transformed into an effective system by replacing the army of officials with an incorruptible cloud service for legitimate and open resource distribution based on metric monitoring of social projects. This means the end of corruption.

The new model tends toward a zero number of officials, turning corrupt-in-essence statehood into a powerful territorial corporation, free from administrative redundancy. It should be noted that the state monopoly creates no alternative to low-quality service for citizens. Instead of this, qualitative multiverse services from private companies are coming into the market, created, among other things, on the basis of an army of competent officials who care about their reputation among the citizens, affecting the profit and quality of life of company employees. The question remains how to initiate the elimination of the redundant link creating corruption and negatively affecting the effectiveness of the social system if there is a link—power.

12.5 Conclusion

A cyberculture of social computing aimed at moral metric management of society based on exhaustive digital monitoring is proposed in order to eliminate corruption, ensure a high quality of life for citizens, and preserve the ecology of the planet. The law of additivity of the nation or social group intellect is formulated, defined by the metric distance between the competencies of the community members.

Prospective directions for creating smart e-infrastructures, states, cities, universities, companies, and houses with a high level of capitalization in the world are considered. Electronic technologies for paperless workflow are proposed, to save hundreds of millions of dollars and foster green ecology on the scale of the state. The issues of human-free moral management of social groups are considered, based on the creation of cyber democracy, cyber governance, and a cyber parliament, to destroy corruption on the scale of the state. Blind faith in power is the sworn enemy of truth (Albert Einstein). Nazism/nationalism is the enemy of science and education.

The capitalization level of the proposed cyber social innovations on the scale of even a weak developing state is billions of dollars. But an even more significant possible achievement is the creation of moral relationships in society and the preservation of ecology for future generations. In addition, cyberculture, cyber finance, and cyber technologies will destroy the criminality associated with the theft of money and cars, corruption and bribery, apartment theft and banditry.

Currently there is a stable dominance of innovative cyberculture, including nanoadditive, bioinformation, cyber physical, and socio–cognitive technologies in the markets of science, education, industry, transport, and quality of life. In this regard, the following areas of innovation are attractive:

1. Creation of a cloud service for monitoring and management of scientific and educational processes, titled the "smart cyber university."
2. Development of intelligent digital monitoring of cars and streetlight-free cloud transport control.
3. Creation of an e-infrastructure and cloud monitoring services for human-free management of social groups.
4. Development of triangular 2D–3D topologies for creation of optimal cyberspace infrastructures: triangle-driven cities, computing architectures, and the cyber ecosystem of the planet.
5. Creation of an e-infrastructure for authenticating a person on the basis of primary signs for banking, travel, and access to rooms, excluding passports, electronic cards, metal keys, plastic keys, and all types of paper documents.
6. Development of an integrated e-infrastructure of the house and apartment for online digital monitoring and cloud management of all household appliances and services.
7. Creation of an e-infrastructure for online e-voting based on human primary sign authentication.
8. Creation of an e-infrastructure for pure and completely transparent electronic money circulation by total exclusion from physical circulation of metal and paper currency, which will bring additional billions of dollars to countries' economies around the world. E-money will be a moral cyber physical metric for the social significance of people, goods, and services.
9. Creation of electronic cyber statehood, when a citizen of the planet will have the right not only to choose his leaders but also to choose his state, where he will pay taxes to obtain moral and qualitative services of legal and social protection.

Municipal services from medical and law enforcement agencies have already become paid services. As for protection from enemies, considering the administrative actions in separate states, there is no greater enemy for the people than their own government. State institutions discredit themselves with their incompetence; therefore, the service market creates effective private parallel structures with the functionalities of courts, police, army, health, science, and education. Financial flows directly go from the consumers of services to their providers, bypassing the state as an intermediary. There are no fundamental restrictions to transforming statehood into an effective private territorial corporation that is morally and service attractive to its citizens.

References

1. Gaol, F. L., & Hutagalung, F. D. (2017). *Social interactions and networking in cyber society*. Singapore: Springer International Publishing.
2. Meiselwitz, G. (2016). *Social computing and social media*. Cham: Springer International Publishing.
3. Koch, F., Koster, A., & Tiago, P. (2016). *Social computing in digital education*. Cham: Springer International Publishing.
4. Barnaghi, P., Sheth, A., Singh, V., & Hauswirth, M. (2015). Physical–cyber–social computing: Looking back, looking forward. *IEEE Internet Computing, 19*(3), 7–11.
5. Hahanov, V. I., Litvinova, E. I., Chumachenko, S. V., & Mishchenko, A. S. (2016). Cyber social system—smart cyber university. *Radioelectronic and Computer Systems Journal, 5*(79), 187–194.
6. http://fakty.ictv.ua/ru/ukraine/20170410-rejtyng-biznes-koruptsiyi-ukrayina

Chapter 13
Practical Conclusion

Vladimir Hahanov

According to quantum theory, a particle (electron) can be in two places simultaneously. This statement does not contradict the Schrödinger equation, where the coordinate of a (cyber) space can have at the same time two particles (electrons) or two stable states. Qubit $Q = (Q_1, Q_2, \ldots, Q_i, \ldots, Q_n)$ is the discrete state of a point in a cyberspace defined by the superposition of a finite number of primitives (carriers) of the universe represented by a vector. A quantum of functionality (XQY) is a universal elementary structure (particle) focused to create any complex memory-driven computer architecture, free of logic. The quantum of functionality is a qubit with identifiers of input and output variables that form the addresses of primitives (memory cells) for purpose-driven execution of transactions (read–write).

A memory structure that uses Manhattan or triangular topology creates quasioptimal conditions for increasing the speed of address transactions between the qubit vectors of functionalities. The sequential processing of the quantum of functionalities for the computing architecture is specified by a qubit of the algorithm, which describes the oriented graph, and is implemented in the memory.

Two memory-driven mechanisms for managing and executing the computer system are created, using only the qubit vector form of the functions and the structure description. The access address to the qubit cell is formed by concatenating the input variable (X) states of the functionality quantum. The content of the Q ($M(X)$) qubit bit, according to the concatenated address $M(X)$, determines the state of the quantum functionality that is written to the cell (Y) of the state vector (M) of the computer system.

Thus, computing (Fig. 13.1)—globally and locally, virtually and physically, software and hardware—is an implementation in memory of a trivial transaction:

V. Hahanov (✉)
Design Automation Department, Kharkov National University of Radio Electronics, Kharkiv, Ukraine
e-mail: hahanov@icloud.com

V. Hahanov, *Cyber Physical Computing for IoT-driven Services*,
DOI 10.1007/978-3-319-54825-8_13

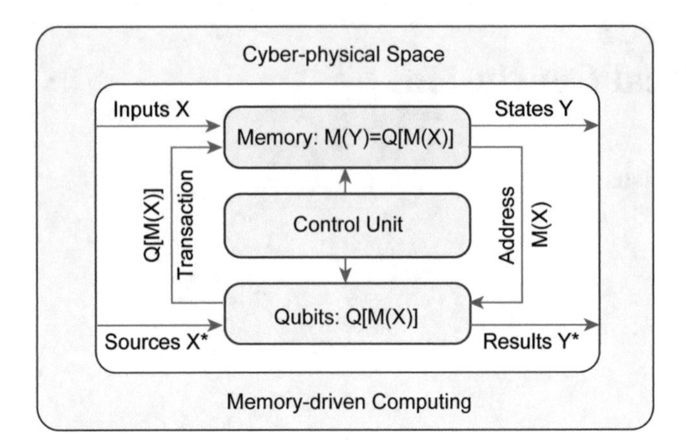

Fig. 13.1 Memory–address–transaction computing

$M(Y) = Q[M(X)]$, called the memory–address–transaction (MAT) equation of computing, and nothing more.

In other words, any computational process is reduced to addressing read–write operations on memory. The practical significance of this paradigm lies in the exclusion of hardware logic from computing, which makes it possible to (1) create a regular computer architecture based on leverage of memory matrices; (2) perform all arithmetic, logical, and specific operations on the matrix memory structure, without accessing the external ALU module, which significantly reduces the performance; (3) technologically solve all the problems of reliability of computer systems based on built-in online testing and repair of failed memory modules by redirecting them to spare cells; (4) have almost unlimited opportunities within a system on a chip (SoC) for parallel execution of logical operations of any register dimension; (5) essentially reduce the time-to-market design of specialized processors by eliminating complex irregular blocks of ALU with information exchange buses; and (6) reprogram the logic in the memory elements online for performing other operations.

There has been a sustainable dominance of innovative cyberculture, which contains nanoadditive, bioinformation, cyber physical, and social–cognitive technologies, creating the new market of science, education, industry, transport, and quality of life. The main attractors connected with the chapters are covered by the keywords qubit function, qubit structure, qubit coverage, qubit vector, quantum of functionality, memory–address–transaction, quantum simulation, qubit design and testing, memory-driven computing, cloud computing, IoT computing, big data computing, cyber physical computing, and cosmological computing.

This book is aimed at the problem of research that consists of eliminating the contradiction between the two-dimensional classical description of functions and structures and the one-dimensional representation of register variables in hardware computing, by reducing functions and structures to a one-dimensional qubit vector metric for technological solution of problems of synthesis and analysis of digital systems based on parallel logical operations that create memory-driven computing.

The substance of the presented scientific and technological research consists of creation of vector data structures and cubic methods of synthesis, testing, modeling, and simulation integrated into the cloud service infrastructure of SoC components in order to improve the quality of computing and the product yield due to addressability of all computing processes and phenomena.

The main innovative idea of the proposed MAT model of computing is the synthesis and analysis of qubit vector digital structures based on addressable memory elements that exclude the leverage of reusable or new logic.

The aim of this research is focused on significantly increasing the yield of software and hardware products and the quality of computing by creating an infrastructure for cloud modeling, simulation, testing, and repair services based on qubit vector data structures describing the functional black box and increasing the performance of the qubit synthesis and analysis methods.

The attractive directions of the further research are: (1) creation of the theory of qubit design and testing of digital systems; (2) development of cloud services for qubit synthesis and analysis of digital systems; (3) creation of cloud services for monitoring and management of scientific and educational processes—a "smart cyber university"; (4) development of cloud monitoring services and streetlight-free transport control; (5) creation of e-infrastructure, cloud monitoring services, and human-free management of social groups; (6) development of triangular 2D–3D topologies for creation of an optimal cyberspace infrastructure (triangle-driven cities, computer architecture, the cyber ecosystem of the planet); (7) creation of the theory of cosmological quantum expansion of humanity in the universe; and (8) quantum computing for intelligent analysis of cyber physical space.

Computing is a binary union. A computing system always consists of two interacting components that perform (intelligent) control and (quantum) execution functions to achieve a specified goal of operating in the cyber physical–biosocial space (Fig. 13.2). Any other system can be represented by a computing interaction of two components. The presence of only one component or the breaking of connections cannot achieve the goal, and leads to the destruction of the system. Any change in the structural relationship or emergence of different goals for the two components of the system lead to collisions and its self-destruction.

Examples of computing systems are:

1. Energy and matter, the interaction of which creates a computing couple, where the goal is a harmonic change in the matter and energy substance in time and space.
2. Space and time, the interaction of which creates a computing couple, where the goal is a harmonic change in the space–time form to create material and energy diversity.
3. Soul (brain) and body, the interaction of which creates a computing couple, where the goal is survival and knowledge of the world.

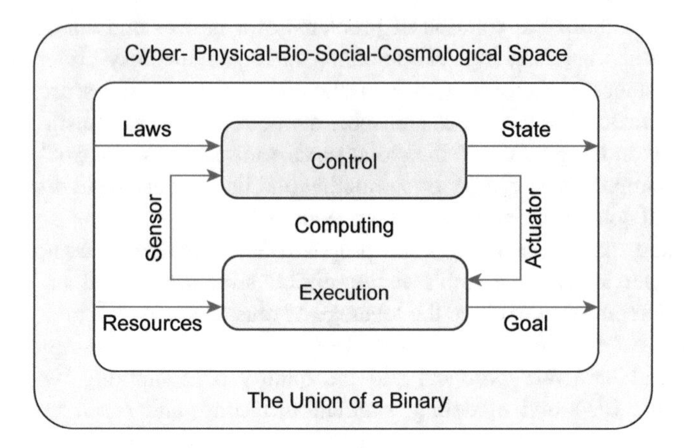

Fig. 13.2 Computing is a binary union

4. A man and a woman, whose interaction creates a computing couple, where the goal is survival and continuation of humanity.
5. Electricity and a silicon crystal, the interaction of which creates a computing couple, where the goal is to create functionality: information storage or computational procedures.
6. A person and a computer, the interaction of which creates a computing couple, where the goal is to improve ecology and the quality of a person's life.
7. Humanity and computing (the Internet), the interaction of which creates a computing couple, where the goal is immortality and expansion of humanity in the universe.
8. The person and an avatar, as his digital copy, the interaction of which creates a computing couple, where the goal is cyber information immortality, in demand for future generations.
9. A social group (state) and the leading elite, whose interaction creates a computing couple, where the overall goal is the quality of life of citizens and the marketability of the community. The existence of different goals for the leadership and the people (the community) leads to social conflicts, revolutions, and riots, and then to the self-destruction of the society or state. Paradoxically, it is a fact that university management in developing countries ignores the theory of system management. The insolvency of the administration of higher education is the absence of strict personnel separation between the rectorate (executive power) and the scientific council (legislative power), which is elevated to the rank of law. This creates, as usual, a corrupt management system of the university, when the rectorate by its majority actually dictates its will to the remnants of the academic council while making ignorant or incompetent decisions that destroy science and education.
10. Humanity and God, whose interaction creates a computing couple, where the goal is moral and creative perfection of people in harmony with nature.

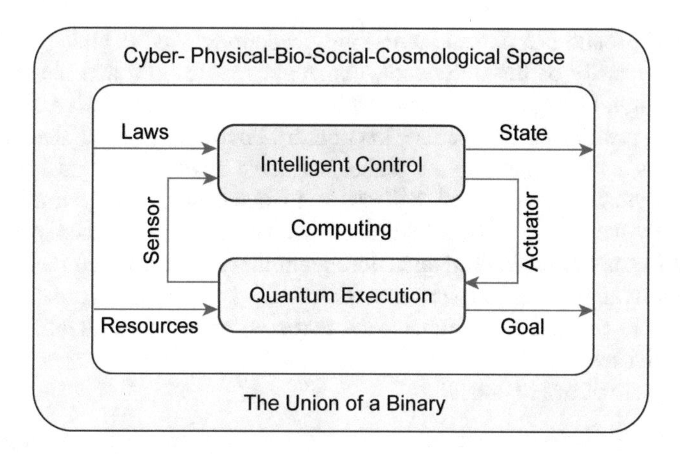

Fig. 13.3 The attractive market-driven computing of the future

The cosmological, cyber physical, and cyber social marriages of matter and energy, space and time, essence and form, control and execution are infinite in their diversity, which shapes the universe. Naturally, the computing couples interact with other couples, which together create a competitive game for survival according to the existing physical laws.

The pair of matter and energy forms a systemic, equitable interaction, where the components provide each other with what they have. The result is a memory genome that contains information about the past, present, and future of the couple's life cycle.

A primitive system containing two interacting and mutually complementary components is the simplest, most reliable, and most vital.

One of the most attractive market-driven computing developments in the next 20 years will be the union of an intelligent control mechanism interacting with quantum execution mechanisms (Fig. 13.3) to achieve the goals of increasing performance, reducing energy consumption and materials, and saving the planet's ecology.

For traveling in space and time, it is necessary to create primitive pairs of matter and energy substances that form a memory for storing information about the structure and algorithms of the system's life cycle in order to reproduce it anywhere in the universe with acceptable conditions. All we need are the abilities to transfer information over any distance, to structure the available matter, and then to breathe the energy of algorithmic life into a new teleported system.

The following possible start-up—"Cyber Human Distance"—completely fits into the scheme of market-driven computing.

The goal is to determine the metric distance between objects or processes in cyberspace to make an exact decision. The service can complement search engines such as Google, Yahoo, IEEE Xplore, Yandex, and Facebook.

The background is a person increasingly becoming a cyber biological phenomenon. The activity of the individual, his social recognition, and his significance create his digital image in cyberspace. Moreover, every constructive citizen of the planet has a positive desire to create his immortal portrait in cyberspace. During the next 5 years, every person on the planet will have a cyber biological image. This also means that the individual will strive to make his cyber portrait open and attractive by the metric of social significance. Thus, a person becomes potentially available for metric evaluation and subsequent moral and material incentives.

The motivation is cloud services for metric analysis of big data becoming marketable to extract, from cyberspace garbage, a pure portrait of each citizen and his activities.

The objectives are to create:

1. A cloud service—"Where have we met?"—for a metric definition of the space–time coordinate of possible intersection with a specific person
2. A cloud service for a metric search for a person who can create the best solution to a specific problem
3. A cloud service for a metric comparison of the achievements of several applicants for a vacant position
4. A cloud service for a metric search for the minimal chain of people providing contact with a specific person
5. A cloud-based service planning a minimum number of activities to achieve a metrically formulated goal
6. A cloud-based metric search service for like-minded people in relation to a particular problem
7. A cloud service to determine the metric parameters of a significant (constructive) effect on the behavior of a particular person
8. A cloud service for formation of a social image of a particular person according to an ordered metric
9. A cloud service for formation of a social portrait of a particular team according to an ordered metric
10. A cloud service for metric analysis of big data in order to search for socially dangerous citizens and prevent crime

The algorithm for solving the abovementioned problems leverages a logical *xor* operation comparing the metric vector m and the possible solutions (A), calculating the quality criteria Q of the cyber physical object's interactions:

$$P(m, A) = \min Q_i \left[m \mathop{\oplus}_{i=1}^{n} A_i \right].$$

Therefore, the methodology for creating cloud search services for people is reduced to the synthesis of a minimal metric of essential parameters.

The market attractiveness of these cloud services is determined by the demand for technologies of precise digital management by social groups, communities, and individuals from state structures and private companies to improve the quality of a citizen's life, goods, and services. In addition, the services will undoubtedly be useful for every person who wants to compare themselves to the surrounding people in order to improve their social status and quality of life. The level of capitalization of the proposed cloud service could reach $0.5 billion.

Chapter 14
Cyber-Physical Technologies: Hype Cycle 2017

Vladimir Hahanov, Wajeb Gharibi, Ka Lok Man, Igor Iemelianov, Mykhailo Liubarskyi, Vugar Abdullayev, Eugenia Litvinova, and Svetlana Chumachenko

14.1 Introduction

Gartner Inc., which is creating a global technology, cyber-fashion, added eight new trends to its Hype Emerging Technologies Cycle brand in 2017 (Fig. 14.1): 5G, Artificial General Intelligence, Deep Learning, Deep Reinforcement Learning, Digital Twin, Edge Computing, Serverless Platform as a Service (PaaS), and Cognitive Computing [1, 2].

Edge computing technology aims to improve the performance of cloud services by leveraging special intelligent computational procedures at the location of the mobile user or embedded microsystem. Digital Twin creates cyber images of physical processes and phenomena. As in a mirror, if a digitized company (university) has no reflection in cyberspace, then it is not in the physical space. Serverless PaaS is hardware-free architecture for organizing cloud-computing processes.

How can we understand the phases of the Gartner cycle? (1) "Innovation trigger" means the launch of innovation, where potential breakthrough technologies that are

V. Hahanov (✉) • I. Iemelianov • M. Liubarskyi • E. Litvinova • S. Chumachenko
Kharkov National University of Radio Electronics, Kharkiv, Ukraine
e-mail: hahanov@icloud.com; igor@itdelight.com; mlyubarskyy@gmail.com; litvinova_eugenia@icloud.com; svetachumachenko@icloud.com

W. Gharibi
Jazan University, Jazan, Saudi Arabia
e-mail: gharibiw2002@yahoo.com

K.L. Man
Xian Jiaotong-Liverpool University, Suzhou Shi, China
e-mail: ka.man@xjtlu.edu.cn

V. Abdullayev
Azerbaijan State Oil and Industry University, Baku, Azerbaijan
e-mail: abdulvugar@mail.ru

V. Hahanov, *Cyber Physical Computing for IoT-driven Services*,
DOI 10.1007/978-3-319-54825-8_14

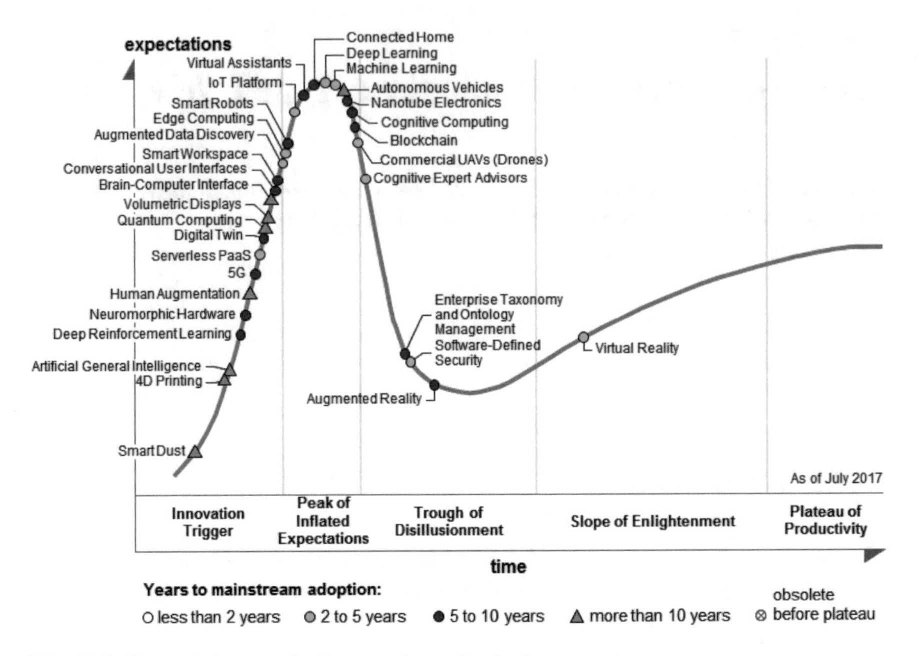

Fig. 14.1 Gartner's hype cycle for emerging technologies

interesting for the market but have unproven commercial consistency replace existing cyber-physical structures. (2) The "peak of inflated expectations" is the highest level of bloated market expectations, where timely advertising creates successful precedents for implementing innovative technologies on the failure field. (3) A "trough of disillusionment" refers to the disappointment that occurs when interest in technology goes down, experiments do not confirm the expected market attractiveness, and individual developers improve their products and learn about investments. (4) "Slope of enlightenment" means an insight when examples exist of technologies that benefit the enterprise and finances are available for pilot projects. (5) "Plateau of productivity" means an area of sustainable enhanced productivity, when the technologies, products, and services that are created find their consumers in the market.

14.2 Three Main Areas of the Cyber Culture

Hype Cycle for Emerging Technologies 2017 forms the planet's cyber culture for the next 5–10 years—and beyond—by expert analysis of more than 1800 possible technologies performed by leading research and consulting companies. A list of 33 + 2 top Gartner-table technologies, shown in Fig. 14.2, creates a cyber culture and competitive advantages for the markets in the fields of science, education, industry, and transport.

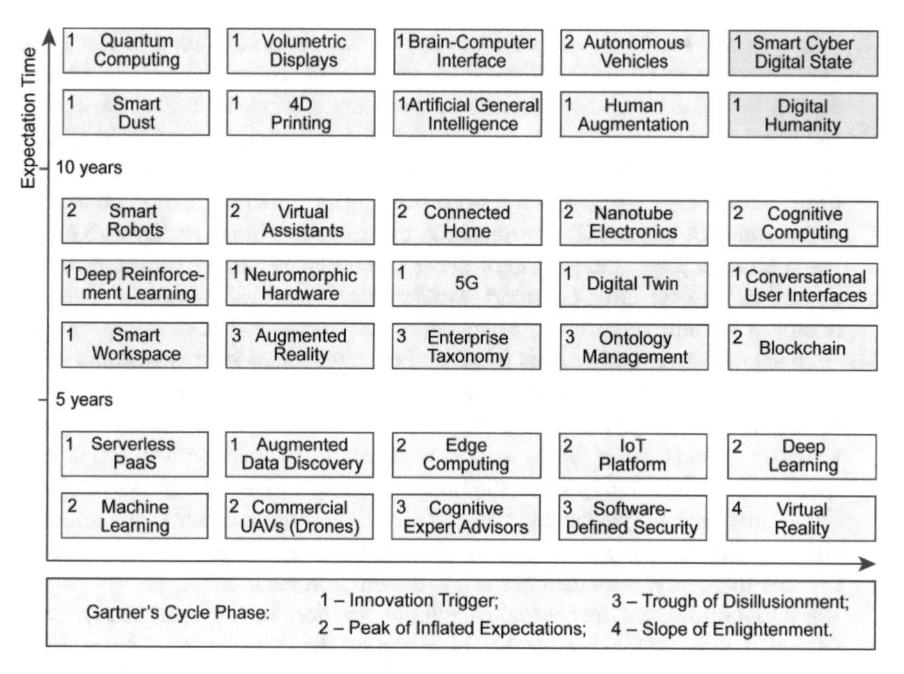

Fig. 14.2 Gartner's table for emerging technologies

The first three positions in the Gartner cycle are assigned to the following strategic directions: artificial intelligence everywhere, transparently immersive experiences, and digital platforms.

1. Artificial intelligence everywhere will become the most disruptive technology in the next 10 years because of the availability of computing power, infinite volumes of big data, and improvements in the ability of neural networks to adapt to new situations that no one has encountered before. Enterprises that are interested in leveraging artificial intelligence consider the following technologies to be useful: deep learning, deep reinforcement learning, artificial general intelligence, autonomous vehicles, cognitive computing, commercial unmanned aerial vehicles (drones), conversational user interfaces, enterprise taxonomy and ontology management, machine learning, smart dust, smart robots, and smart workspaces. Thus, in the next 10 years artificial general intelligence will penetrate all spheres of human activity as a technological service immersed in the cyber space, including 30% of high-tech and transport companies.

   Smart workspaces will be connected 24 hours a day, 7 days a week to infrastructure for solving production problems in space and time. At the same time, virtual private networks are used—metrics for measuring potential and performance results, the presence of a certain cyber culture, and the most convenient places for doing business. High self-motivation for successful and efficient performance of a task stipulates leveraging dynamically changing cyber-physical workspaces for creativity, which are invariably the office, home, vehicles, and places of rest and sports.

2. Transparently immersive experiences are becoming more human-oriented and provide transparency of relations between people, business, and things; and flexibility and adaptability of links between the workplace, home, enterprise, and other people. Gartner Inc. also predicts the following critical technologies will be introduced into practice: autonomous vehicles, brain–computer interfaces, smart dust, 4-dimensional (4D) printing, augmented reality, connected homes, human augmentation, nanotube electronics, virtual reality (VR), and volumetric displays. Integration of cyber technologies aims to ensure humans' quality of life by creating smart workspaces, connected homes, augmented reality, VR, and the growing brain–computer interface. For example, human augmentation technology aims to expand or supplement human capabilities to improve health and quality of life through the harmonious use of cognitive and biotechnical improvements as parts of the human body. Volumetric displays visualize objects using active voxels in three dimensions with a 360-degree spherical viewing angle, where the image changes as the viewer moves. Four-dimensional printing technology is an innovation of three-dimensional printing, whereby structural materials can be transformed after manufacturing a product in order to adapt it to human needs and the environment.
3. Digital platforms of technological culture are formed by components: 5G, Digital Twin, Edge Computing, blockchain, the Internet of Things IoT), neuromorphic hardware, quantum computing, serverless PaaS, and software-defined security (SDS). Technologies such as quantum computing and blockchain will create the most unpredictable and disruptive breakthroughs for humans in the next 5–10 years. Neuromorphic hardware is considered as the future of artificial intelligence; it aims to create a neuromorphic computing chip that can replace the cloud computing power of the Apple Siri Data Center in solving complex machine learning problems (Chris Eliasmith, a theoretical neuroscientist and co-CEO of the Canadian artificial intelligence startup Applied Brain Research) [3]. Inside the iPhone will be a digital brain in the form of a neuromorphic intellectual property core, which solves in real time all the tasks during which a gadget interacts with the outside world. Because of spike asynchronism, IBM's neuromorphic universal chip consumes three orders of magnitude less energy, with a fivefold increase in the number of transistors, exceeding Intel's existing hardware solutions. Compilers such as Nengo and Python are used to program hardware-oriented algorithms. At present, the following chips are already implemented using the Nengo compiler: vision systems, speech systems, motion control, adaptive robotic controllers, and the Spaun chip for offline interactive communication between computers and the environment. SDS, or Catbird, is designed to protect system objects or logical structures in virtual space. This is necessary because network security no longer has physical boundaries within the framework of the logical architecture existence of cloud services. Therefore, an accurate and flexible SDS is created as a complement to infrastructures and data centers without the presence of special security devices. Scaling SDS makes it possible to create or acquire the minimum necessary security conditions in a certain place and time, which significantly reduces the material costs of forming a quality SDS service.

14.3 Examples of Leveraging Top Technologies

High-level costs for research and development from Amazon, Apple, Baidu, Google, IBM, Microsoft, and Facebook stimulate the creation of original, patentable solutions in the fields of deep learning and machine learning; among these technologies are Amazon's Alexa, Apple's Siri, Google's Now, and Microsoft's Cortana. Gartner Inc. is sure that tools for in-depth training will account for 80% of standard funds for scientists by 2018. Today, technologies and data on scientific research are becoming available on companies' websites: Amazon Machine Learning, Apple Machine Learning Journal, Baidu Research, Google Research, IBM AI and Cognitive Computing, and Facebook Research.

The introduction of 5G telecommunication technology (Fig. 14.3) in the coming decade will provide the market with expected innovative solutions for the security, scalability, and performance of global networks and connections in transport, IoT, industry, and healthcare.

Gartner Inc. predicts that by 2020, 3% of mobile network providers will launch commercial networks in 5G format, which will provide qualitatively new conditions for the widespread introduction of telecommunications for scalable globalization: IoT, cloud-transport control, and ultrahigh-definition television. Leaders of 5G implementation in 2017–2018 are AT&T, NTT Docomo, Sprint USA, Telstra, T-Mobile, and Verizon. The 5G technology is an ultra-wideband mobile connection

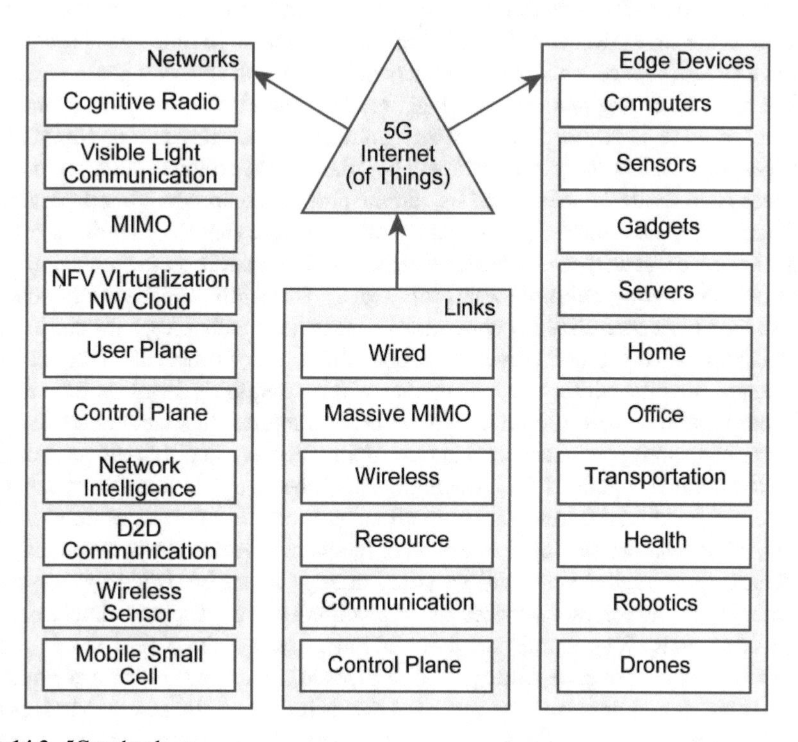

Fig. 14.3 5G technology

in the millimeter range for massive real-time machine-to-machine transactions with permissible delay control (1 ms), while simultaneously connecting about 10 million devices per 1 km^2. The 5G uses beam division multiple access technology to interface a base station with mobile devices. The 5G wireless cellular architecture provides 10–50 Gbps throughput in the 30- to 300-GHz range of ultrahigh-definition video applications and VR creation [4]. The innovative 5G technology features the use of the massive multiple input and multiple output array, cognitive radio network, direct device-to-device connection for IoT, radio access network as a service, and network function virtualization cloud.

The hype cycle does not consider two technologies related to cyber social computing that are essential for humanity: digital humanity and smart cyber digital state. They have been added to the Gartner's Table of Emerging Technologies as the launch of the innovation trigger, the implementation of which is expected in 10 years.

Digital humanity suggests an accurate digital identification of a person based on natural biometric parameters (fingerprints, retinal scanning, and DNA), excluding paper data carriers, plastic cards, certificates, diplomas, and passports on a planetary scale. The digital identifier makes it possible to delegate the positioning of each person in time and space to the cloud service, which removes all problems associated with the cyber-physical analysis of the illegitimate actions by humans. The consequence of sustainable development of digital humanity technology within the green IoT to create a smart world [5] represents multibillion-dollar savings in the costs of producing and using paper products, thereby preserving forests and planetary ecology. The payment for receiving these dividends is the cost of creating an electronic infrastructure for digital authentication of each individual in time and space. Green IoT is a cyber-physical culture of human activity aimed at ensuring people's quality of life and preserving planetary ecology, energy, materials, money, and time. IoT components are identification, sensing, controlling, communication, computation services, and intelligent infrastructure. The smart world provides each person with services from smart devices (watches, mobile phones, computers), intelligent transport (aircrafts, cars, buses, trains), smart infrastructure (homes, hospitals, offices, factories, cities, states), and intelligent remote online education (schools, universities). An interesting solution was proposed by Christidis and Devetsikiotis [6]: blockchain (a decentralized cryptographic transparent technology for storing and exchanging data on completing transactions) allows two businessmen, for example, to interact directly within the IoT service without requiring legal intermediaries. Without legitimate recognition by government structures, this technology is still a resource-consuming, experimental anticorruption enclave in the cyberspace of data and transactions legalized by governments. If the people's trust in blockchain technology increases, then a cyber revolution may occur that will overthrow existing state institutions that are now guarantors of mediators in relations between citizens.

Smart cyber digital state will certainly emerge as an innovative cyber social culture, providing tolerant services based on digital monitoring and cloud management of citizens. Why should a cyber state necessarily take place? (1) A citizen must have a choice of acceptable public services in exchange for the taxes he or she pays. There should be no monopoly on the provision of traditional state services to a citizen, especially if those provisions become logically or emotionally unacceptable. (2) The care of citizens turns from depending on residential location (place) in extraterritorial cloud service, regardless of location, because of the presence of cyberspace, which connects citizens with state organizations online [7]. (3) Conflicts between the authorities and the people should be resolved by online monitoring of public opinion on substantive issues. (4) Like social networks, new states will emerge and old, uncompetitive social entities will collapse. Whoever becomes the leader in cyberspace will rule in the physical world as well. (5) The struggle for the souls of people in the virtual world from the side of cyber-states becomes the main argument of the authorities about the consistency of market attractiveness. (6) In the cyber-physical world, physical prototypes and digital images gradually change places. The "master–slave" pair, or who is the leader and who is the follower, is formed in the sequence "cyber image–physical prototype." This is especially true for states that have existed throughout the history of mankind as long as citizens supported the basic doctrine – the immaterial substance put forward by the ruling elite. Today it has become more important to make personal decisions not as a citizen, but as a metric cyber-image of their social significance. (7) Statehood becomes an integral cyber-service available in the market worldwide, which can be bought by a citizen in exchange for taxes. (8) The main function of the cyber-state is the formation of the moral, emotional intelligence of society by means of a tolerant integrating association of languages, histories, cultures, religions, and traditions, which is dominant when making strategic decisions, in comparison with the logical intelligence of a social group.

The world is becoming more and more intelligent, digitized, and strongly connected (networked) among people, things, and services. Figure 14.4 presents a picture of the top 10 strategic trends in the information technology industry in

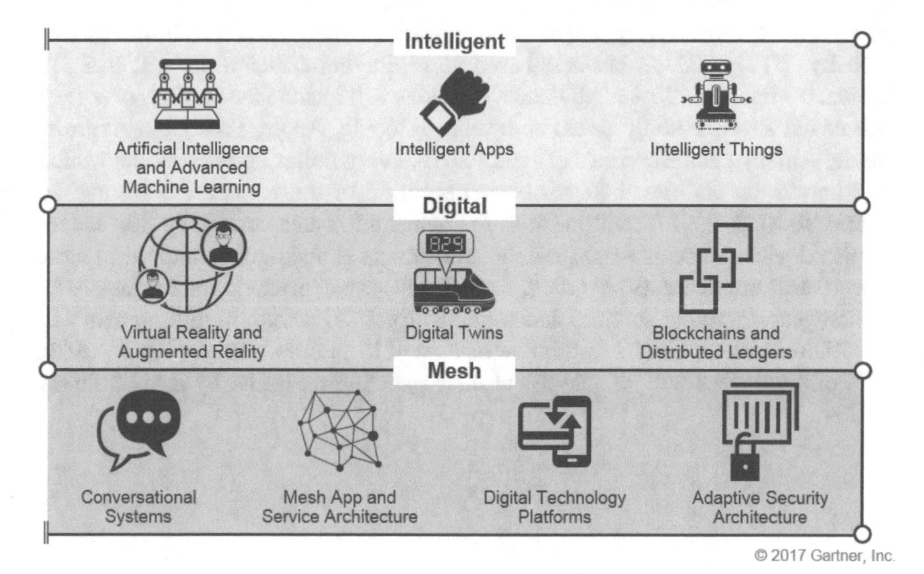

Fig. 14.4 Top 10 disruptive directions in the information technology industry

2017 [8], which should be adopted by all companies and universities that wish to form new business projects on the NASDAQ market of goods and services.

Today, a new, more sophisticated, and intelligent digital cyber-physical and intellectual world has appeared to allow the harmonious existence of humans in an environment of people-friendly goods and services. Thus, the general picture of the friendly world consists of the desires to (1) digitize all objects and processes on the planet (spatial, biological, technical, social, virtual); (2) extend scalable artificial intelligence into all digitized cyber-physical processes and phenomena; and (3) connect all intelligent objects and processes to a smart scalable network within the framework of a single digitized cyberspace. The goal of creating a digital intelligent cyber-physical world is to ensure the quality of life of moral people, eliminate social collisions, and provide green ecology on the planet.

Gartner Inc. considers it necessary to withdraw from the current market the following technologies, which do not meet the expectations of information technology business: affective computing, micro data centers, natural-language question answering, personal analytics, smart data discovery, and virtual personal assistants.

To create successful businesses and new educational courses, Gartner Inc. recommends taking into account assumptions about strategic planning, which include 10 points: (1) By 2020, 100 million consumers will buy into expanded reality, including using head-mounted displays. (2) By 2020, 30% of web browsing sessions will be performed without a screen. More than 5 million of the 550 million Apple iPhone owners will use AirPods to exchange voice messages. Five percent of consumer-oriented websites will be equipped with audio interfaces (including voice chats with live support). (3) By 2019, 20% of brands will abandon their mobile applications (in favor of the mesh app and service architecture). (4) By 2020, smart algorithms will positively affect the behavior of more than 1 billion workers globally. (5) By 2022, a business based on leveraging detachments will cost $10 billion. (6) By 2021, 20% of all human activities will include the services of at least one of the seven leading global companies (Google, Apple, Facebook, Amazon, Baidu, Alibaba, and Tencent). (7) Until 2019, every dollar invested in innovation will require an additional $7 for basic execution of the project. (8) During the course of 2020, the IoT will increase the demand for data centers by 3%. Indoor display devices, such as Amazon Echo and Google Home, will be located in more than 10 million homes. (9) By 2022, the IoT will save consumers and businesses $1 trillion/year, targeting services and supplies. By 2020, about 40 million cars will use Android Auto, and 37 million vehicles will leverage CarPlay. (10) By 2020, 40% of employees will be able to cut their spending on health by using a fitness tracker.

14.4 Innovation for the Architecture of Quantum Computing

The physical basis of classical quantum computing (Fig. 14.5) is to leverage superposition and entanglement operations over electron states, which are quite sufficient for the computational process [10–12]. The electron performs a memory function, storing bits of information. The low and high electron orbits correspond to the values of 0 and 1. A functionally complete basis for creating a quantum computing architecture is represented by the operations of superposition and entanglement, which can correspond with the traditional logical basis or not. In set theory, an isomorphism in the form of a union–complement pair is set as a given. Based on this pair of primitives, a more complex system of logical elements is constructed to organize and optimize computational processes. Disadvantages of quantum classical computing are (1) the high cost of maintaining the temperature conditions (−270 °C) required to operate quantum atomic structures and (2) the observability of the results of computational processes, leading to data being destroyed after they are read.

Innovation in the architecture of quantum computing is determined by the elimination of logic, associated with superposition and entanglement. The analogy is the memory-driven architecture of the classic computer, free of reusable logic. In such a computer, there is nothing but a memory; a transaction (write-read operation) is performed on the address memory. Transactions are sufficient to organize any computational process by using a unique characteristic equation [10, 12]:

$$M_i = Q_i[M(X_i)].$$

Here is the memory for the vector state of the computational process; vector qubit of a logical primitive; vector address of the Q-logic coverage cell. Q-logic is implemented on addressable memory, where all primitives are also integrated under

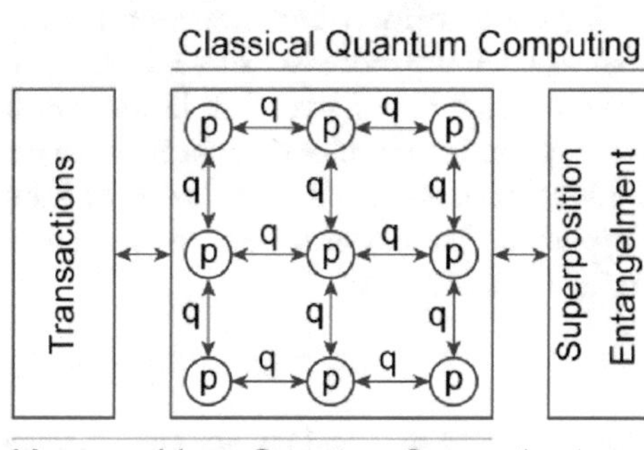

Fig. 14.5 Two types of quantum computing

an M-state vector, which forms the binary addresses of M(X) from the array of variable X.

An innovative proposal is to create quantum memory–driven computing without superposition and entanglement operations, based on the use of the above characteristic equation, which realizes read-write transactions on the electron structure (see Fig. 14.5). To exclude the two aforementioned operations from quantum computing means to simplify significantly the architecture based on the memory of electrons, performing transactions between them using quanta or photons.

Several recent publications that focus on stable trends to create quantum computing, based on the atomic structure of memory with the transmission of information by means of photons or quanta, serve as confirmation of the validity of the proposed innovative quantum architecture.

Scientists from the California Institute of Technology created an optical quantum memory in which information is transmitted by encoding data by leveraging the quantum state of photons [13]. Memory is realized on rare earth elements and is capable of storing photon states with the help of intermediate resonators between atoms and light. The dimension of quantum memory is 1000 times smaller than that of traditional solutions. It is implemented in a nanotube, which allows information to be stored in a very small volume.

Practical realization of the idea of replacing electrons with photons led to the creation of computing with a performance close to the speed of light [14]. Korean researchers have taken one more step toward quantum optical computing. They created a photon-triggered nanowire transistor based on crystal and porous silicon, in which the current is switched and amplified under the influence of a photon. The use of photons in logical AND, OR, and NAND gates leads to ultracompact nanoprocessors and nanoscale photodetectors for high-resolution imaging.

Scientists from Columbia University are on the way to creating a molecular electronics transistor from one atom [15]. They created a geometrically ordered cluster of inorganic atoms with a central nucleus consisting of 14 atoms; this was connected to gold electrodes, which allowed the transistor to be controlled under the influence of one electron at room temperature.

What are the formal differences between classical and quantum computing? The first sequentially processes addressable or ordered heterogeneous data, spending on the procedure $Q = n$ cycles. It is also capable of processing homogeneous data in parallel and in one automatic cycle. If the data are not ordered and are sets, then the limiting computational complexity of their processing on a classical computer depends on the power of the two sets and is defined as $Q = n \times m$. For example, to intersect two sets:

$$M_1 \cap M_2 = \{Q, E, H\} \cap \{E, H, J\} = \{E, H\}$$

it is necessary to spend six automatic cycles. Quantum computing eliminates this drawback associated with the quadratic or multiplicative computational complexity of the intersection procedure on a classical computer. He solves the problem of simultaneous and parallel processing of set theoretical data. An example of this is the parallel execution of the aforementioned operation of intersection over sets in one automatic cycle. To do this requires the operation of superposition (union) of primitive symbols, entering into sets:

$$M_1 \cap M_2 = \{V\} \cap \{C\} = \{P\} = \{E, H\},$$

taking into account a closed set theoretical alphabet [10]:
B*(Y)={Q, E, H, J, O={Q, H}, I={E, J}, A={Q, E}, B={H, J}, S={Q, J}, P={E, H}, C={E, H, J}, F={Q, H, J}, L={Q, E, J}, V={Q, E, H}, Y={Q, E, H, J}, U=∅}.

The symbols of the alphabet represent the set of all subsets on the universe Y, which are composed by superposition of primitives. Quantum superposition makes it possible to concentrate several discrete states at one point in the Hilbert space. Similarly, the join operation also creates a symbolic image in a single point in the discrete space containing several states.

Based on the mentioned, it is enough to just use a multivalued closed alphabet to simulate quantum computing on a classical one. But for this it is necessary to first create a symbolic system (set theory algebra) to encode states. The simplest is Cantor algebra, which operates with two discrete states and creates four symbols: $A^k = \{0, 1, X = \{0, 1\}, \varnothing\}$. The symbols of this alphabet are a set theoretic interpretation and isomorphism of the qubit. Otherwise, the superposition of two states of one qubit creates four symbols. Naturally, two qubits are able to generate 16 states, and three qubits, 64 states. In general, the number of states Q has a dependence on the number of qubits n, which is represented by the formula $Q = 2^{2^n}$.

For parallel execution but already logical operations on qubits, it is necessary to encode the primitive symbols of the alphabet with a unary binary code. The remaining symbols are obtained by superpositioning the primitive codes. The exception is the character code of the empty set, which is obtained by applying the logical AND operation. For Cantor algebra the correspondence table "Symbol-Code" has the following form:

$a_i \in A^k$	0	1	X	∅
$C(a_i)$	10	01	11	00

The cost of the parallelism of logical operations on sets in a classical computer is a significant increase in the bit length (register, memory) for encoding symbols of the alphabet. A similar correspondence table for encoding the hexadecimal alphabet B*(Y) has the form

$a_i \in B^*$	Q	E	H	J	O	I	A	B	S	P	C	F	L	V	Y	∅
$C(a_i)$	1	0	0	0	1	0	1	0	1	0	0	1	1	1	1	0
	0	1	0	0	0	1	1	0	0	1	1	0	1	1	1	0
	0	0	1	0	1	0	0	1	0	1	1	1	0	1	1	0
	0	0	0	1	0	1	0	1	1	0	1	1	1	0	1	0

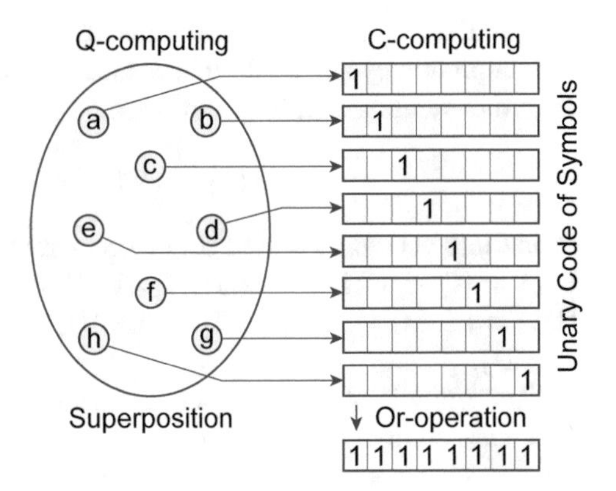

Fig. 14.6 Superposition of primitive elements and logical union of vectors

Thus, the expansion of the power of the set theoretical alphabet can be matched with the buildup of qubits in a classical quantum computer. This makes it possible to perform computational procedures in parallel and in one automatic clock cycle based on the leverage of logical (set theoretic) operations.

The tabular model of a logical element is initially represented by a set of rows or by a set of discrete relations between input and output variables. Instead of such a set number amount, a qubit vector of output states is proposed, oriented toward addressable parallel simulation of digital logic circuits. Replacement of an unordered row set of a truth table by an ordered vector of addressable states makes it possible to create parallel computation on classical computers by increasing the memory for unary encoding of each state. Otherwise, the superposition of n elements of a finite set in a quantum (Q) computer has a one-to-one correspondence to the n-dimensional vector in the classical (C) addressable computer (Fig. 14.6). This vector is obtained by performing an OR operation on unary codes of primitive elements of the original set. Naturally, any intersection (AND), union (OR), or addition (NOT) in C-computer of unary data codes is performed in parallel in one automatic cycle in a C-computer, as in a Q-computer. The pay for the speed is the increase in the memory (the number of bits) for unary encoding of symbols, relative to positional encoding, which is determined by the following expression: $Q = n/\log n$.

14.5 Conclusion

1. Cyber trends from Gartner Inc. give the opportunity for corporate architecture and university leaders to keep up with digital business processes in science, education, and industry; provide timely responses to cyber threats; use business innovations; and determine an effective digital business strategy for the sustainable development of states.
2. In fact, the hype cycle is a deep spatiotemporal 4D analytics of the modern market state related to sustainable cyber-physical development of smart technologies for the next 10–15 years.

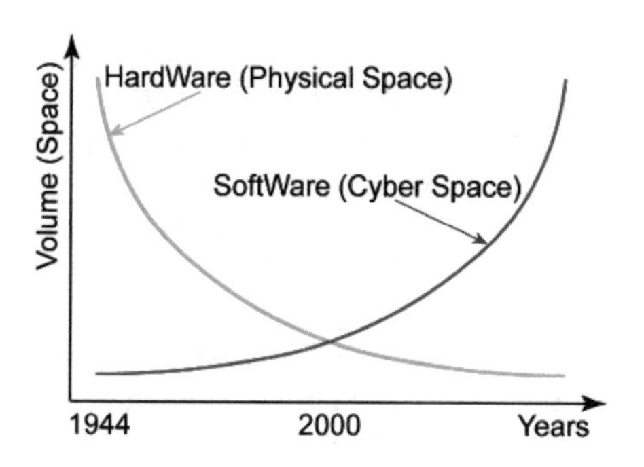

Fig. 14.7 Interaction of volumes of cyber-physical components

3. For universities, the hype cycle determines the vital need to invest in innovative technologies to enhance students' knowledge, in order to obtain within 5–10 years an army of creative experts capable of raising the state from the ruins of modern cyber ignorance. Otherwise, the Gartner cycle for the university is a strategy of its cyber-physical sustainable development in time and space. Any strategy developed without knowledge of the pace and direction of technological change will suffer from incorrectly planned actions and the destruction of business, science, and education. For instance, it should be borne in mind that in 2018 robobosses will accurately monitor and remotely manage 3 million workers worldwide, with the goal of metrically assessing employees' potential; distributing tasks and logically routing their implementation, regardless of the location of the workplace; and assessing quality and productivity.
4. The hype cycle implicitly differentiates all top technologies as either master or slave, which means that making hardware (physical space) platforms more compact is always given priority, since the rest of the virtual world (cyberspace) will always be striving for unlimited expansion of software applications. The interaction of two worlds associated with the steady development of volumes of hardware and software that form a cyberspace is shown in Fig. 14.7.
5. Nevertheless, hardware and software technologies are represented in the hype cycle (on the market) in practically the same proportions (50:50):

   Hardware-driven technologies are 4D printing, volumetric displays, nanotube electronics, brain–computer interface, human augmentation, autonomous vehicles, cognitive computing, commercial unmanned aerial vehicles (drones), smart dust, smart robots, smart workspace, connected home, 5G, IoT platform, Edge computing, neuromorphic hardware, and quantum computing.
   Software-driven technologies are deep learning, deep reinforcement learning, artificial general intelligence, enterprise taxonomy, ontology management, machine learning, virtual assistants, cognitive expert advisors, digital twin, blockchain, serverless PaaS, SDS, VR, augmented reality, augmented data discovery, conversational user interfaces, digital humanity, and the smart cyber digital state.

6. The identical ratio of hardware and software technologies in Gartner Inc.'s forecast means that the levels of their capitalization of the NASDAQ market tend toward parity, a clear example of which are Apple and Google. These manufacturers are significantly different in that they rely on the wisdom of their

teams (experts) armed with the doctrine "consumers cannot predict their own needs" [9]. An alternative policy is that of Microsoft, which conducts extensive research before launching a product, such as the Windows Phone. According to Gartner Inc., Apple's share in the global mobile phone market is 14.2%, compared with 3.3% for Microsoft. Whom should these companies trust, experts or consumers? The answer is unambiguous to the experts: 4D format (always, everywhere, and on all issues).

7. The memory-driven innovation architecture of quantum computing is determined by the ability to eliminate logic associated with superposition and entanglement of states, based on the use of the characteristic equation that realizes read-write transactions on the structure of electrons. Eliminating logical operations from quantum computing will greatly simplify the architecture to the level of the memory structure of electrons, allowing transactions between them using quanta or photons. The formal difference between quantum computing and classical computing consists of the possibility of parallel and simultaneous execution of logical operations on sets.

References

1. https://www.forbes.com/sites/louiscolumbus/2017/08/15/gartners-hype-cycle-for-emerging-technologies-2017-adds-5g-and-deep-learning-for-first-time/#646a4cf34be2
2. http://www.gartner.com/newsroom/id/3784363
3. http://www.wired.co.uk/article/ai-neuromorphic-chips-brains
4. Gupta, A., & Jha, R. K. (2015). A survey of 5G network: Architecture and emerging technologies. *IEEE Access, 3*, 1206–1232.
5. Zhu, C., Leung, V. C. M., Shu, L., & Ngai, E. C. H. (2015). Green internet of things for smart world. *IEEE Access, 3*, 2151–2162.
6. Christidis, K., & Devetsikiotis, M. (2016). Blockchains and smart contracts for the internet of things. *IEEE Access, 4*, 2292–2303.
7. Zanella, A., Bui, N., Castellani, A., Vangelista, L., & Zorzi, M. (2014, Feb). Internet of things for smart cities. *in IEEE Internet of Things Journal, 1*(1), 22–32.
8. https://www.gartner.com/doc/3471559?srcId=1-7578984202&utm_campaign=RM_GB_2017_TRENDS_QC_E2_What&utm_medium=email&utm_source=Eloqua&cm_mmc=Eloqua-_-Email-_-LM_RM_GB_2017_TRENDS_QC_E2_What-_-0000
9. http://www.gartner.com/smarterwithgartner/three-digital-marketing-habits-to-break-2/
10. Hahanov, V. (2017). *Cyber physical computing for IoT-driven services* (p. 243). New York: Springer.
11. Hahanov, V. I., Tamer, B. A., Chumachenko, S. V., & Litvinova, E. I. (2015). Qubit technology for analysis and diagnosis of digital devices. *Electronic Modeling Journal, 37* (3), 17–40.
12. Vladimir Hahanov, Wajeb Gharibi, Eugenia Litvinova, Mykhailo Liubarskyi, & Anastasia Hahanova. (2017). Quantum memory-driven computing for test synthesis. *IEEE East-West Design and Test Symposium*. Novi Sad, Serbia. p. 63–68.
13. https://spectrum.ieee.org/nanoclast/semiconductors/nanotechnology/quantum-optical-memory-device-one-thousand-times-smaller-than-previous-options
14. https://spectrum.ieee.org/nanoclast/semiconductors/devices/nanowire-transisitors-triggered-by-photons-combine-photonics-and-electronics
15. https://spectrum.ieee.org/nanoclast/semiconductors/devices/single-molecule-transistors-get-reproducibility-and-roomtemperature-operation

Index

A
Actuators, 204
Amazon's Alexa, 263
Apple's Siri, 263
Assertion-based verification, 161
Assertion-driven verification, 149, 150
Attractive market-driven computing, 255
Automaton computing model, 17
Automaton structure, 13

B
Big Bang theory, 229
Big data, 183
Big data quantum computing
  Boolean metric, 48
  circuit implementation, 48
  computational procedure, 48
  cyberspace, 58
  multiprocessor operation, 55
  multivalued alphabet, 51
  multivalued logic space, 57
  quality criterion, 44
  query–component, 51
  structures, 43
  topological cyber physical structure, 61
  triangle structure row, 58
  variables, 45
  vector logical processor, 50
  vector operation, 44
  vector quality criterion, 46
Big data–based cloud computing, 6
Binary logical vector, 35
Binary union, 253, 254
Bolognese metric, 53
Boolean Cantor algebra, 98
Boolean derivative, 109, 112
Boolean function, 114, 115
Boolean variables, 47
Built-in self-test (BIST), 95

C
Cantor algebra, 74, 98
Cartesian product, 79
Cloud and infrastructure interactions, 215
Cloud EDA system, 159
Cloud micro services, 207
Cloud–mobile management, 199
Cloud monitoring, 199
Cloud service, 145, 146
Cloud traffic control
  advantages, 214, 216
  classical automaton computing model, 201
  computing model, 202
  crossroad, 205
  cyber culture, 202
  drawback, 216
  goals, 203, 204
  planet's ecosystems, 202
  register-based structure, 205, 206
  scalable computing model, 201
  technological culture, 201
  virtual cyber computing, 202
Cloud traffic control system, 216
Code-Flow Transaction (CFT), 151
Computing systems, 253
Corruption, 247
Corruption-free society, 248
Cosmological computing

V. Hahanov, *Cyber Physical Computing for IoT-driven Services*,
DOI 10.1007/978-3-319-54825-8

Cosmological computing (*cont.*)
arbitrary phase slot, 223
automaton model, 219
definitions, 220
energy, 220
energy transformation, 222
galaxies' expansion, 222
genome universe program, 222
goals, 220
harmonic function, 222, 223
intergalactic processes, 222
matter, 220
microcosm and macrocosm, 219
space, 221
timing, 220
universe, 222
Cyber and physical technology integration, 215
Cyber culture, 2, 5, 184
EDC, 244
internal and external enemies, 242
post-Soviet, 242
principles, 244–246
win–win methods, 242
Cyber culture, Gartner cycle
artificial intelligence everywhere, 261
digital platforms, 262
transparently immersive experiences, 262
Cyber democracy, 196–198
Cyber Human Distance, 255
Cyber image–physical prototype, 265
Cyber physical computing
artificial intelligence, 6
automaton determinism, 4
automaton model, 4, 10
automaton structure, 14
components, 10
computing model, 4
control, 7
creation, 7
cyber physical space, 13
development, 1
e-citizenship, 6
humanity, 5
information technology, 3
innovations and discoveries, 2
IT engineering, 1
levels, 7
market-attractive features, 2
MAT, 9
measurement processes, 4
mechanisms, 4
metric assessment, 5
physical and social processes, 2
reflection, 7
self-learning and self-evolution, 7
structures, 9
Cyber physical monitoring, 206, 207
Cyber physical structure, 61
Cyber physical system (CPS), 185, 198
Cyber physical–biosocial space, 253
Cyber-physical technologies, 267
businesses and educational courses, 266
cyber culture, 260–262
digital humanity, 264
Gartner's hype cycle, 259, 260
people-friendly goods and services, 266
quantum computing (*see* Quantum computing)
smart cyber digital state, 264, 265
telecommunications, for scalable globalization, 263
Cyber Physical Topology, 60–63
Cyber service market, 187, 188
Cyber social computing
attractive statehood, 233
cloud metric human-free management, 236
cyber community management, 233
cyber democracy, 238
e-infrastructure, 238
harmonic genome, 236
incompetent persons, 236
intelligence additivity, 235
IoT, 237
logical operations, 235
reliable metric management, 236
reliable system, 236
social significance, 234
social system, 234, 235, 237
statehood and corruption, 234
structure, 233
Cyber social governance
authoritarian rule, 240
economically efficient system, 240
EDC, 239
humanity, 241
humanity's development, 239
immoral democracy, 240
incompetent leaders and uneducated political elite, 239
physical monitoring and control, 241
scalability, 241
scientist's/employee's achievements, 240
social protection mechanism, 240
teacher–student communication, 241
US presidential election system, 241
Cybersecurity, 184
Cyberspace, 56–59, 184, 256, 264–266, 271

D
Deductive parallel fault simulator, 117
Design under verification (DUV), 150
Digital functional elements, 99
Digital humanity, 264
Digital systems, 131
Digital Twin, 259

E
e-citizenship, 6
Edge computing technology, 259, 262, 271
Einstein equation, 229
Einstein's theory, 222
Electronic design automation (EDA), 160
Electronic document circulation (EDC), 239, 245
Electronic document management system (EDMS), 185
Electronic infrastructure, 243
Electronic technologies, 249
ERA-GLONASS, 212
Eternity, 230

F
Fault detection matrices, 121
Fault detection table, 165, 166
Fault detection test, 102
Fault detection vectors, 128
Fault-free quantum simulation method, 138
Flip-flop structure visualization, 144
FPGA Family Xilinx®, 95

G
Gartner cycle, 259, 261
Gartner Inc., 263, 266, 270
Genome algorithm, 221
Global cyber physical system, 204, 205
Global e-infrastructure, 6
Global navigation satellite systems (GNSS), 212–214
Google's Now, 263
Graph structures, 79–81
Green IoT, 264

H
Hamming distance, 23, 49
Hardware description language (HDL)
  diagnosis, 155, 156
  digital system, 150, 157
  TABA graph diagnosability, 160
  transaction-level TABA graph, 160
  verification methods, 159
Hardware implementation, 75
Hardware logic computing, 252
Hardware–software system, 157–160
Harmonic cycling, 229
Harmonic evolution model, 228
Hasse multiprocessor, 170
Hasse processor, 63, 65
Hilbert space, 74
Human-free moral management, 249
Hype Cycle for Emerging Technologies 2017, 260

I
Image transactions processor, 56
Inertial navigation system (INS), 213
Innovative cloud–mobile services, 199
Innovative cyber culture, 252
Intelligent cyber physical transport system (ICTS), 206
Internet of Nature Computing, 1
Internet of Things (IoT), 32, 188, 236

L
Library Control module, 143
Logic Circuit Quantum Vector, 85–88
Logic circuits
  addressable and spare elements, 177
  combinational circuit, 178, 180
  control algorithm, simulation, 179
  digital product functionality, 177
  element-addressable combinational circuits, 178
  functional elements/structural primitives, 178
  multiplexers, 176
  operational structure, 179
  PLD chips, 180
  PLD memory elements/software modules, 177
  primitive types, 176
  procedures, 178
  SoC infrastructure IP, 178
  software simulation, 177
Look-up tables (LUTs), 95

M
Machine learning, 261–263, 271
Manhattan/triangular topology, 251
Manhattan topology, 59

Massive open online courses (MOOCs), 247
Matter–energy universal set, 227
Memory, 9, 17, 38, 72
Memory devices, 97
Memory–address–transaction (MAT), 9, 96, 252
Memory-based digital circuit, 139
Memory-based models, 89
Memory-driven quantum computing, 268
Metric cyber relationships, 246–248
Microelectromechanical systems (MEMS), 213
Microsoft's Cortana, 263
Mnemonic description, 144
Modeling digital systems, 84
Multiprocessor architecture
  analytical model, 30
  components, 33
  cyberspace component, 24
  devectorization, 29
  discrete space, 22
  functional elements, 31
  logical processor, 34
  membership functions, 27
  memory cell, 31
  minimum number, 28
  null vector, 23
  operations, 29
  prototype, 39
  quality metric, 30
  shifting and compacting, 28
  symmetric difference, 24
  theoretic vector operation, 25
  vector logic criterion, 26
  vector logical computing, 22
  vector quality criterion, 27
  vectors' interaction, 25
  VLMP, 32, 33
Multitree traversal, 158

N
Nonarithmetic metrics, 22–30

O
On-chip distribution, 96

P
Parallel logical operations, 253
Pareto principle, 239–241
Physical computing, 12
Physical computing equation, 10
Post's theorem, 76
Power consumption, 136
Programmable logic devices (PLDs), 95

Q
Q-logic, 267
Quantum computing, 16–18
  adjacent qubit parts, 106
  components, 120
  data organization, 98
  deductive simulation, 117
  digital devices, 120
  digital functional elements, 99
  disadvantages, 267
  FinFET transistors, 96
  functional components, 97
  innovation, 267
  logical register operations, 124
  market feasibility, 97
  MAT, 96
  memory, 95, 268
  operation of superposition, 269
  qubit vector cells, 105, 114, 118
  Schneider scheme, 123
  sequential algorithm, 106
  test synthesis algorithm, 100, 101
  two-input element, 116
  types, 267
  variables, 110, 112
  vector representation, 98
  zero signals, 117
  zero value, 110
Quantum computing architecture, 63
Quantum computing methods, 71, 73
Quantum data structures, 73–78
Quantum execution mechanisms, 255
Quantum Hasse device, 66
Quantum Hasse structure, 64
Quantum of functionality, 251
Quantum processor, 136
Quantum Sequential Primitives, 88–91
Quantum teleportation computing, 14
QuaSim, 142
QuaSim cloud service, 144, 145
  digital device, 137
  fault-free interpretative simulation method, 141
  memory components, 137
  objectives, 143
  parameter, 142
  Q coverage–driven software, 137

Q coverages, 141
QuaSim cloud service, 143
qubit elements, 136
sync signals, 140
Quasioptimal coverage, 36
Qubit computing, 165–176
binary variables, 164
cloud Internet services, 163
cyberspace, 163
data structures, 164
digital system diagnosis
Cartesian product, 165
components, 165
digital circuit and fault detection table, 166
digital unit, discrete process/ phenomenon, 165
disjoint subsets, 165
expression, 169
expression and test patterns, 167
fault detection table, 168
multiple stuck-at faults, 170
output response vector, 170
parallel logical operations, 165
repair cycle, logic blocks, 171
single stuck-at faults, 169
single vector operation, 169
structural activation matrix, 168
structural fault activation, 166, 167
test vector, 167
theorems, 169
transducers implementation, 170, 171
truth table, 168
digital systems
circuit structure, 172
components, 173
computational complexity, 174
digital circuit, 171–174
fault-free behavior, 172, 173
fault-free interpretative modeling, 171
fault-free simulation, 176
noninput variables, 174
preprocessor procedure, 174
quasioptimal data structures, 175
simple iterations/Seidel iterations, 174
single coordinate value, 173
FinFET transistors and 3D technology, 163
nanoelectronic technologies, 163
nondeterministic quantum interactions, 164
parallel computation, 180, 181
quantum computing, 164
time-consuming models, 163
unitary positional coding states, 164
Qubit coverage method, 114
Qubit data structures, 11, 136
Qubit deductive method, 126
Qubit deductive simulation, 128
Qubit description
automaton model, 71
binary vector, 75
digital circuits, 83
digital system, 76
electronic technology, 75
fault-free behavior, 76
graph structures, 82
logical elements, 72
memory element, 89
memory-driven processor, 72
node, 80, 81
power set, 74
primitive graphs, 80
quantum-like data structures, 72
qubit vector, 76
SoC, 72, 81
software applications, 73
unitary and binary codes, 76
vector, 76
vector data structures, 73
Qubit fault simulation method, 120
Qubit method, 116
Qubit *Q* coverage, 173
Qubit simulation method, 138
Qubit simulation processor, 121
Qubit Test Generation Method, 99–109
Qubit vector, 102
Qubit vector digital structures, 253

R
Road user (RU), 209, 210
Row analysis architecture, 35
RTL-level scheme, 67

S
Schneider circuit, 122–132
Schrödinger equation, 251
Search engines, 255
Seidel's method, 175
Sensors and actuators, 16
Smart cyber physical system, 185
Smart cyber university (SCU)
big data, 183
cause–effect cycle, 189
competence, 184
computational methods, 189

Smart cyber university (SCU) (*cont.*)
  CPS, 185
  cyber physical system, 189
  cyber social relationships, 183
  cyberculture, 184
  cybersecurity, 184
  cyberspace, 184
  description, 185
  digital monitoring and active cloud cyber management, 189
  digitization, 189
  EDMS, 185
  education, 184
  entities, 193
  human-free cyber management service, 190
  humanity, 188
  implementation, rating system, 195
  innovative services, 191, 192
  integral parameters, 194
  integration, 189
  international cooperation, 186
  legitimate relationships, 186, 187
  management, 189
  metric of a valid scientist, 186
  metric of relationships, 186
  metrics, 184, 193, 194
  parameters, 193
  personal achievements, 195
  principles, 187
  public relations, 186
  quality, 184
  science, 184
  scientific and educational achievements, 195
  scientific and educational processes, 183
  scientific–educational and socioeconomics, 195
  sensor–actuator fog-network monitoring, 188
  smart cyber physical system, 185
  smart, 185
  social system, 189, 190
  social–metric relations, 187
  structural university components, 189
  system, 185
  university, 184
Smart digital state, 264, 271
Smart workspaces, 261
Social computing, 248
Socio–cognitive technologies, 249
Software implementation, 75
Space and time digitalization, 5
Space–time universal set, 227
Speed of light, 231
Structural–analytical memory–address–transaction model, 136
SyncD flip-flops, 140
Sync signals, 140
System-on-a-chip (SoC), 95, 135

T

Teleportation property, 18
Test Assertion Blocks Activated Matrix (TABA)
  components, 154
  faulty functional blocks, 152, 153
  functional coverage, 152
  HDL code, 151, 154
  math expressions, 153
  matrix structure and diagnosis method, 150
  software–hardware blocks, 151
  temporal assertion, 151
  test quality, 152
  test segments, 152, 154
  transaction-level graph, 150
Test synthesis method, 99
Test synthesis sequencer, 108
Through-silicon vias (TSVs), 163
Topological connections, 62
Traffic control system (TCS)
  analytical and automaton form, 209
  basic and additional expenses, 212
  cloud cyber system technology, 208
  cloud monitoring, 209
  components, 210
  computer control, 208
  cyber physical system, 207
  innovative proposal, 208
  road infrastructure, 211
  RU, 210
  software–hardware mobile systems, 208
  streetlight switching cycle, 211
  time interval, 211
  traffic jams, 211
  transport control signals, 209
  visualization, 210
  windshield, 210
Transaction-driven verification (TDV), 150
Transaction-Level Test Assertion Blocks Activated Graph (TABA graph), 150
Transport monitoring, 213
Triangle metric advantages, 62
Triangular cyberspace, 59
Triangular urban topology, 62

U
UltraScale™, 95
Unified transport system, 203
Unit under test (UUT), 37
Universal engine algorithm, 157
Universe's genome
  acceleration and deceleration synchronicity, 225
  cycling, 225
  Einstein's formula, 227
  entities and forms, 226
  entropy, 224
  four-component computing model, 223
  harmonics, 224, 226
  humanity, 225, 227
  Lorentz factor, 223
  phases, 225
  primitive types, 227
  singularity, 227
  substance and form, 228
  tetrahedron constraints, 226
  transitions, 224

V
Vector logical analysis infrastructure, 35–39
Vector logical criteria, 29
Vector logical multiprocessor (VLMP), 21, 32, 49
Vector Logical Multiprocessor Architecture, 32–35
Vector logical operations, 44
Vector logical space, 22, 33
Vector representation, 98
Verification template library (VTL), 160
Virtual traffic lights (VTLs), 208

Printed in the United States
By Bookmasters